12.95

Foundations of
GENETICS
A Science for Society

Foundations of GENETICS
A Science for Society

Anna C. Pai
ASSISTANT PROFESSOR OF BIOLOGY
MONTCLAIR STATE COLLEGE

McGRAW-HILL BOOK COMPANY

New York / St. Louis / San Francisco
Düsseldorf / Johannesburg / Kuala Lumpur
London / Mexico / Montreal
New Delhi / Panama / Paris / São Paulo
Singapore / Sydney / Tokyo / Toronto

FOUNDATIONS OF GENETICS
A Science for Society

Copyright © 1974 by McGraw-Hill, Inc. All rights reserved. Printed in the United States of America. No part of this publication may be reproduced, stored in a retrieval system, or transmitted, in any form or by any means, electronic, mechanical, photocopying, recording, or otherwise, without the prior written permission of the publisher.

34567890 MUMU 798765

This book was set in **Trade Gothic** by **Progressive Typographers.** The editors were **William J. Willey** and **Richard S. Laufer;** the designer was **J. Paul Kirouac;** cover illustration by **Edward A. Butler;** the production supervisor was **Joe Campanella.** The drawings were done by **Vantage Art, Inc. The Murray Printing Company** was printer and binder.

Library of Congress Cataloging in Publication Data

Pai, Anna C   date
    Foundations of genetics.

    Includes bibliographical references.
    1. Genetics. I. Title. [DNLM: 1. Genetics. QH431 P142f 1974]
QH430.P34   575.1   74-1372
ISBN 0-07-048093-1
ISBN 0-07-048092-3 (pbk.)

TO MY FAMILY

Contents

Preface   xi

Chapter 1   Patterns of Inheritance   3
Genetics: The Study of Life / Historical Theories on the Origin of Life / Gregor Mendel and the Laws of Genetics / The Prediction of Offspring by Using Laws of Probabilities / The Rediscovery of Mendel's Laws

Chapter 2   The Physical Basis of Inheritance   27
The Cell / Mitosis / The Role of Chromosomes / Meiosis / The Physical Basis for the Law of Segregation / The Physical Basis for the Law of Random Assortment / The Completion of Gametogenesis / The Effect of Age on Egg Formation / Cell Division in the Zygote: Mitosis and Genetic Continuity

Chapter 3   The Chromosomal Determination of Sex   51
The Sex Chromosomes / Genetics of the Y Chromosome / X-linked Traits / Sex-limited Traits / Sex Determination in Other Organisms / Why Only Two Sexes?

Chapter 4   Mendelian Heredity in Man   63
Some Mendelian Traits in Humans / The Pedigree Chart / Genetic Disadvantages of Marriage Between Related Individuals / Genetic Counseling

Chapter 5   Beyond Mendelian Genetics   89
Incomplete Dominance and Codominance / Polygenic Inheritance / Epistasis / Modifiers / Linkage / Crossing-over / Lethal Genes

Chapter 6 What Is a Gene? 109
Proteins and Nucleic Acids in Chromosomes / Identification of the Genetic Substance / The Structure of DNA / The Significance of the Molecular Nature of the Gene

Chapter 7 What Does a Gene Do? 125
Genes and Enzymes / The Mechanism of Gene-determined Protein Synthesis / How Does a Cell Produce the Correct Protein? / The Genetic Code / Some Problems of Gene Action in Complex Organisms

Chapter 8 The Regulation of Gene Action 145
Regulation in *E. coli:* The Lac Operon / A Redefinition of "Gene" / Developmental Genetics / Gene Regulation in Complex Organisms / Levels of Organization in Multicellular Organisms / The Present Status of Gene Regulation

Chapter 9 The Genetics of Immune Reactions 171
What Is an Immune Reaction? / Circulating Antibodies: Blood Groups and Resistance to Disease / Immune Reactions: Problems of Cell-mediated Transplantation / The Physical Nature of the Antibody Molecule / The Recognition of "Self" / Other Problems in Immunology

Chapter 10 The Genetics of Viruses and Cancer 193
General Characteristics of Viruses / Bacteriophages / Viruslike Genetic Particles / Animal Viruses / Transformation in Animal Cells: Cancer / Cancer-causing Agents and Innate Resistance / Present Theories on the Origin of Cancer / Chromosome Abnormalities and Cancer / A Cure for Cancer?

Chapter 11 Chromosomal or Gross Mutations 213
Germinal and Somatic Mutation / Chromosomal Mutations / Aberrations of Chromosome Number / Structural Abnormalities of Chromosomes / New Techniques for Studying Chromosomes

Chapter 12 Point Mutations and Population Genetics   **235**

Point Mutations / Hemoglobin Abnormalities in Man / Repetitive DNA and Gene Duplication / Determination of Mutation Rates / Population Genetics

Chapter 13 The Genetic Basis of Evolution   **255**

The Role of Mutations / Charles Darwin and the Theory of Natural Selection / The Origin of Species and the Descent of Man / Evidence for Darwin's Theory of Evolution / Factors Influencing the Course of Evolution / Some Unanswered Questions

Chapter 14 Man and Evolution   **283**

Resistance to Darwinism / Origins of Man / Man and His Environment / The Formation of Human Races / IQ / Genetics and Behavior

Chapter 15 Radiation and Chemical Mutagenesis   **305**

Radiation Mutagenesis / Man-made Radioactivity / Chemical Mutagenesis / The Determination of Mutagenicity / Other Biological Effects of Chemicals / Long-Range Effects of Chemical Mutagenesis

Chapter 16 Now and to Come   **329**

The Control of Life? / Euphenics / Eugenics / Genetic Engineering / The Future—with Imagination

Appendix A Answers to Chapter Problems   **349**

Appendix B The Chi-square Test   **351**

Appendix C Determination of Linkage and Chromosome Mapping Using Crossover Frequencies   **355**

Appendix D The Chemical Nature of Mutation   **359**

Appendix E Glossary of Terms   **361**

Index   **373**

Preface

Recent developments in the field of genetics have been prominently discussed not only in scientific journals but in every type of mass media from front-page stories in newspapers to feature reports on television. The extent of this popularization of achievements in a life science is virtually unique to the field of genetics, and for good reason. More than any other, genetics is a science whose subdisciplines have the potential to affect man and his society directly in countless ways, from contributing toward the solution of major medical problems such as cancer to possibly creating a "brave new world" of test tube babies with controlled genetic constitutions, or even to influencing the evolutionary future of our species.

Because the potential impact of genetic study is so great, it is absolutely essential that everyone have some basic knowledge of the principles of this science. Only then can the real significance of the science be grasped and our role as part of the natural environment of a threatened planet be understood.

Towards this end I have been conducting a course for a number of years at Montclair State College in New Jersey, called Genetics for the Layman. The enthusiasm with which non-science majors have registered for the course and their ability to cope with basic technical concepts have convinced me that not only *should* nonscientists approach this subject, they *can* digest the material with little strain.

However, I have found that the books now available for non-scientists in this relatively young science are inadequate. Many of them omit basic genetic principles that form the foundation for a true understanding of some of the recent "hot topics" receiving wide attention. Underlying problems such as those encountered in the rejection of organ transplants, for example, are basic genetic phenomena. Without knowledge of the basic principles, you only know that rejection occurs without understanding why.

The first portion of this book, then, devotes itself to the classic principles of heredity and to the molecular nature of genetic information, because a solid understanding of these principles is necessary for understanding the more complex phenomena dealt with later. Since the discussion is directed at nonscientists, technical aspects have been simplified whenever possible without detracting from the validity of the information. For example, although problem-solving and data analysis are standard methods of genetic study (the geneticist analyzes data obtained from breeding experiments to establish the manner in which the traits are transmitted), the number of practice problems has been intentionally limited. Some of the problems listed at the end of the first few chapters are solved; answers to those that are not solved are listed at the end of the book in Appendix A.

It must be emphasized here again that the book does not pretend to be comprehensive; the reader is not expected to be a trained geneticist by the completion of the book. Because of the simplification of some of the technical aspects of genetics, there may be readers who desire more details. For those who have the background and interest, some of the concepts that are merely mentioned in the text are discussed in greater depth in other appendixes, to which reference is made where appropriate in the text. Also, a list of references can be found at the end of each chapter, annotated to indicate the level for each selection.

While I have found fault with books on genetics for nonscientists for not being sufficiently technical, on the other hand introductory texts that can be used by biology majors are often too technical and lack discussions which adequately relate the possible effects of new developments to man and society. Since I have found the main interest of the nonscientist to be in the relevance of genetic principles to himself, illustrations of genetic principles have been drawn from conditions found in humans whenever possible.

Furthermore, most traditional texts are written in a somewhat arid style, which those not accustomed to reading scientific literature frequently find stifling and even intimidating. From this has developed a problem of communication for which we in science are largely responsible. We have not always tried to make ourselves understandable to the layman. For this reason, the style of this book is intentionally informal.

On the other hand, the nonscientist must also realize that it is simply not possible to discuss a field of science without using *some* technical terminology which one would not normally encounter in everyday life. The necessity for learning such terms, however, should not be a deterrent from approaching basic books such as this one. It is far easier than picking up a foreign language, for example, because the number of new words is far smaller, and one need not worry about grammar! As an aid in learning terminology, an extensive glossary is given in Appendix E.

In recent years, the number of basic texts dealing with various aspects of biology has increased. This development reflects both laymen's growing awareness of the importance of biological principles to man and scientists' increasing awareness of the necessity to inform nonscientists of new developments. Although it is the responsibility of geneticists to inform the public of implications of their research, it is the responsibility, in turn, of the informed public to apply their knowledge to help determine the direction of research and the manner in which new developments may be applied for the benefit, and not the destruction, of man and his society. Should we continue developing techniques for test tube babies? Do we want to be able to control heredity? These are typical of the questions that face us now and in the future, and which, it is to be devoutly wished, those who finish this book will be able to handle with greater insight and wisdom.

ACKNOWLEDGMENT

I would like to acknowledge with deep gratitude the amused tolerance and encouragement offered me by my family, especially my husband, David, who had to suffer the brunt of a fledgling author's agonies.

A number of people contributed their expertise towards improving the quality of the content of this book. Among them, I am especially grateful to Dr. Robert Edgar of the University of California at Santa Cruz, Dr. Ernest H. Y. Chu of The University of Michigan, Dr. Elizabeth L. Russell of The Jackson Laboratory, Dr. Elizabeth Jones of Case Western Reserve University, and Miss Lillian Eoyang of the Albert Einstein College of Medicine.

To my students I owe a vote of thanks for their enthusiastic support and interest in this project, which gave me the confidence to make the attempt. In particular, I appreciate the efforts of Mrs. Barbara Zeiler and Miss JoAnn Carollo for their suggestions and their assistance in proofreading the manuscript.

For their expert typing skills I am indebted to Mrs. Victoria Berutti and Miss Karen Durkin. Finally, and certainly not the least, I am grateful to the staff of the McGraw-Hill Book Company for their professional guidance and expert advice.

Anna C. Pai

Foundations of
GENETICS
A Science for Society

Chapter 1

Patterns of Inheritance

GENETICS: THE STUDY OF LIFE

The science of genetics is basically the study of life. Although most people associate genetics primarily with the transmission of traits from one generation to another, or what we would call heredity, we now know it reaches much further, encompassing every biological process.

Consider the criteria commonly used by biologists to distinguish living organisms from nonliving matter: the capacity to reproduce, to evolve, to respond to stimuli, and to carry out internal biochemical (i.e., metabolic) reactions. All these phenomena are in fact under genetic control.

As a result of recently acquired knowledge about the genetic basis of almost all biological functions, the science of genetics has been subdivided into many active and interrelated fields such as biochemical genetics, immunogenetics, medical genetics, human genetics, developmental genetics, and behavioral genetics. Paradoxically, while it has been necessary to subdivide the study of genetics, there is simultaneously a growing awareness among scientists that cooperation among the different fields in areas that overlap is essential if they are to unlock the many secrets of life that remain beyond their knowledge.

A family tree by Norman Rockwell. [*Reprinted with permission from The Saturday Evening Post,* © *1959, The Curtis Publishing Company.*]

Many of these specialized branches of genetics will be discussed in subsequent chapters of this book.

Not unnaturally, the inheritance of characteristics was the aspect of genetics that first drew man's attention, and probably still arouses the greatest curiosity in laymen. It is obvious that, inherent in the ability of living organisms to reproduce, there is an invariable rule that like begets only like. What, then, in the nature of reproduction, guarantees this perpetuation of the species? What are the processes by which the features of the young resemble in many fine details those of either or both parents, in some cases unmistakably, in others remotely? Why are some traits invariably passed on from one generation to another, while others are seemingly lost, only to reappear many generations removed from the last individual in which they were present?

These questions regarding the nature of inheritance have occupied the attention of scholars for many centuries. The earliest theories proposed to explain such mysteries of nature reflected the lack of scientific equipment by which the classical scientists could explore their world in fine detail, as well as the strength of religious doctrines in the society in which they lived.

In view of both these restrictions, it is amazing how often our predecessors were able to approach the truth. Many of their theories appear ludicrous today, but each nonetheless served to advance knowledge a step, if not by its own contribution, then by stimulating studies in the right direction.

HISTORICAL THEORIES ON THE ORIGIN OF LIFE

Prior to the seventeenth century, it was thought that life was created spontaneously. Common observations supported this theory of "spontaneous generation." For example, scientists without microscopes observed that maggots suddenly appeared on raw meat that showed no previous evidence of infestation. Had microscopes been available, they would have revealed the presence of tiny insect eggs.

With the development of the first crude microscopes in the seventeenth century, innumerable minute structures, the sperms, could be seen in the fertilizing fluid, or semen, of men. Evidence from lower forms of life indicated that females produced ova for procreation. The theory of "preformation" then arose as a natural consequence of these observations and the theory of spontaneous generation. Preformationists proposed that an entire miniature individual resided in the germ cell (and there was a controversy over which germ cell—egg or sperm) and had only to enlarge in the female's womb. How the parents passed traits to this embryo was not understood, only that somehow there was a transmission of some aspects of the parents' bodies to the "homunculus," as it was called (Figure 1.1).

In the eighteenth and nineteenth centuries, with the improvement of microscopes and scientific techniques, it was realized that there is no spontaneous generation, that all life comes from preexisting life. It was also established that germ cells were united in fertilization, to produce a single cell which then underwent divisions and gradual developmental changes, to become the embryo—no preformation, no homunculus!*

But how traits were inherited was still not understood. Where, for example, is the trait for eye color embodied in the

* The reader interested in the historical development of scientific thought will be referred to several sources at the end of this chapter.

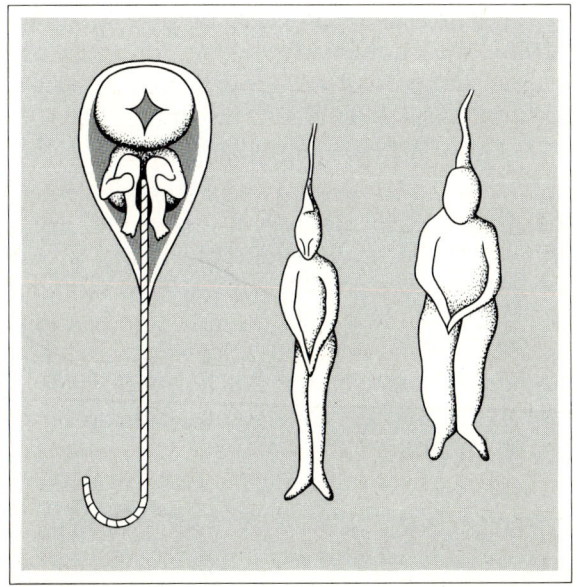

FIGURE 1.1
Drawings depicting the homunculus as envisioned by early biologists. [*Reprinted with permission of Macmillan Publishing Co., Inc., from* Genetics *by Monroe Strickberger. Copyright © Monroe W. Strickberger, 1968.*]

FIGURE 1.2 Portraits of some members of the Hapsburg family, showing the famous "Hapsburg lip": (*A*) Rudolph I, (*B*) Karl V, and (*C*) Ferdinand I. [*Photos of Rudolph I and Ferdinand I from the Bettmann Archive, Inc., photo of Karl V from the Austrian National Archives.*]

parent? Is it in the cells of the parents? Which cells? Is it some elusive factor that cannot be localized? How is it introduced into the germ cells, and where is it stored in them?

One widely held belief was the "blood theory of heredity," which stated that trait-carrying factors floating in the blood of both parents were somehow passed into the fertilized egg in sexual reproduction. To this day, our language retains meaningless idioms that reflect this idea of trait transmission by way of the blood, such as "It's in the blood" and "blueblood" and "blood brothers" and even "He has bad blood."

Implicit in all these ideas was a lack of precision in defining the processes of heredity. Indeed, there does not appear offhand to be any measurable entity by which one could predict the combinations of parental features that would be found in the offspring. Certainly some families are known by distinctive traits that appeared frequently, such as the Hapsburg lip (Figure 1.2), but most other traits in even these individuals showed no consistency in occurrence from generation to generation.

GREGOR MENDEL AND THE LAWS OF GENETICS

The first scientist to show that the transmission of traits is not always ambiguous but can have predictable patterns was not from any academic or scientific institution, as you might suppose, but from the quiet surroundings of a monastery in Austria. In the mid-nineteenth century, Gregor Mendel (Figure 1.3), a monk of the Augustinian order, combined a logical mind, an interest in plant hybridization (the crossing of different varieties), and a talent for statistical analysis to arrive at conclusions known as the classical *laws of genetics*.

How did he manage to succeed where others before had failed? For one thing, heredity is a quantitative science, involving the accumulation of data which must be sorted out, and only a man with Mendel's mathematical training would recognize certain trends in the data. (With Mendel's laws for guidance, however, students of genetics fortunately need not be mathematicians in order to understand the basic pattern of inheritance.) Another reason for Mendel's success was a fortuitous choice of traits for study. Why the choice was fortuitous will be discussed in a later chapter, but as is the case in many monumental scientific discoveries, luck was on Mendel's side. This certainly does not demean the enormity of his accomplishment, as it still requires a great mind to capitalize on such opportunities.

Mendel had more than a bent for mathematics and luck. He succeeded primarily because he had the insight to establish a set of criteria for experimentation that to this day must be met in genetic studies if meaningful results are to be obtained. It was the painstaking application of these criteria that led to the formulation of the Mendelian laws of genetics, which earned for this brilliant but unassuming man the sobriquet "Father of Genetics."

Mendel recognized that the choice of experimental material is of utmost importance. In his own words:

The experimental plants must necessarily—1. Possess constant differentiating characters. 2. The hybrids of such plants must, during the flowering period, be protected from the influence of all foreign pollen, or easily capable of such protection. 3. The hybrids and their offspring should suffer no marked disturbance in their fertility in the successive generations.*

These criteria may seem obvious. If the characteristics chosen for study were not "constant differentiating characters" and changed from generation to generation, obviously

FIGURE 1.3
Gregor Mendel. [*From The American Museum of Natural History, New York City.*]

* G. Mendel, Verh. Naturf. Ver. in Brunn, Albhandlungen, 1865. Translated by the Royal Horticultural Society of London, *Experiments in Plant Hybridization*.

their transmission could not be traced. If the plants used could be pollinated by other unknown plants, then obviously genetic material other than what was under study would be introduced, and the results would be meaningless. And certainly, if a cross were made that caused sterility in the offspring, this would immediately terminate the experiment in one generation, and one generation does *not* a pattern make!

Obvious? Yet centuries of work directed at the elucidation of inheritance failed because no one before Mendel recognized the importance of using simple, unchanging characters in systems that can be completely controlled by the experimenter. Time and again in the history of scientific endeavors, it has taken the genius to recognize some aspect of a problem that afterward seems so obvious.

The plant that satisfied all Mendel's criteria was the common garden pea, the breeding of which he could closely control, as it is capable of both self-pollination and cross-pollination. Given space on the monastery grounds, Mendel made crosses and recorded the characteristics of tens of thousands of plants over a period of eight years.

He chose seven different characteristics for study. We shall discuss two in detail: differences in the length of the stems and differences in the color of the immature pods. For each of these traits, he bred plants that became uniform in contrasting ways, thus satisfying his first criterion. For example, he crossed tall plants, around 6 feet in height, with other tall plants until all offspring were uniformly tall, either by cross-pollination or by self-pollination. Similarly, short plants were bred with each other until their offspring all bred uniformly short, about $1\frac{1}{2}$ feet.

THE LAW OF DOMINANCE AND RECESSIVENESS

Mendel's first hybridization experiment was to cross the so-called true-breeding short plants with true-breeding tall plants. The result of this cross was that all the offspring in the first generation were between 6 and $7\frac{1}{2}$ feet; that is, they all resembled the tall members of the parent generation.

When the experiment was repeated for the six other traits under study, the results were the same. All the first-genera-

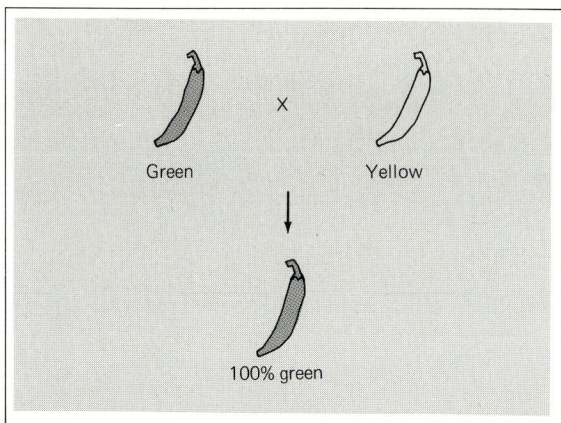

FIGURE 1.4
Mendel's experimental cross of two true-breeding plants differing in one particular characteristic, in this case the color of immature pods. Typically, the offspring resemble only one of the parental types.

tion plants resembled only one of the parental types. In the color of the pods, for example, a cross of true-breeding green-podded plants with true-breeding yellow-podded plants produced first-generation offspring that were all plants with green pods (Figure 1.4).

Mendel then crossed the first-generation plants with each other. [For simplicity, the symbol F_1 (for filial 1) will be used to denote the first generation of offspring; F_2 to denote the second generation; and so on. The parental generation with which the experiment first began is traditionally referred to as the P generation.] When the F_1 (tall) × F_1 (tall) crosses were made, he found that *both* parental types were represented in the F_2 generation. This indicated that the factors determining shortness were still present in the F_1 plants but that their expression was somehow masked.

From these results, Mendel concluded that of two contrasting alternatives for a certain trait such as height, one appears to be dominant over the other. This finding is one of the Mendelian laws of genetics, that of *dominance and recessiveness*. With regard to height in garden peas, tallness is dominant to shortness; with regard to pod color, green is dominant to yellow.

Further, Mendel detected a consistency to the numbers of parental types in the F_2 generation. There always seemed to

FIGURE 1.5
Mendel's cross of F_1 plants in his study of the transmission of seven different characteristics in the garden pea. [*Redrawn from Hickman, Cleveland P., and Hickman, Cleveland P., Jr.:* Biology of Animals, *St. Louis, 1972, The C. V. Mosby Company, p. 293.*]

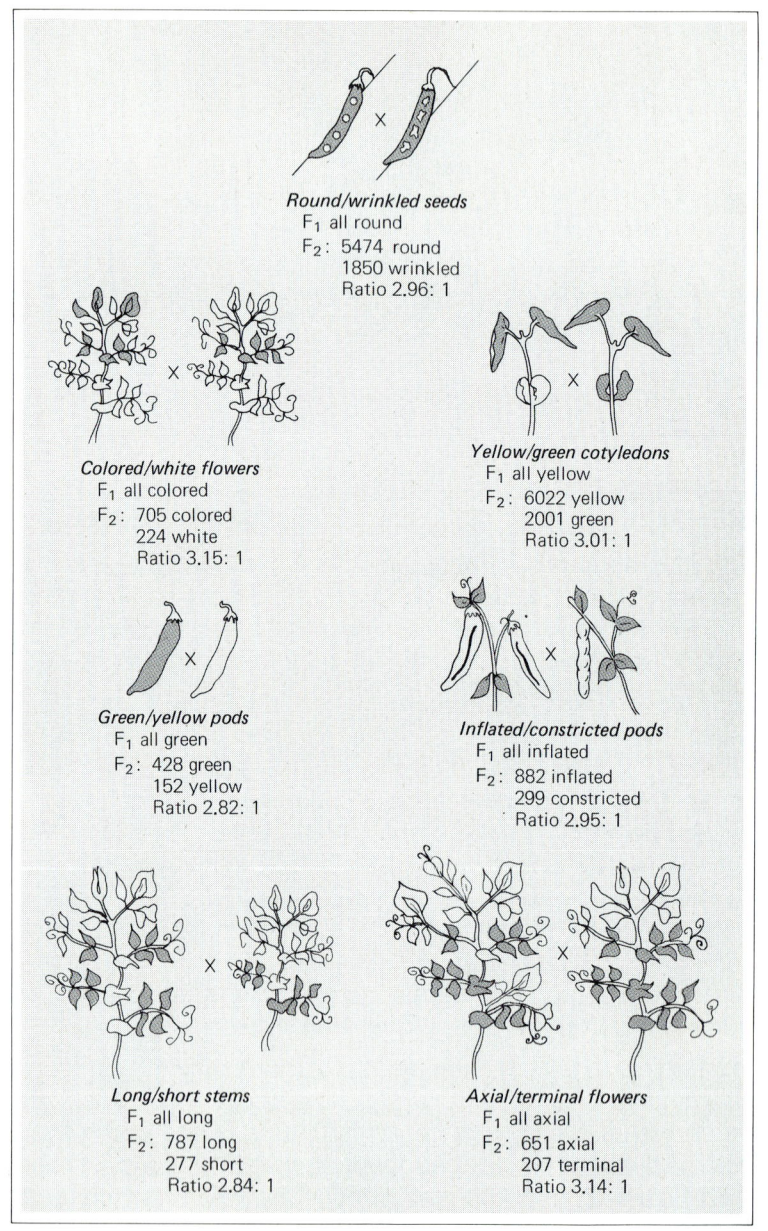

be a ratio of dominant types to recessive types on the order of 3:1, regardless of the trait studied, as shown in Figure 1.5.

When the F_2 plants were allowed to self-pollinate, the recessive F_2 plants bred true. Of the dominant F_2 plants, one-third bred true, and the other two-thirds produced offspring of dominant to recessive types in a 3:1 ratio, just as had the F_1 plants (Figure 1.6).

By these experiments, Mendel established that there is indeed a *pattern* to the transmission of traits. But Mendel's contribution went beyond just recognition of ratios. Although

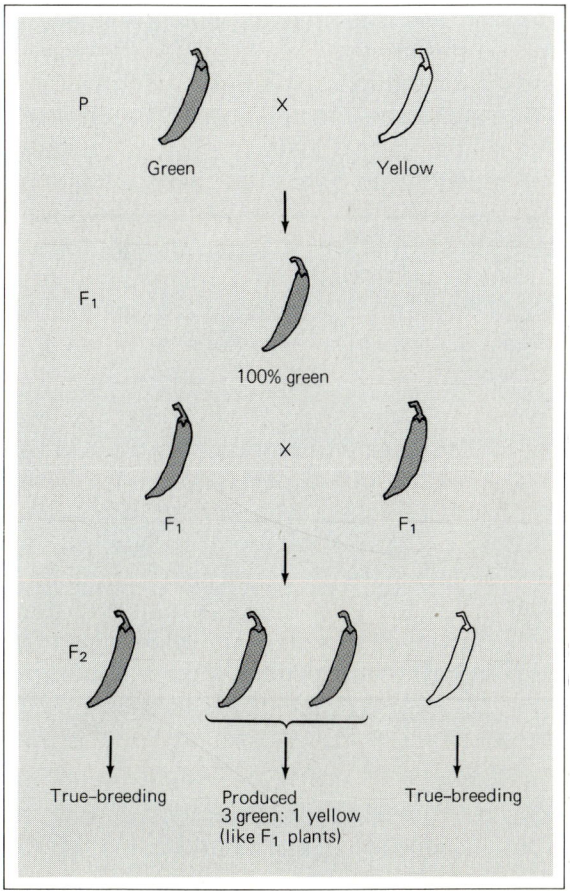

FIGURE 1.6
The results of Mendel's hybridization experiments with plants possessing contrasting colors of immature pods. The F_2 ratios obtained in these plants were characteristic of ratios in all experiments on seven different traits.

a lesser analyst might have stopped there, Mendel hypothesized that one could explain the numbers if it is assumed that traits are determined by *a pair of units,* only one of which is passed to the offspring by each parent. The units which determined traits that Mendel studied were, as already noted, either dominant or recessive.

Let us denote the dominant unit with capital letters and the recessive unit with small letters, as Mendel proposed and as geneticists have since done traditionally. Mendel used *A* and *a*, and *B* and *b*, but for clarity let us use the letter *T* for the unit determining tallness and *t* for the unit determining shortness. Similarly, we shall use *G* for the unit determining green pods and *g* for that determining yellow pods.

The preceding crosses can be explained in the following manner: The true-breeding plants in the parental generation can be designated *TT* and *GG* for the dominant types and *tt* and *gg* for the recessive types. If the assumption is correct, it is easy to see why they bred true, having only one kind of unit in their germ cells to pass on to their offspring when mated with their own kind. Such individuals are said to be *homozygotes,* or individuals *homozygous* for a particular unit. This means that they possess a pair of identical units determining a particular trait. The original hybridizing crosses would then be *TT* × *tt* and *GG* × *gg*. Individuals possessing different units in a pair, such as *Tt* or *Gg,* are called *heterozygotes,* or *hybrids*.

To work out what kinds of offspring may result from a simple Mendelian cross of homozygotes, we can use what is known as a *Punnett square,* devised by one of the classical geneticists, R. C. Punnett, who worked in the first decades of this century. To make a Punnett square, place the genetic content of one parent's germ cells (i.e., the unit contributed) across the top of a square, and that of the second parent's cells down the left side. Subdivide the square into four parts and fill in the units from both parents in the appropriate squares. As each parent would pass only one of its units to the offspring, a Punnett square for the hybridizing crosses would appear as shown in the margin.

As you can see, the F_1 generation, having inherited a *T* or *G* from the dominant plants and a *t* or *g* from the recessive

	T	T
t	Tt	Tt
t	tT	tT

	G	G
g	gG	gG
g	Gg	Gg

plants, would all have combinations of the dominant and recessive units, or *Tt* and *Gg*, and therefore would have the same appearance as the dominant homozygote. (Note that the order in which the letters are written, *Tt* or *tT*, does not represent any genetic difference. Traditionally, however, heterozygotes are denoted with the dominant-gene letter first: *Tt* or *Gg*, for example.)

Since each F_1 parent has two different units to transmit, the F_1 heterozygotes when crossed would give rise to four possible combinations of units in the F_2 generation, as shown in the margin.

	T	t
T	TT	Tt
t	Tt	tt

	G	g
G	GG	Gg
g	Gg	gg

Mendel did not use the Punnett square, of course, as this method was not devised until well after his death, but his reasoning followed the same lines. With this analysis, Mendel swept away much of the mysticism previously associated with the process of inheritance. And he gave the scientific world something specific to search for: discrete units whose expression may be masked in a generation, but which apparently are not lost or changed as they are transmitted from generation to generation, units which must therefore be located in the germ cells of sexually reproducing organisms.

What Mendel referred to as units of inheritance were given the designation *genes* by W. Johanssen, another of the classical geneticists who did so much at the turn of the century to propel us into the Age of Genetics. Mendel's astute observation that every individual possesses only one pair of genes for each trait meant, of course, that each parent contributed only one of the pair to the offspring; otherwise, if the offspring inherited both genes from each parent, he would have two pairs, and there would be a geometric increase in the number of genes per generation.

THE CONCEPT OF ALLELISM

The pair of "partner" genes determining any trait are referred to in modern terminology as <u>alleles</u>. Although the alleles which determined traits that Mendel studied in the garden pea plants appeared to exist in only two forms, the dominant and recessive, we now know that some traits can be determined by genes which can exist in many different forms. This

14
Patterns of Inheritance

phenomenon, known as *multiple allelism,* is illustrated in the inheritance of ABO blood types.

As you probably know, there are people with blood types A, B, AB, and O. These are determined by a series of three alleles, designated I^A, I^B, and i. The genetic constitution of individuals in the different blood groups may be summarized as shown in the accompanying table.

BLOOD TYPE	POSSIBLE GENE COMBINATIONS
A	$I^A I^A$, $I^A i$
B	$I^B I^B$, $I^B i$
AB	$I^A I^B$
O	ii

It appears from this information that alleles I^A and I^B are dominant over i, and that when both dominant alleles are present, both are expressed so that the individual is type AB. Such alleles are said to be *codominant.* Both codominance and the ABO blood groups will be discussed in greater detail later, but here the genes involved serve as a good example of multiple allelism.

We consider I^A, I^B, and i alleles because they determine the same trait and exist only as pairs of partner genes. In other words, regardless of how many different forms a series of alleles may have, only two alleles may be present in an individual. This is an unvarying fact, which Mendel recognized in his analysis of "units." In Chapter 2 we shall explore the reasons why only two alleles exist in an individual for any inherited trait.

Use of the same letter for the different blood-type alleles is intentional. In cases of multiple allelism, superscripts are frequently used to denote the specific allele, such as I^A and I^B. We could not, for example, refer to the blood-type genes as *A, B,* or *O.* To a geneticist, this would indicate three different, nonallelic genes.

THE LAW OF SEGREGATION

In his analysis, Mendel noted that the pair of alleles determining a particular trait seem to be transmitted separately. With rare exceptions in nature, no pair of alleles are normally transmitted together from one generation to another. This phenomenon, termed the *law of segregation,* is illustrated by the F_1 (tall) × F_1 (tall) cross. Only if the two genes *T* and *t* from each F_1 plant segregate (are transmitted separately) during reproduction is it possible to obtain the combinations diagrammed previously, namely, *TT, Tt,* and *tt* (see page 13).

One mark of a good scientist is thoroughness, in the preparation of his experiments, in the recording of data, and in the testing of his hypotheses. We have already discussed the methodical manner in which Mendel carried out his first experiments and dissected the results for some meaning.

To test the validity of his assumption of unit inheritance further—and remember that it was only an assumption, as genes were not identified or named until the twentieth century—Mendel combined different traits into a line of plants, then hybridized these, allowed the F_1 generation to self-pollinate, and again recorded ratios.

Let us stay with the traits of stem length and pod color to illustrate the results of this second group of experiments. Mendel first bred plants that expressed only the dominant traits, tallness and green pods, and plants that combined the recessive traits, shortness and yellow pods. Using these lines, he crossed tall green plants with short yellow ones (Figure 1.7).

The F_1 generation contained plants that were all like the dominant parent. This indicated that combining traits did not in any way affect the dominance and recessiveness of the individual traits. What was dominant when studied singly remains dominant when combined with other known dominant genes.

When the F_1 tall green plants were allowed to self-pollinate (Figure 1.8), both parental types appeared in the F_2 generation, as may be expected from previous results. But in addition, two new classes of plants appeared: tall yellow and short green, representing *new combinations* of parental traits. And again, these four classes appeared consistently in a ratio in the following manner: 9 (*T–G–*):3 (*T–gg*): 3 (*ttG–*):1 (*ttgg*). A dash following a gene symbol, as in *T–*, means that the second allele could be either dominant or recessive and would not change the appearance since *TT* is identical to *Tt*.

Repeating this kind of cross with different combinations of characteristics, Mendel invariably obtained the same results. These observations led to the formulation of a third Mendelian law of genetics, that of *random assortment,* which states that the genes determining *different* traits are transmitted *in-*

THE LAW OF RANDOM ASSORTMENT

FIGURE 1.7
Mendel's experimental cross of two true-breeding plants differing in two characteristics, the height of the stem and the color of the immature pods. Again, the offspring resemble only one of the parental types, the dominant one.

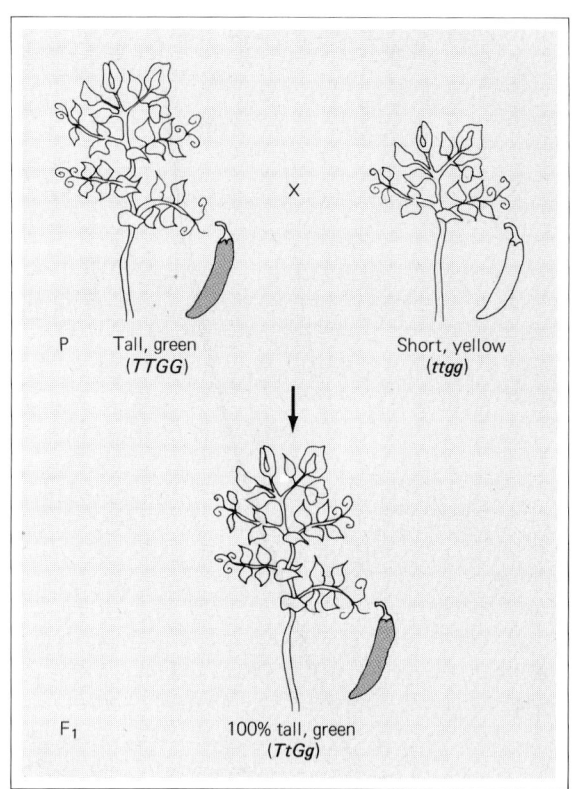

dependently of each other and therefore will reappear in random assortments in the offspring.

This becomes quite clear if we diagram the above experiments, starting with the parental generation, which is designated *TTGG* and *ttgg*. Since the F_1 generation all inherit *one of each pair* of genes present in the parent, they would have *TtGg* as their genetic makeup. Such heterozygous individuals are called <u>*dihybrids,*</u> or <u>hybrids</u> for two different pairs of genes.

Now, what about the F_2 generation? The situation becomes a bit more complicated — complicated but not different, as we are not dealing with new principles but only with an additional pair of genes.

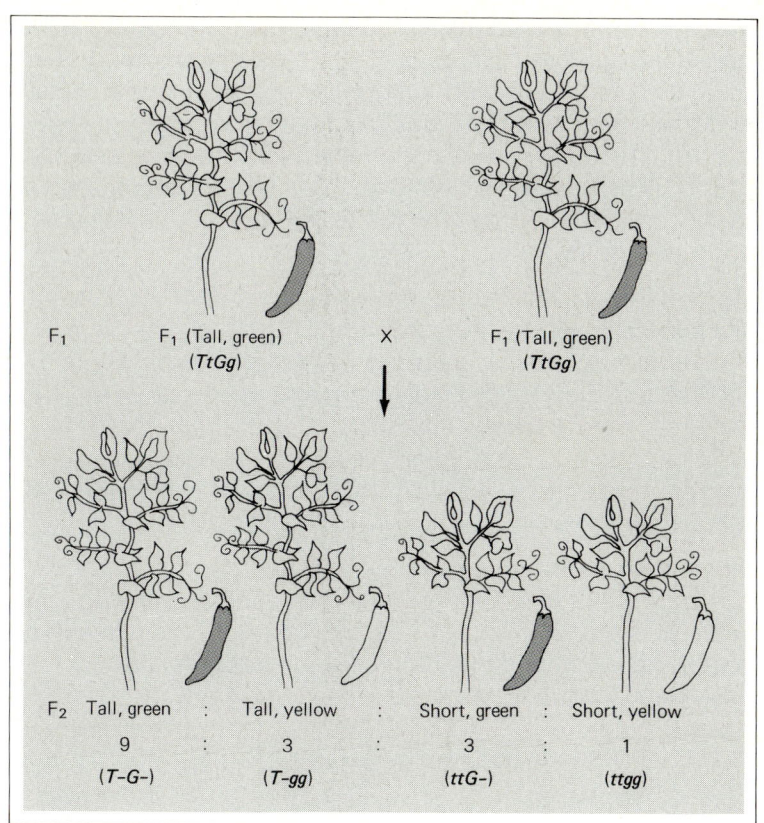

FIGURE 1.8
The cross of F_1 *TtGg* pea plants resulted in parental types in the offspring as well as new combinations of traits, and always in a 9:3:3:1 ratio. These results led to the formulation of Mendel's law of random assortment.

Because of the laws of segregation and random assortment, the F_1 plants have two pairs of alleles, but only *one of each pair* (two genes) will be transmitted in the parental germ cells. Thus the offspring will receive two from each parent, and the total number of genes in the offspring for these two traits will be four. The possible combinations of genes in germ cells from one dihybrid are the following:

F_1 *TtGg*

Genes of F_1 T   t   G   g

Germ cells of F_1 TG   Tg   tG   tg

	TG	Tg	tG	tg
TG	TTGG	TTGg	TtGG	TtGg
Tg	TTGg	TTgg	TtGg	Ttgg
tG	TtGG	TtGg	ttGG	ttGg
tg	TtGg	Ttgg	ttGg	ttgg

When two such dihybrids are crossed, each will have the four genetically different kinds of germ cells diagrammed above to contribute. A Punnett square for such a dihybrid cross is shown in the margin. Since dominance and recessiveness are not affected by combining these traits, it can be seen that of the 16 possible combinations in the F_2 generation, 9 would be tall and green, 3 short and green, 3 tall and yellow, and 1 short and yellow.

An important point can be brought out here. If you examine the genetic makeup of the individuals that look alike in all but the doubly recessive class, you will notice that where two individuals may look alike in expressing a dominant trait, they may not be genetically alike. One may be homozygous dominant, one heterozygous. On the other hand, individuals who express a recessive trait must of necessity be genetically alike, that is, homozygous for the recessive gene.

Geneticists have coined two terms that differentiate between an individual's appearance and his genetic makeup. We refer to the appearance of an individual as the *phenotype*, and to his genetic constitution as his *genotype*. As you will see, these two terms are used repeatedly in any discussion of a genetic nature.

THE PREDICTION OF OFFSPRING BY USING LAWS OF PROBABILITIES

The method of the Punnett square is actually useful only in monohybrid and dihybrid crosses, since crosses of hybrids heterozygous for more than two pairs of genes would entail squares with so many subdivisions as to be unmanageable. For example, even a trihybrid cross, *AaBbCc* × *AaBbCc*, would entail 64 subdivisions in a Punnett square, as each trihybrid can produce eight different genetic combinations in its gametes. How, then, can we easily determine the proportion of a particular genotype, such as *aabbCc*, among the offspring of such a cross?

What we do is apply a shortcut, using a law of probability which states that *the probability of independent events occurring simultaneously is the product of their separate probabilities.* We can use this law of probability because we know that every individual possesses a pair of alleles determining any trait and, by the law of segregation, which of these two is passed into a gamete is a matter of equal chance, of equal probability.

We can make an analogy between the transmission of one of a pair of genes and the tossing of a coin which has a head and a tail side, either of which may turn up with equal probability, and this probability is $\frac{1}{2}$ (Figure 1.9). In other words, the chance of either allele being passed to a gamete is $\frac{1}{2}$, the same as the chance of a head or a tail turning up when a coin is tossed.

If three coins are tossed at once, the probability of getting all three heads would be $\frac{1}{2} \times \frac{1}{2} \times \frac{1}{2}$. This is because what turns up in one coin is totally independent of what turns up in the other two. The same principle applies if several pairs of genes are involved in a problem, such as our trihybrid cross. Whichever allele of each pair is transmitted, the result is independent of any of the other gene pairs because of the law of random assortment.

How, then, do we apply this technique to the solution of problems such as finding the proportion of offspring of a trihybrid cross having the genotype *aabbCc?* The answer is $\frac{1}{4}$ (the probability of obtaining an *aa* offspring from *Aa* × *Aa*

FIGURE 1.9
An analogy may be made between the turning up of heads and tails in tossing two coins and the transmission of either gene of a pair of alleles from two parents. What one coin does is independent of the other coin; which allele is transmitted by one parent is independent of the other parent. Therefore, a law of probability may be applied to both cases: the probability of two independent events occurring simultaneously is the product of their individual probabilities.

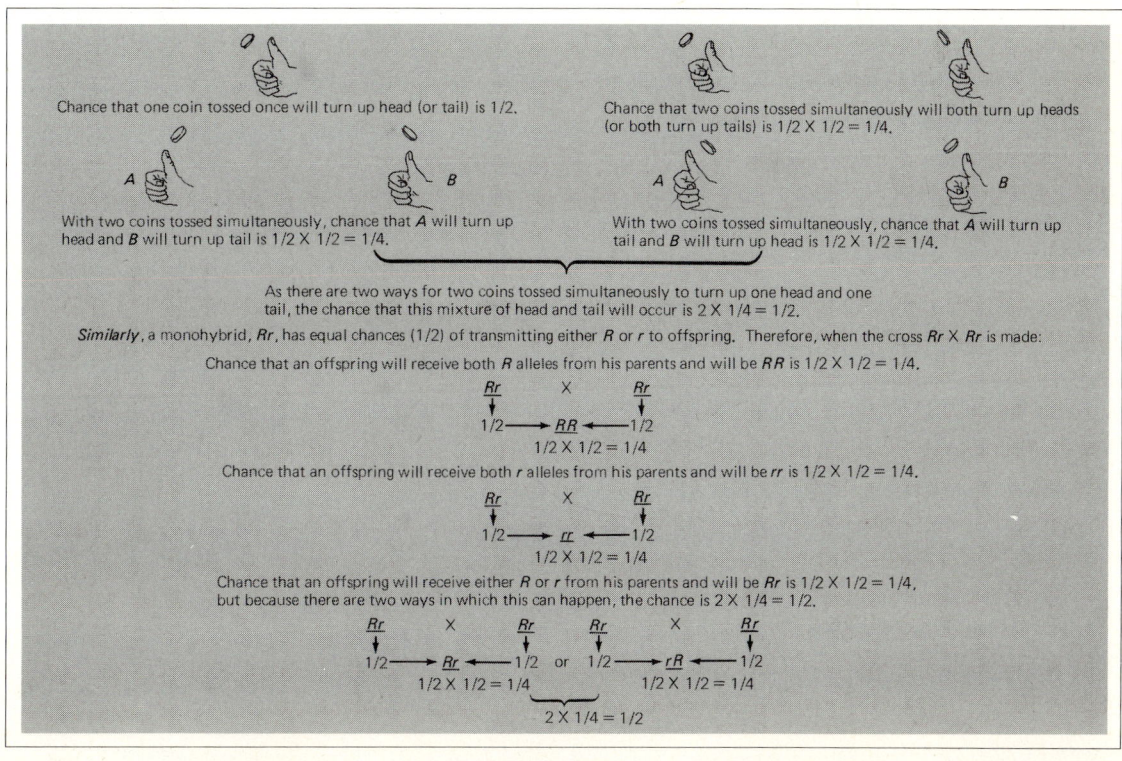

crosses) $\times \frac{1}{4}$ (the probability of obtaining a *bb* offspring from *Bb* × *Bb*) × $\frac{1}{2}$ (the probability of obtaining a *Cc* offspring from *Cc* × *Cc*) = $\frac{1}{32}$.

If you wish to prove the accuracy of this answer, make out the correct Punnett square, and you will find that 2 of the 64 subdivisions have the genotype under consideration ($\frac{2}{64} = \frac{1}{32}$). But how much quicker and easier it is to use probabilities!

This method is applicable regardless of the number of genes involved, or the genotypes of the parents. Suppose you are studying five pairs of randomly assorting genes, and the cross is of hybrids: *AaBbCcDdEe* × *AaBbCcDdEe*. What is the probability of obtaining offspring with the genotype *AABbccDdEE*? The answer is $\frac{1}{4} \times \frac{1}{2} \times \frac{1}{4} \times \frac{1}{2} \times \frac{1}{4} = \frac{1}{256}$.

Consider another example. In man, albinism is determined by a recessive gene which we shall designate *a*. Brachydactyly, a condition in which the fingers are abnormally short, is determined by a dominant gene, *B*. Suppose that a couple know their genotypes to be *BbAa* and *bbAa* and wish to learn what the chances are that one or both conditions will be passed to their children. The chance of brachydactyly occurring in their offspring would be $\frac{1}{2}$, because half the offspring from a *Bb* × *bb* cross would receive the dominant gene from the affected parent. Similarly, one could expect one-fourth of this couple's offspring to be albino, as both parents are heterozygous for the albinism gene. The chances that their children will be doubly affected would be $\frac{1}{2} \times \frac{1}{4} = \frac{1}{8}$.

Note that these are only probabilities. Just as one could toss three coins and get three heads just by chance, so the above couple could have children that are all perfectly normal—or, alas, all affected. This can happen in human families because relatively small numbers are involved. It is only when a large number of coins are flipped or a large number of progeny are produced by the organism under study that the expected ratios are approached, as in Mendel's experiments.

For the same reason, the absence of affected children does not prove that a couple is homozygous normal and could not produce albino offspring. They could be heterozygotes who, by chance, did not have albino offspring. In other words, whereas the presence of an albino child proves the heterozygosity of phenotypically normal parents, the absence of such children does not disprove this possibility.

Unfortunately, for couples who have family histories of an abnormal condition, the best a genetic counselor can do in most cases is to predict the proportion of offspring that might be affected, since there are only a few conditions in humans that can be detected with certainty in the early embryo. Whether the chance of having an affected child is taken or not is the decision of the parents. We shall discuss this problem further in subsequent chapters.

Because we are dealing with chance events, geneticists must rely on large numbers: the larger the number, the smaller the error through chance. There are methods known to statisticians by which deviations from expected ratios can be tested for validity; for these and the application of other laws of probability, the interested reader should consult advanced genetics texts and/or books on probability and statistics. Some titles are given at the end of this chapter.

We shall not discuss these methods because it is improbable that the average layman or non-science major will have to analyze genetic data in such depth. But recognize that geneticists must be able to distinguish between *chance* deviations from expected ratios and deviations caused by other factors. For example, if a dihybrid cross is made and the offspring do not exhibit a 9:3:3:1 phenotype ratio as expected, has the deviation occurred because the numbers are insufficient or because there has been some genetic interference with, say, random assortment?

Appendix B contains a discussion of one of the methods used to determine the "goodness of fit" of data from genetic experiments. The expression *goodness of fit* refers to the degree of deviation of obtained ratios from the expected ratios, and whether the deviation is significant. The technique, called chi square, requires only simple algebra.

In fact, as later chapters will reveal, many genetic phenomena have been discovered because of the geneticists' ability to analyze unexpected results from experimental crosses. This aspect of genetic analysis is challenging, certainly, but also can be stimulating and just plain fun.

THE REDISCOVERY OF MENDEL'S LAWS

Mendel performed other experiments which will not be detailed here. The three laws discussed in this chapter—dominance and recessiveness, segregation, and random assortment—are the main fruits of his labor. They form the basis for inheritance.

One would have thought that the presentation of his work in 1865 would have evoked a storm of excitement. But unfortunately for Mendel and the scientific world, no one recognized the implications of his findings and he was totally ignored.

It was not until the turn of the century that three biologists—DeVries of Holland, Tschermak of Austria, and Correns of Germany—simultaneously "rediscovered" Mendel's laws and the science of genetics was born. Mendel did not live to see this recognition, however. He had died in 1884.

PROBLEMS

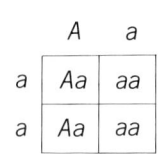

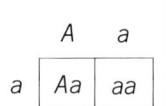

1 A normal woman whose father was albino marries a man who is albino. What phenotypic proportions can be expected among their offspring? (Let A = the gene for normal pigmentation and a = the recessive gene for albinism.)

	A	a
a	Aa	aa
a	Aa	aa

	A	a
a	Aa	aa

SOLUTION *Both the father and the husband of the woman are known to be aa. The normal woman must be Aa, having inherited a recessive gene from her father. The Punnett squares are shown in the margin.*

ANSWER 1 *normal:* 1 *albino.*

2 What proportions may be expected in the offspring of two heterozygotes for albinism?

3 Dark hair (M) in man is dominant over blond hair (m). Freckles (F) are dominant over no freckles (f). If a blond, freckled man whose mother had no freckles marries a dark-haired nonfreckled woman whose mother was blond, what proportions of their children would have the following phenotypes: (a) dark-haired and freckled; (b) blond and nonfreckled?

SOLUTION *First determine the genotypes of the man and woman. Being blond, the man is mm; having a mother homozygous for no freckling, he must therefore also be Ff. He is* therefore *Ffmm*. By the same reasoning, the woman must be *ffMm*.

You may make a Punnett square here, because each is homozygous for one pair of genes and the number of squares is reduced.

	fM	fm
Fm	FfMm	Ffmm
fm	ffMm	ffmm

ANSWER (*a*) One-fourth (*the progeny whose genotype is FfMm*).
(*b*) One-fourth (*ffmm*).

4 If the woman in problem 3 were homozygous recessive for both genes, how would this change your answer?

5 In probabilities, what proportion of the offspring would be (*a*) dark-haired and freckled and (*b*) blond and nonfreckled if both parents were dihybrids (*MmFf* × *MmFf*)?

SOLUTION *All offspring that are M–F– would be dark-haired and freckled. From a cross of monohybrids, three-fourths would be expected to inherit at least one dominant gene (see Figure 1.6). Therefore the chance of the offspring's being M– from the cross is $\frac{3}{4}$; the chance of the offspring's being F– is also $\frac{3}{4}$.*

ANSWER (*a*) The chance of the offspring's being M–F– is $\frac{3}{4} \times \frac{3}{4} = \frac{9}{16}$.
(*b*) See Appendix A.

6 If the parents in problem 5 were also heterozygous for short fingers (brachydactyly, caused by a dominant gene) and thus were *MmFfBb* × *MmFfBb*, what proportions of their children would be (*a*) blond, freckled, and short-fingered; (*b*) blond and nonfreckled, with normal fingers?

SOLUTION *The children of the phenotype in problem 6a must have the genotype mmF–B–. By the same reasoning used in problem 5, the probabilities would be $\frac{1}{4} \times \frac{3}{4} \times \frac{3}{4}$.*

ANSWER (*a*) $\frac{9}{64}$.
(*b*) See Appendix A.

REFERENCES

Almost any introductory biology textbook will have in it a section on Mendelian genetics. The following titles are good:

1971: *Biology Today,* CRM Books, DelMar, Calif.

Hardin, G., 1966: *Biology: Its Principles and Implications,* W. H. Freeman and Company, San Francisco.

Keeton, W. T., 1972: *Elements of Biological Science,* 2d ed., W. W. Norton & Company, Inc., New York.

Simpson, G. G., and W. S. Beck, 1969: *Life: An Introduction to Biology,* Harcourt Brace Jovanovich, Inc., New York.

A classical genetics textbook which contains lucid discussions of basic principles, though it is woefully out of date, is:

Sinnott, E. W., L. C. Dunn, and T. Dobzhansky, 1958: *Principles of Genetics,* McGraw-Hill Book Company, New York.

For historical development of ideas:

Butterfield, H., 1953: *The Origins of Modern Science: 1300–1800,* The Macmillan Company, New York.

Dunn, L. C., 1966: *A Short History of Genetics,* McGraw-Hill Book Company, New York.

For a collection of original papers, including Mendel's, with comments by the editor on their significance:

Peters, J. (ed.), 1959: *Classic Papers in Genetics,* Prentice-Hall, Inc., Englewood Cliffs, N.J.

Mendel's paper was translated by the Royal Horticultural Society of London, entitled *Experiments in Plant Hybridization.* It was originally published in 1865 in the Verh. Naturf. Ver. in Brunn, Albhandlungen, IV, 1865.

Three genetics texts written on a fairly elementary level are the following:

Lerner, I. M., 1968: *Heredity, Evolution and Society,* W. H. Freeman and Company, San Francisco.

Winchester, A. M., 1972: *Genetics: A Survey of the Principles of Heredity,* Houghton Mifflin Company, Boston.

_____, 1971: *Human Genetics,* Charles E. Merrill, Inc., Englewood Cliffs, N.J.

Three good genetics texts intended for science majors are:

Burns, G. W., 1972: *The Science of Genetics,* 2d ed., The Macmillan Company, New York.

Srb, A. M., R. D. Owen, and R. S. Edgar, 1965: *General Genetics,* 2d ed., W. H. Freeman and Company, San Francisco.

Strickberger, M. W., 1968: *Genetics,* The Macmillan Company, New York.

For those interested in exploring the application of statistical methods and probabilities to genetics, the texts by Srb, Owen, and Edgar, by Strickberger, and by Burns, mentioned above, have much more extensive discussions than we have gone into here. In addition, for those with a mathematical bent, any introductory text on statistics and probabilities may be of interest. The following titles may also be helpful:

Bishop, O. N., 1967: *Statistics for Biology,* Houghton Mifflin Company, Boston.

Hill, A. B., 1961: *Principles of Medical Statistics,* Oxford University Press, New York.

Mosimann, J. E., 1968: *Elementary Probability for the Biological Sciences,* Appleton Century Crofts, New York.

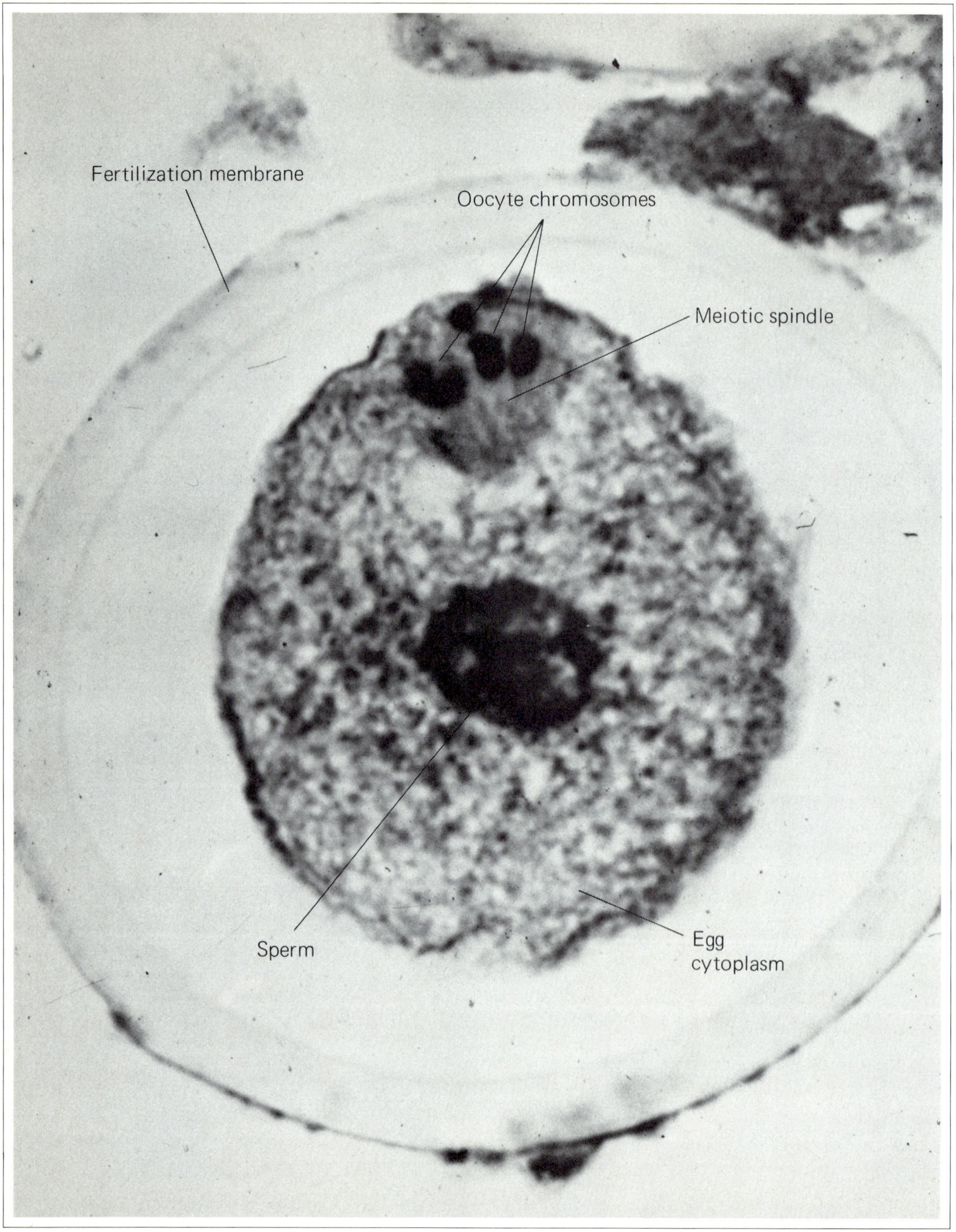
Fertilization membrane
Oocyte chromosomes
Meiotic spindle
Sperm
Egg
cytoplasm

Chapter 2

The Physical Basis of Inheritance

Let us turn now to the physical basis of heredity: how genes are passed to the next generation via the germ cells. The major work toward clarification of this process was done in the latter half of the nineteenth century, following the development of the cell theory. By this time, use of improved microscopes had revealed that all living organisms are composed of subunits called *cells* and that all cells come from preexisting cells.

The great biologists of that age reasoned that if this is true, then all cells of an organism must trace back to a single cell—the fertilized egg, or *zygote,* as it is often called. Since the zygote undergoes cell divisions, to give rise to the embryo and finally to the adult form, and since all *daughter cells* (the two cells that result from division of an original cell) are known to be alike genetically, then cell division itself must in some way be the basis for genetic continuity.

THE CELL

There is no such thing as a "typical" cell, for all cells have their own shape, size, and function, as illustrated by the vari-

An *Ascaris* (roundworm) egg cell undergoing maturation division following fertilization. [*Carolina Biological Supply Company.*]

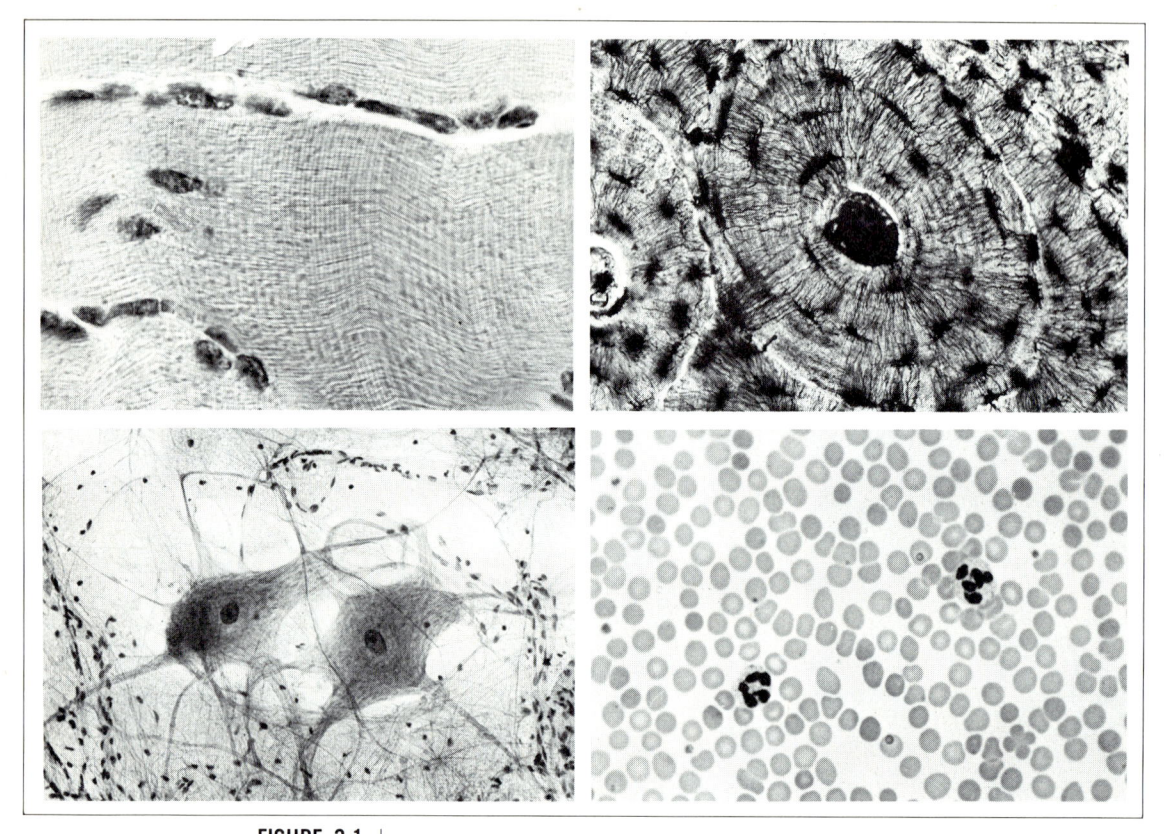

FIGURE 2.1
Cell types as they appear through the microscope: (A) muscle, (B) bone, (C) nerve, and (D) blood smear showing circular biconcave red blood cells, and white blood cells with darkly staining, irregularly shaped nuclei. [*Carolina Biological Supply Company.*]

ety of cell types found in man (Figure 2.1). All cells, however, do have certain features in common. Let us consider those which appear to have the most direct bearing on genetic processes.

Figure 2.2, a diagram of a generalized animal cell under high magnification, shows the two major portions of a cell, the *nucleus* and the *cytoplasm*. Within the nucleus is a network of fine threads called *chromatin*. When the cell is ready to divide, the fine threads condense to form *chromosomes*, which are rods of varying shapes and lengths. There is also a nuclear structure known as the *nucleolus*, which disappears during division and reappears when division is completed. All the structures of the nucleus are bounded by the *nuclear membrane*.

In the cytoplasm, near the nucleus, are two *centrioles*. These structures are fascinating in that they are capable of movement and self-duplication, and play an essential role in all kinds of cell division. Smaller particles, the *ribosomes*, which are sometimes also attached to membranes forming the *endoplasmic reticulum*, will be discussed later, as they are important in the molecular aspects of gene function. The cell as a whole is bounded by a *cell membrane*, which is now known to play an active role in several essential cell functions, such as the transport of materials into and out of the cell, and is also involved in the process of immune reactions.

Experiments have indicated that there is nothing precise about the redistribution of cytoplasmic structures during cell division. Whatever happens to be located in one-half of a

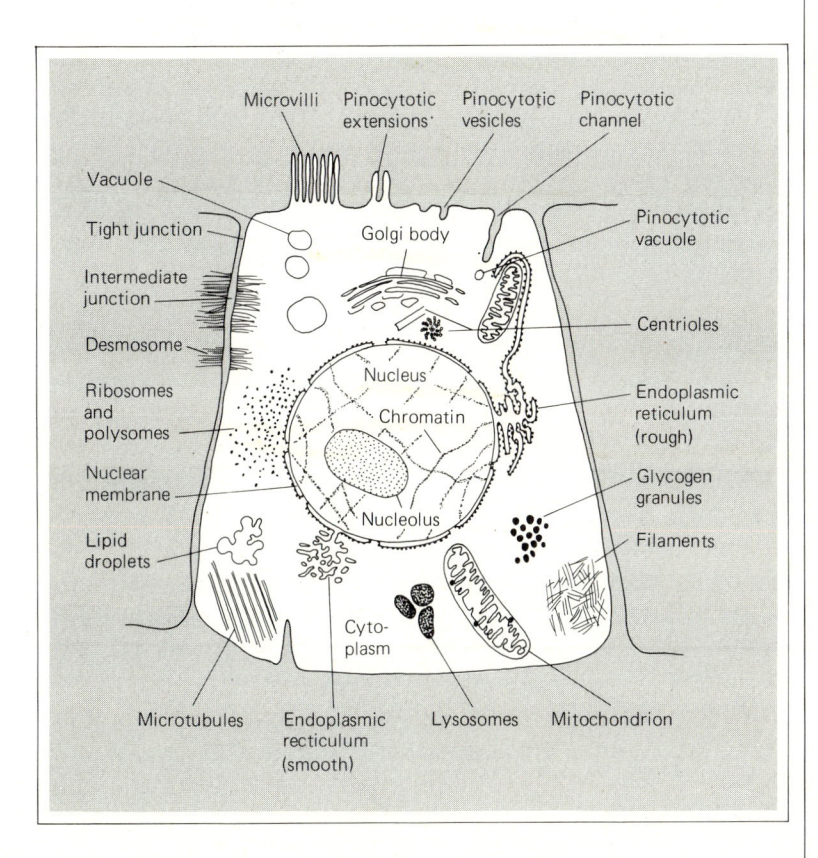

FIGURE 2.2
A generalized animal cell. [*After Cell Structure and Function, second edition, by Ariel G. Loewy and Philip Siekevitz. Copyright* (c) *1963, 1969 by Holt, Rinehart and Winston, Inc. Adapted by permission of Holt, Rinehart and Winston, Inc.*]

cell that is dividing remains in that daughter cell when halves pull apart. What *is* certain is that the chromosomes double in number, half of them passing into one daughter cell and half into the other, so that all cells of an organism have *exactly* the same number of chromosomes.

MITOSIS

Let us turn now to the details of cell division, known as *mitosis,* to see how the constancy on which the growth and normal development of all organisms depend is maintained in spite of the physical separation of the cells. Mitosis is a continuous process, but for convenience of reference, biologists have divided its events into five stages: interphase, prophase, metaphase, anaphase, and telophase.

INTERPHASE

The stage of *interphase* has sometimes been referred to, erroneously, as the "resting stage." Indeed, visual observation seems to indicate that nothing much is happening in the cell. By using special techniques, however, biologists have found that the cell is doing anything but resting. It actually is undergoing great biochemical change. Most significantly for our discussion, the chromosomes are being increased in quantity. Specifically, they are doubling. Figure 2.2 shows, in fact, a generalized animal cell in interphase.

PROPHASE

The beginning of the *prophase* stage of mitosis is marked by a condensation of the very thin chromatin threads to form rodlike structures which are the duplicated chromosomes. Figure 2.3 illustrates a cell with four chromosomes, during prophase.

In each chromosome there is a constriction, sometimes depicted in diagrams as a dot, called the *centromere.* Different chromosomes have centromeres at different locations along their length, and by recognition of this characteristic as well as differences in length, we can distinguish one chromosome from another (Figure 2.3*B* and *C*).

The centromere is also important in that it is the point of attachment of the *sister chromatids,* the two threads of a doubled chromosome, and it largely determines the movement of the chromosomes during division. In fact, if you focus your attention on what happens to the centromere at the different stages of division, you will have no difficulty understanding the distribution of chromosomes in dividing cells.

Sister chromatids are considered identical in their chemical content. *How* the genetic material duplicates to form identical sister chromatids will be discussed in Chapter 6.

Once the chromosomes coil after duplication, they appear as discrete structures and begin to migrate to the center of the cell.

A number of complex cellular changes take place at this time. The nuclear membrane and the nucleolus disintegrate, and the centrioles self-duplicate, separate, and move to opposite sides of the cell. There also appear in the cytoplasm some translucent fibers that eventually stretch between, and connect with, the two sets of centrioles, forming a structure named, for its shape, the mitotic *spindle*. The origin of the spindle fibers is still unclear. In fact, although the events of mitosis have been known since the nineteenth century, we still do not understand what factors cause the components of a cell to undergo the fascinating changes required by cell division.

Formation of the spindle gives the cell an orientation: the centrioles form the poles, and a plane bisecting the spindle is the *equator,* which marks the plane of division, or cleavage, that will result in the separation of the daughter halves of the original cell (Figure 2.4). Prophase ends as the chromosomes migrate to the equator of the spindle.

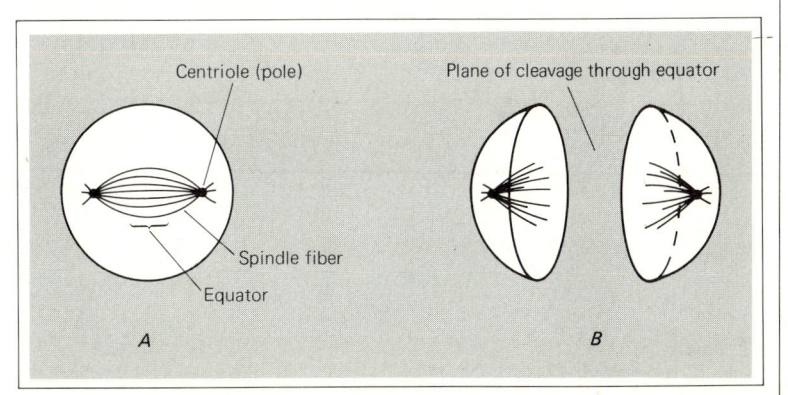

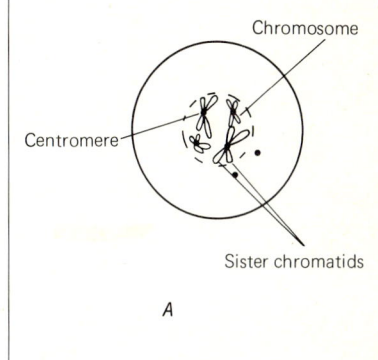

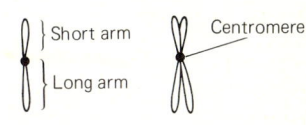

FIGURE 2.3
(*A*) A cell with four chromosomes in the prophase stage of mitosis. The nuclear membrane is beginning to disintegrate, and the centrioles (shown as black dots) begin to separate. (Other cell structures not visibly involved with mitotic events have been omitted.) (*B*) A single chromosome showing arms of unequal length. (*C*) A duplicated chromosome.

FIGURE 2.4
(*A*) Spindle formation. (*B*) The plane of cleavage of a dividing cell.

FIGURE 2.5 (left)
The metaphase stage of mitosis, in which chromosomes are attached to the spindle fibers at the equator by their centromeres.

FIGURE 2.6 (right)
The anaphase stage of mitosis. Arrows indicate direction of movement of the sister chromatids, now chromosomes.

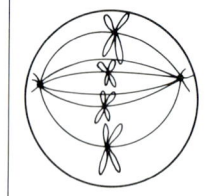

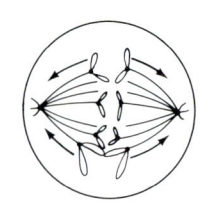

METAPHASE

When the chromosomes reach the equator of the spindle, they each attach to a spindle fiber by the centromere in a random manner (Figure 2.5). This is the *metaphase* stage of mitosis.

ANAPHASE

Shortly after metaphase, in the *anaphase* stage, the spindle fibers appear to break at the attachment of the chromosomes, causing the centromeres to divide. The spindle fibers appear to contract toward the poles; as they do so, they drag one sister chromatid of each original chromosome toward opposite ends of the cell by the duplicated centromeres, which are still attached to the fibers (Figure 2.6). Because they are now single threads with their own centromeres, the chromatids are now referred to as chromosomes.

TELOPHASE

In *telophase,* the last stage of mitosis (Figure 2.7), the chromosomes continue to their respective poles. Characteristic of telophase, then, is a cell in which there are two clusters of chromosomes, each cluster containing *exactly the same chromosomes as the other.* Moreover, each cluster is identical in content to the original cell, having received one sister chromatid from each original chromosome. A constriction of the cell appears in the plane of the equator.

FIGURE 2.7
The telophase stage of mitosis. The constriction will deepen, and two cells will pinch off from each other. The cells resulting from division are identical in chromosome content.

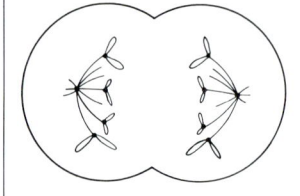

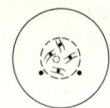

Mitosis is completed when the cell constriction deepens and eventually pinches off two cells, identical in their chromosome makeup. The nuclear membrane and nucleolus reform in each of the daughter cells, and the chromosomes become attenuated to form chromatin.

DIVISION

From the precise nature of the duplication and redistribution of chromosomes during cell division, it was a logical conclusion that if some kind of genetic continuity is to be preserved, the chromosomes must play an important role. But what? If they contain genetic information, does each chromosome contain the entire set of genes, or is each chromosome different?

The answer came from several investigators, but one of the most definitive experiments on the subject, carried out at the turn of the century, was the brilliant work of an embryologist, Theodor Boveri. He studied the development of sea urchin embryos, which are easily manipulated since they develop externally. If the cells resulting from the first divisions following fertilization are separated from each other, each cell or group of cells can develop normally into a whole adult.

By altering environmental conditions, Boveri was able to induce the fertilization of one egg by two sperms, thereby producing a zygote with an extra set of chromosomes. Upon division, the daughter cells of the zygote would receive abnormal numbers and combinations of chromosomes because of the odd set.

Boveri found that when he separated these cells, only some of those possessing the correct number of chromosomes developed normally. Because other cells with the correct number showed abnormalities, he deduced that not only the *correct number of chromosomes* but also the *correct combination of chromosomes* is essential for normal development.*

These findings and other studies showed that chromosomes are indeed the carriers of genetic information and that each chromosome contains different information from others. In order for the offspring to develop normally, it must receive the entire normal set of chromosomes from each parent.

THE ROLE OF CHROMOSOMES
QUALITATIVE DIFFERENCES IN CHROMOSOMES

* See the reference to his work at the end of this chapter.

THE SPECIES-SPECIFIC NUMBER OF CHROMOSOMES

Another essential fact discovered at this time had bearing on studies on the role of chromosomes. It became apparent that every species of organism studied had a specific number of chromosomes. In other words, every member of a species has the same number of chromosomes. For example, we now know that all genetically normal human beings have 46 chromosomes in each body cell.

But in sexually reproducing organisms, if a zygote is formed by the union of two cells, would it not receive double the number of chromosomes? Upon further study this problem was resolved: all functional germ cells, or *gametes,* of humans have but 23 chromosomes. Upon fertilization, therefore, the zygote is restored to the species number.

The species number of chromosomes is called the *diploid number,* and the number of chromosomes in mature germ cells is referred to as the *haploid number.* Sometimes these numbers are designated *2n* and *n*, respectively.

Does this seem reminiscent of anything we have already discussed? You are right if you thought about Mendel's units, which he theorized occurred in pairs, only one of which is passed to the offspring. The work of Boveri and of the classical geneticist W. S. Sutton resulted in the recognition that the physical basis of Mendel's laws is the movement of chromosomes during germ-cell maturation and fertilization.

MEIOSIS

Meiosis is the very special kind of cell division, occurring during the maturation of germ cells, which reduces the chromosome number in half. No other cells of the body are capable of this kind of division.

There are actually two divisions, or separation events, in meiosis, and the steps leading to each have been divided into the same five stages as was mitosis. The stages preceding the first division are called interphase I, prophase I, etc., and those following the first division are called interphase II, prophase II, etc. Let us now follow the changes that occur during the maturation of a sperm in a species that has four chromosomes.

INTERPHASE I

Interphase I is the same in appearance in meiosis as it is in mitosis.

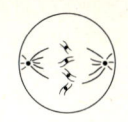

Prophase I is a very important stage in meiosis, and perhaps the most complicated. Just as in mitosis, the chromosomes appear after condensation of the chromatin threads. In meiosis, however, because the sister chromatids remain tightly wound up with each other, the chromosomes look like single structures. Figure 2.8 diagrams a cell with four chromosomes in early prophase I.

Once the chromosomes appear as discrete structures, they begin to move within the nucleus. When the movement finally ceases, it becomes apparent that chromosomes actually exist in pairs which are identical in appearance; such physically identical chromosomes are said to be *homologous* to each other and are frequently referred to as *homologs*. Each homologous pair seems to be drawn together so that the identical chromosomes line up side by side. This process of chromosome pairing is called *synapsis,* and a pair of homologs that have completed synapsis are known as *bivalents* (Figure 2.9). Although the process of synapsis has been known for many decades, we are unable to say what causes this movement to take place.

Following synapsis, the same cellular changes occur as described in the prophase stage of mitosis, such as the duplication of centrioles and the disappearance of the nucleolus. The two sister chromatids of each homolog now unwind to reveal four threads lined up together. This formation is known as a *tetrad* (Figure 2.10).

The tetrads begin to migrate through the cell to line up on, and become attached by their centromeres to, the spindle. This marks the end of prophase I.

The tetrads attach to the spindle fibers at the equator (Figure 2.11). Note that the homologs are *not* randomly dispersed as was the case in the metaphase stage of mitosis.

The next stage in meiosis begins shortly after the tetrads reach the equator of the spindle. During anaphase I, the spindle fibers break in the space between the homologous chromosomes. The sister chromatids remain attached to each other by their centromeres. The fibers contract and simultaneously drag the homologs, which remain attached to them, to opposite ends of the cell. As this results in only two

PROPHASE I

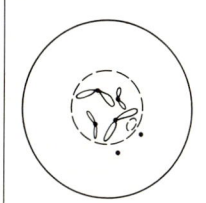

FIGURE 2.8
A cell with four chromosomes in the early prophase I stage of meiosis.

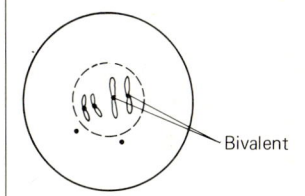

Bivalent

FIGURE 2.9
Synapsis in prophase I and formation of bivalents.

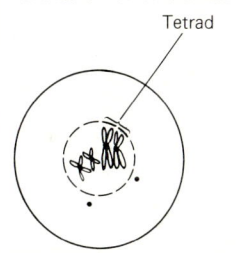

Tetrad

FIGURE 2.10
Appearance of tetrads in later prophase I.

METAPHASE I

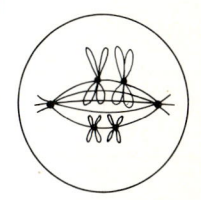

ANAPHASE I

Figure 2.11
Metaphase I in meiosis.

FIGURE 2.12
(A) Separation of homologs to produce dyads in the anaphase I stage of meiosis. (B) A tetrad and a dyad, for comparison.

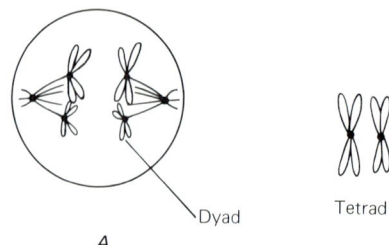

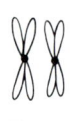

Figure 2.13
The telophase I stage of meiosis. When division is completed, each daughter cell has only half the number of chromosomes as the original cell.

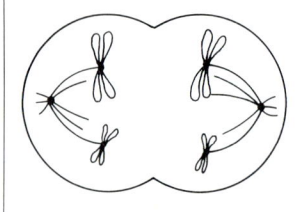

threads together, not four as in prophase I, the halves of the tetrad are called *dyads* (Figure 2.12).

TELOPHASE I

The movement of dyads continues until the homologs reach their respective poles. Characteristic of the telophase I stage of meiosis, then, is a cell in which there are two clusters of chromosomes, but in contrast to mitosis each cluster contains half the original number with *one representative of each of the original pairs*. A constriction appears in the plane of the equator (Figure 2.13).

DIVISION I

The constriction deepens, and eventually the daughter cells are separated. Note that the cells resulting from the first division are haploid. Because of the reduction in chromosome *number* from the diploid to the haploid (although the *amount* of genetic material remains the same), the first meiotic division is frequently referred to as the *reduction division*.

INTERPHASE II

In many organisms, including man, interphase II is merely a pause in the meiotic sequence, and the cells go directly into prophase II.

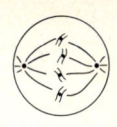

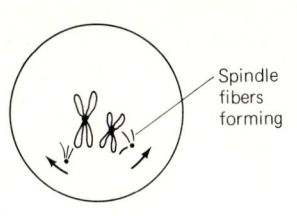

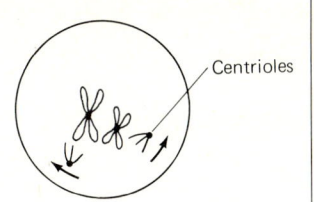

FIGURE 2.14
Prophase II in the daughter cells following reduction division. Arrows indicate the movement of centrioles to form the poles of the new spindles.

The chromosomes are still in dyad formation during prophase II. The centrioles in the daughter cells migrate to opposite poles, and a spindle is formed in each cell (Figure 2.14).

PROPHASE II

In metaphase II, as in metaphase I, the dyads migrate to the equator of the spindle and attach to the spindle fibers at the centromeres (Figure 2.15).

METAPHASE II

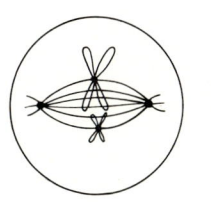

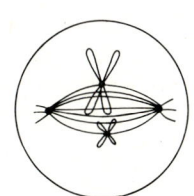

FIGURE 2.15
Metaphase II. One representative of each original pair of chromosomes is present in the daughter cells.

In anaphase II the spindle fibers contract and pull the dyads apart, separating the sister chromatids from each other (Figure 2.16). It is only now, when the centromeres split, that the sister chromatids are considered chromosomes. Temporarily, therefore, the cells are again diploid, but this condition lasts only a short time.

ANAPHASE II

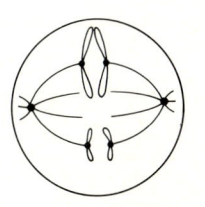

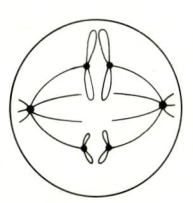

FIGURE 2.16
The anaphase II stage of meiosis. The dyads have split, and the sister chromatids, now chromosomes, move to opposite poles, just as in mitosis.

FIGURE 2.17
Telophase II. The daughter cells will each in turn give rise to two daughter cells, each of which has half the original number of chromosomes and is haploid.

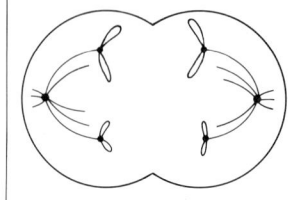

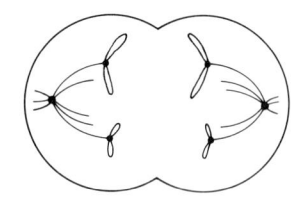

TELOPHASE II

In each daughter cell there are now two clusters of single chromosomes, one representative from each of the two pairs of chromosomes in the original cell (Figure 2.17). Constriction of the cells in telophase II leads to the second and final division.

DIVISION II

When the two halves of each daughter cell pull apart, the second division does not change the *number* of *chromosomes*, although the *quantity* of genetic material is halved. Each daughter cell at prophase II had two duplicated chromosomes; each of the four cells after the second division also has two chromosomes, but two single ones. This division is therefore known as the *equational division*. Meiosis is now completed: the nuclear membrane and nucleolus re-form in each cell.

THE PHYSICAL BASIS FOR THE LAW OF SEGREGATION

The separation of homologous chromosomes during the meiotic events through reduction division (pages 35 to 36) is in fact the physical basis for Mendel's law of segregation. Alleles, or genes which determine a particular trait, exist in pairs because they are located on a pair of homologous chromosomes at the same site or locus. Because the homologs are always separated into different germ cells during meiosis, alleles must therefore also be segregated from each other.

It is necessary to include in our definition of alleles the fact that they are on homologous chromosomes, because we now know of many traits that are determined by more than one pair of genes frequently located on nonhomologous chromosomes.

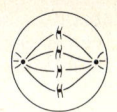

It may now be easier for you to understand why only two alleles can exist in an individual at any one locus, even in multiple allele systems. Homologous chromosomes in sexually reproducing organisms can exist only in pairs, one having been inherited from the mother, and one from the father.

THE PHYSICAL BASIS FOR THE LAW OF RANDOM ASSORTMENT

The physical basis for the law of random assortment can be easily understood if we assign genes to two pairs of chromosomes in a cell undergoing meiosis. Suppose we are following a cell from a pea plant which is dihybrid for genes determining length of stem and pod color (Figure 2.18).

The stage of meiosis which is important for an understanding of random assortment is metaphase I. In cells of a dihybrid, the two pairs of chromosomes may line up in two different ways, either of which will occur with equal frequency among germ cells undergoing maturation. Thus *T* and *G* may be on one side of the equator in one cell, and *t* and *g* on the other side; *or* the combinations may be *T* and *g* on one side and *t* and *G* on the other. With the completion of meiosis, therefore, the dihybrid produces four genetically different gametes in equal numbers, namely, *TG, Tg, tG, tg,* a completely random assortment of genes.

A trihybrid, an individual hybrid for three pairs of genes, would by the same token have eight possible combinations of genes as a result of the four possible random alignments during metaphase I (Figure 2.19).

Knowledge of the meiotic process then allows us to generalize that for *n* hybrid pairs of genes, there are 2^n possible combinations of genes in the gametes (Table 2.1).

Consider that humans have 23 pairs of chromosomes, each of which is believed to contain hundreds if not thousands of gene pairs. If an individual is heterozygous for only one pair of genes on each pair of chromosomes, the genetic combinations that could be found in his gametes is 2^{23}, or over 8 million! It is, therefore, not at all surprising that aside from identical sets of children, no two human beings are genetically the same.

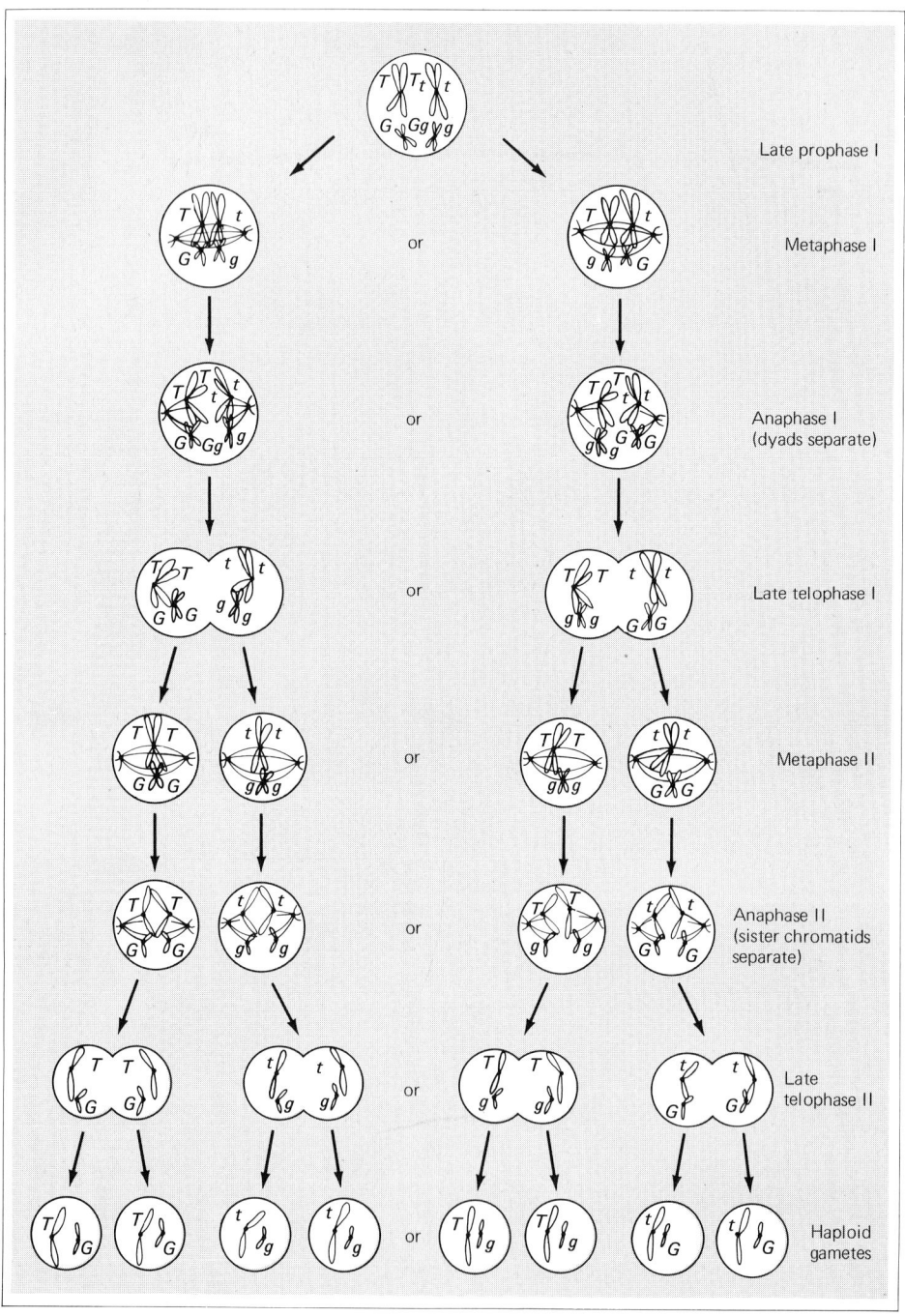

FIGURE 2.18

Meiosis of germ cells from a dihybrid, *TtGg*, illustrating the physical basis for the law of random assortment. For clarity, some of the stages following reduction division have been omitted.

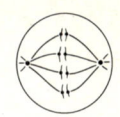

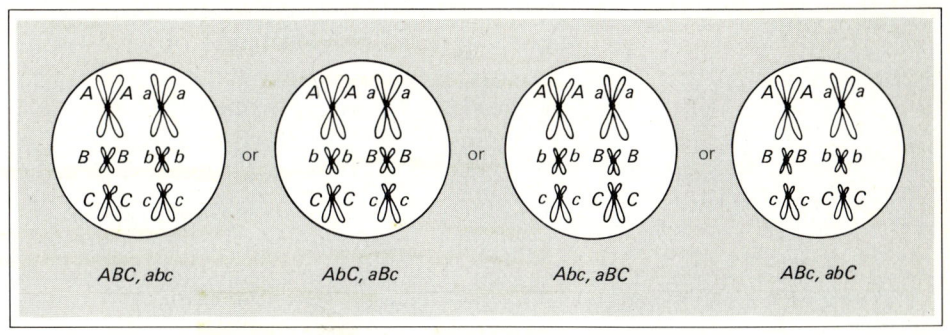

[Figure 2.19 diagram showing four circles with chromosome alignments:]

ABC, abc | AbC, aBc | Abc, aBC | ABc, abC

FIGURE 2.19
Diagrammatic illustration of the possible kinds of chromosome alignment in metaphase I of a cell of a trihybrid, and the resulting gene combinations in the gametes.

An important point which should be brought out here is that all humans must possess those loci determining traits which classify us as members of our species: loci for erect stature, grasping hands, and so forth. Each of us is unique, however, because of the different combinations of different alleles inherited from our parents.

One of the results of sexual reproduction is a tremendous variability of offspring. Later, when we discuss evolution, the importance of genetic variability will be explained. For now, just be aware that each of us is indeed genetically unique, although we share the common bond of membership in the same species.

It may be of interest to note here that *spermatid*s, the male germ cells that have completed meiosis, are not yet functional gametes. A number of complex changes of cell struc-

THE COMPLETION OF GAMETOGENESIS

PAIRS OF HYBRID GENES	NO. OF POSSIBLE GENETIC COMBINATIONS IN GAMETES
1	2
2	2^2
3	2^3
n	2^n

TABLE 2.1
The relationship between the number of hybrid gene pairs and the number of different gametes resulting from meiosis

The Physical Basis of Inheritance

tures must still take place, such as the loss of almost all the cytoplasm and the formation of a tail, or *flagellum*, that is capable of movement (Figure 2.20). Only after these changes occur can the sperm participate in reproduction.

The human egg undergoes the same divisions and reduction of chromosome numbers as does the sperm (Figure 2.21). One difference, however, is that divisions of the cells are not equal, due to the formation of the spindle near the edge of the cell. The first division results in a very small cell called a *polar body*, and a large one, the *egg*.

The second division is again unequal, and so the result of *oogenesis*, or the maturation of the egg, is one functional gamete and three nonfunctional polar bodies that will eventually disintegrate. Because alignment of chromosomes is random during metaphase I, there is an equal chance that any of the four possible combinations of genes in a dihybrid

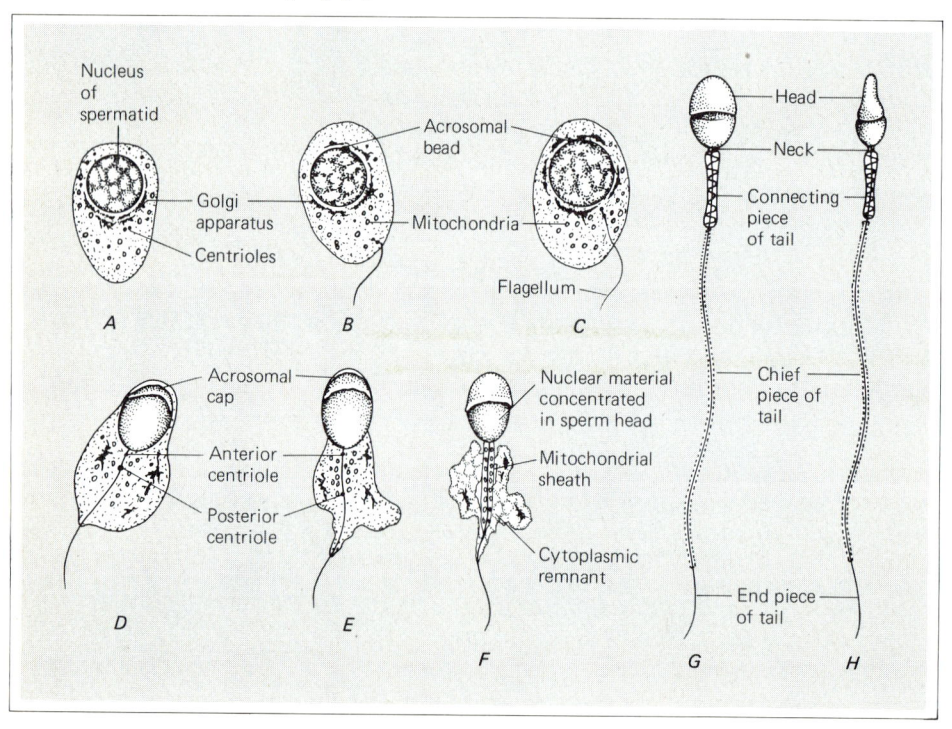

FIGURE 2.20
Stages in the maturation of spermatids. (*A* to *C*) Development of the tail (flagellum) from one of the centrioles and formation of an acrosomal cap from the Golgi body. The acrosomal cap contains enzymes which aid the sperm in reaching the egg cell membrane. (*D* to *F*) Sloughing off of cytoplasm and formation of mitochondrial sheath connecting head and tail. (*G* to *H*) The mature sperm viewed from different angles. [*After B. M. Patten*, Human Embryology, *2d ed. Copyright 1953 by McGraw-Hill, Inc. Used with permission of McGraw-Hill Book Company.*]

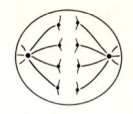

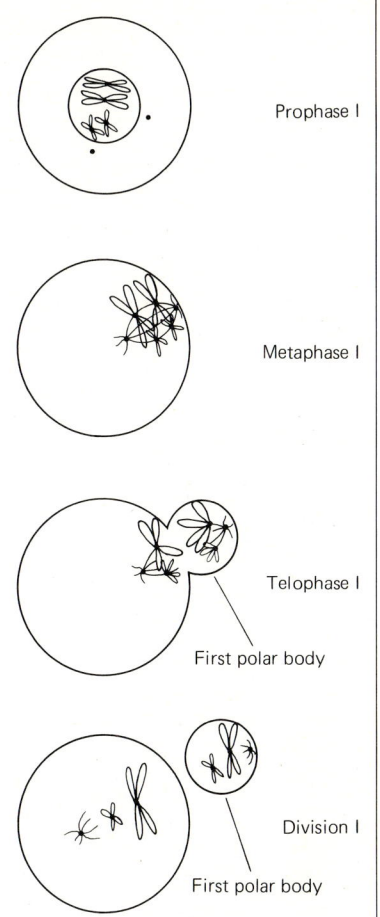

FIGURE 2.21
Reduction division in the maturation of an egg cell with four chromosomes, showing unequal division leading to formation of a polar body.

Prophase I

Metaphase I

Telophase I
First polar body

Division I
First polar body

will be in the functional egg. Consequently, a woman who is a dihybrid may, from four eggs, produce four genetically different gametes, the same as in the male.

One reason for the conservation of cytoplasm in the egg—in direct contrast to the situation in the sperm—seems to be that very important substances stored in the cytoplasm contain essential biochemical information that leads to normal cell division, or mitosis, of the zygote following fertilization. It would appear to be of advantage, therefore, that as much cytoplasm as possible be conserved in the cell to be fertilized.

FIGURE 2.22
A comparison of spermatogenesis and oogenesis. Germ cells in different stages of maturation are designated by special terms such as *primary* and *secondary spermatocytes*; n represents the haploid number, 2n the diploid number.

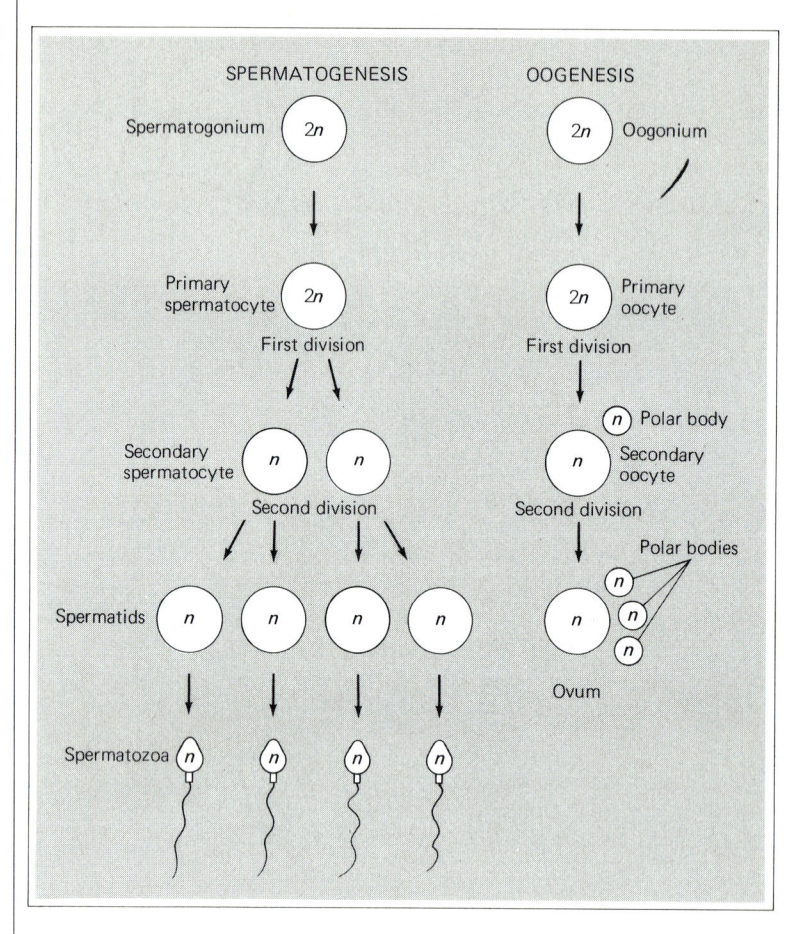

Nature moves not only in mysterious ways, but in practical ones as well! Figure 2.22 diagrams the difference between oogenesis and spermatogenesis.

THE EFFECT OF AGE ON EGG FORMATION

Another important difference between oogenesis and spermatogenesis is that whereas sperms are continuously produced anew from diploid cells arising from tissues of the testes, either to be ejaculated or to degenerate if not used, a human female is born with all the egg cells she will ever have. During her reproductive years, these cells will mature, usually one at a time, reaching maturity on a monthly cycle under hormonal control.

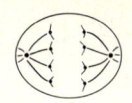

This means that the older a woman is, the older her germ cells are. Experiments on lower forms of life have indicated that age tends to increase the incidence of an aberration of meiosis called *nondisjunction*, which results from the failure of a pair of homologs to separate during the first division, or of sister chromatids in the second division. This leads to the formation of germ cells with too few or too many chromosomes. If fertilized, such eggs would produce either a lethal condition and abortion or an abnormal child.

There is convincing evidence that age results in nondisjunction in human females also. For example, Down's syndrome, or mongoloid idiocy, as it is commonly known, occurs at an overall frequency of 1 in 600 births. In young women, thirty years of age or less, the frequency is roughly 1 in 1400; in women forty-five years of age or older, the frequency seems to be 1 in 50 births.* The most common cause of Down's syndrome is the presence of extra chromosomes, most likely as a result of nondisjunction. We shall discuss the condition in more detail in Chapter 11.

* J. Hamerton, *Human Cytogenetics*, vol. II, Academic Press, New York, 1971, p. 201.

CELL DIVISION IN THE ZYGOTE: MITOSIS AND GENETIC CONTINUITY

Following the intricate maneuvers of meiosis, the act of reproduction is culminated in the union of egg and sperm. Actually, it is an interesting fact that in humans the second meiotic division in the egg does not take place unless there is fertilization.

When ovulated, human eggs are still in the secondary oocyte stage. It is not until a sperm actually penetrates the membrane of the egg cell that a number of cellular changes occur in the egg, including the completion of the second meiotic division. By the time the tiny sperm nucleus migrates through the cytoplasm to reach the egg nucleus, the polar bodies have been extruded, and union of the genetic material from the two gametes to form a single nucleus results in the reconstitution of the diploid number in the zygote.

The single fertilized egg cell then develops into an organism composed of billions of cells by entering into a stage of active mitosis, the kind of division that occurs normally in our bodies millions of times a day for the replacement of cells and tissues or for repair in cases of injury or, as in the embryo, for growth.

For summary and review, Figure 2.23 compares the events of mitosis and meiosis.

The significant differences between mitosis and meiosis are these: In the prophase stage of mitosis there is no synapsis, and so during metaphase the homologs are *not* lying side by side in tetrad formation but are aligned at random along the equator of the spindle. At anaphase, the centromeres split and the sister chromatids separate, and the result in the telophase stage of mitosis is two diploid clusters of chromosomes, identical to each other and to the original cell.

Actually, the equational division in meiosis is the same as mitotic division, except that in the former the cells are already haploid as a result of the first meiotic division and in the latter the cells are diploid.

The occasional production of identical twins is a phenomenon which points out the genetic identity of daughter cells resulting from mitosis. In man, as in many organisms, identical sets of offspring are the result of the separation of two or more embryonic cells into individual cells or clusters which are capable of developing into whole normal beings.

Having been derived from a single zygote, they are said to be *monozygous*, and all evidence indicates that such individuals are genetically identical. The Dionne quintuplets were monozygous, and so we see that separation of early embryonic cells can occur more than once, without affecting the viability of each separated cell.

We do not know very much about what causes such separations in embryos. There may be a familial tendency toward having twins, but as yet no genetic pattern has been established to indicate that predisposition to multiple births is determined by simple Mendelian dominance or recessiveness.

Perhaps this thought has occurred to you: If all cells of an embryo arise from mitosis, they must all be genetically identical. How is it, then, that the newborn baby is composed of not only billions of cells, but a multitude of different cell types: skin, nerve, muscles, and so on?

Apparently it is the activation of certain genes and the deactivation of others in some embryonic cells that starts them on

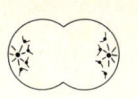

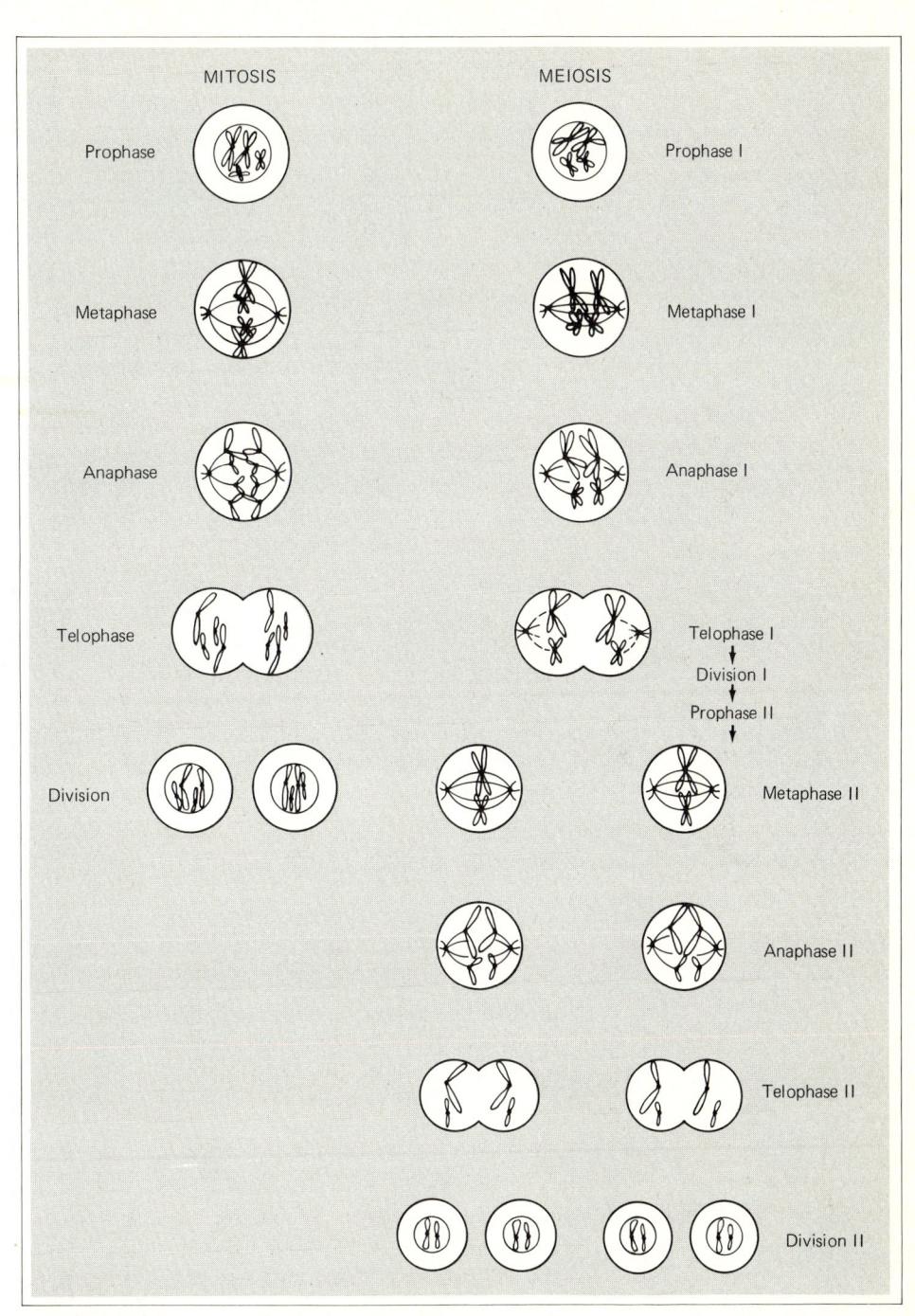

FIGURE 2.23
A comparison of mitosis and meiosis in cells with a diploid number of four.

* This subject is a primary concern of the complex and fascinating area of developmental genetics, which will be discussed in more detail in Chapter 8.

their way to become skin cells, for example, while still other genes in cells perhaps destined to be lung cells will be "turned on and off." What remains the big mystery is the regulation of the turning-on and turning-off process.*

In fact, we know little about the regulation of many cellular processes. In the case of cell division, we are very much in the dark about what factors cause a cell to begin mitosis or to stop it. Certainly such regulatory factors must exist, for no cell or organ or organism grows indefinitely; there is always a limit to size.

Along the same lines, one wonders why only germ cells, of all the diploid cells in the body, undergo meiosis. It is an interesting fact that whereas we can stimulate all kinds of cells to undergo mitosis when we grow them in test tubes and culture dishes outside the body, using techniques of tissue culture, we have never been able to stimulate any of them, even germ cells, to undergo meiosis.

One of the reasons biologists are so keenly interested in cancer research, aside from its medical aspects, is that the cancer cell is characterized by an uncontrolled rate of mitosis because the factors that normally regulate cell division are somehow rendered ineffective. It is a fact of biological research that frequently our understanding of a normal process comes only from the study of some aberration of this process.

PROBLEMS

1. If you were studying cells from an animal whose diploid number is 4, would the cells in the figure on page 49 be indicative of mitosis or meiosis? What stage would each represent?

2. The best way to understand meiosis or mitosis is to draw the stages on a piece of paper. Beginning with the cell of a trihybrid (*AaBbCc*), draw the steps in spermatogenesis that result in the formation of eight genetically different gametes.

REFERENCES

Suggested supplemental reading:

Boveri, T., 1902: "On Multipolar Mitosis as a Means of Analysis of the Cell Nucleus," translated and reprinted in B. H. Willier and J. M. Oppenheimer (eds.), *Foundations of Experimental Embryology*, Prentice-Hall, Inc., Englewood Cliffs, N.J., 1964.

PROBLEM 1

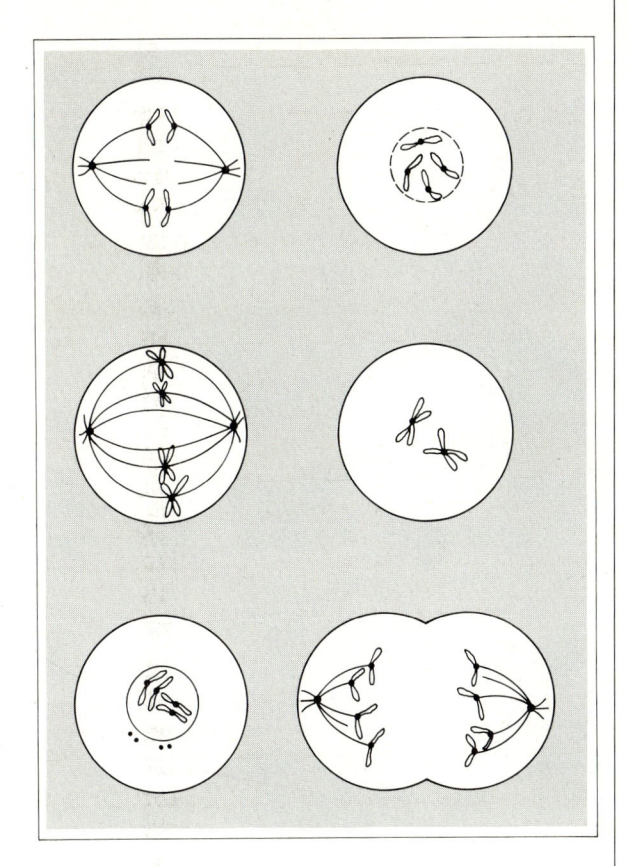

For mitosis and meiosis, any of the general genetics or biology texts mentioned at the end of Chapter 1 would have suitable discussions. In addition:

1965: *The Living Cell, Readings from Scientific American,* W. H. Freeman and Company, San Francisco.

The following two are paperback, part of the EMI Programmed Biology Series, and would be helpful to those who may have difficulties understanding Mendelian inheritance based on the chromosomal changes in cell division:

Parker, G., W. A. Reynolds, and R. Reynolds, 1968: *Mitosis,* Educational Methods, Inc., Chicago.

_____, _____, and _____, 1970: *Heredity,* Educational Methods, Inc., Chicago.

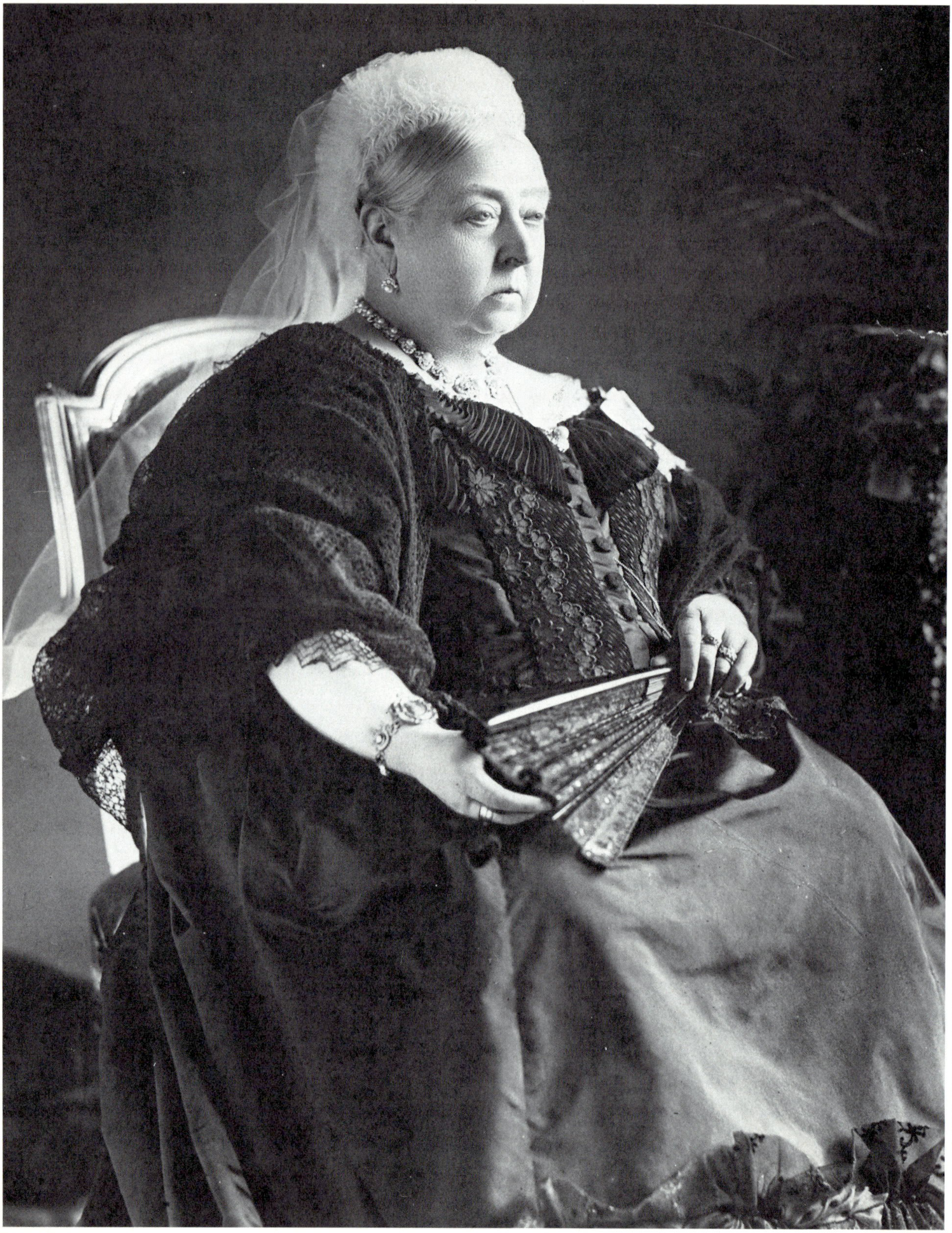

Chapter 3

The Chromosomal Determination of Sex

In Chapter 2 we made mention of the fact that, in all sexually reproducing organisms, chromosomes exist in homologous pairs. Figure 3.1 is a composite picture of all the actual chromosomes found in a cell of a normal human male. Such pictures, known as *karyotypes,* are prepared as follows: Cells grown in tissue culture are stimulated to undergo mitosis. A drug is applied to the cells to stop them in metaphase, when the chromosomes are contracted and duplicated. The cells and their contents are stained and then preserved on a microscope slide. A photograph is taken of the chromosomes under magnification, and the homologs are then cut from the picture and matched; another picture is taken, forming the karyotype.

In many sexually reproducing organisms, including man, one pair of chromosomes found in the male of the species does

THE SEX CHROMOSOMES

Portrait of Queen Victoria, who was a carrier of the X-linked recessive gene for hemophilia (see pp. 58 and 80). [*Radio Times Hulton Picture Library*.]

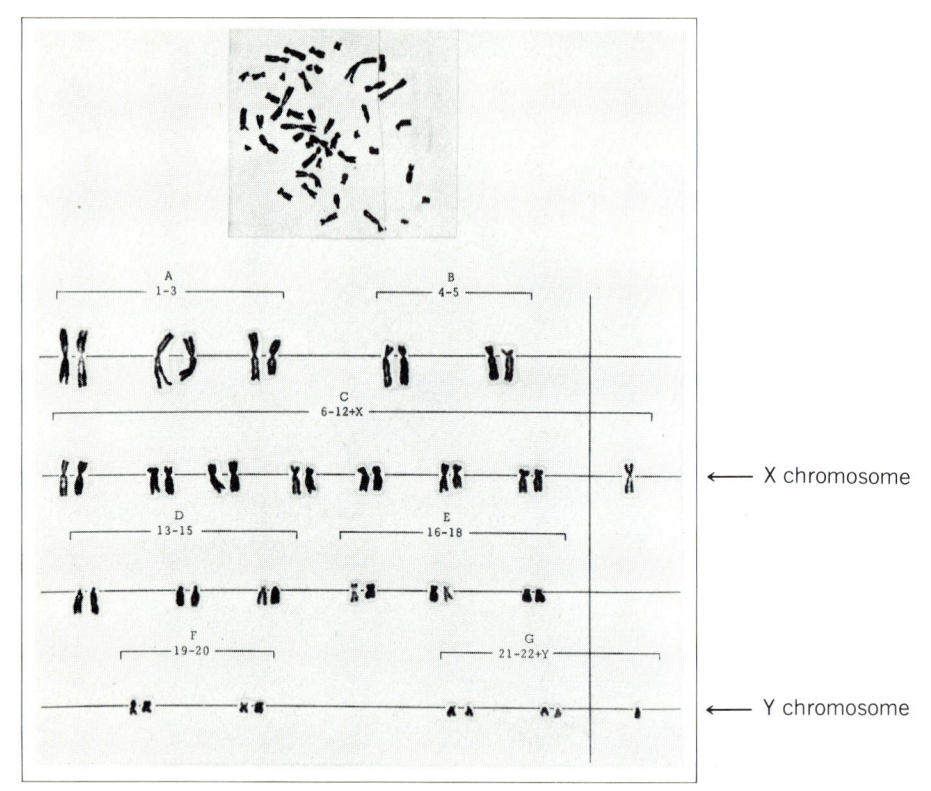

The Chromosomal Determination of Sex

FIGURE 3.1
Karyotype of a normal human male, showing mitotic metaphase chromosomes. [*Courtesy of Emilie Mules and the Chromosome Laboratory, University of Virginia.*]

not fit the description of homologous chromosomes. The members of this unique pair neither look alike nor seem to carry genes that are allelic. But they are of vital importance because the very existence of the species depends on their presence, for they determine the sex of the individual: they are the *sex chromosomes*.

If you examine the karyotype in Figure 3.1 closely, you will note that in the C and G groups there are odd numbers of chromosomes, indicating that one of the chromosomes in each group cannot be matched with a partner. Actually, the C group consists of seven pairs of homologs and one sex chromosome, the *X chromosome,* and the G group consists of two pairs of homologs and another sex chromosome, the *Y chromosome,* which is obviously much smaller than the X.

The chromosomes other than the sex chromosomes are called *autosomes*.

It is the tiny Y chromosome that determines maleness in human beings. When the Y is absent, the individual is female. Figure 3.2 is a karyotype of a normal human female. Note that the C group has eight pairs of chromosomes, including two X chromosomes, and there is no Y chromosome in the G group.

In spite of the fact that they are physically dissimilar, the X and Y chromosomes segregate as a homologous pair during gametogenesis. Following spermatogenesis, half the mature sperm therefore contain the Y chromosome, and half carry the X chromosome. Since the mature ovum, or egg cell, can

FIGURE 3.2
Karyotype of a normal human female. [*Courtesy of Emilie Mules and the Chromosome Laboratory, University of Virginia.*]

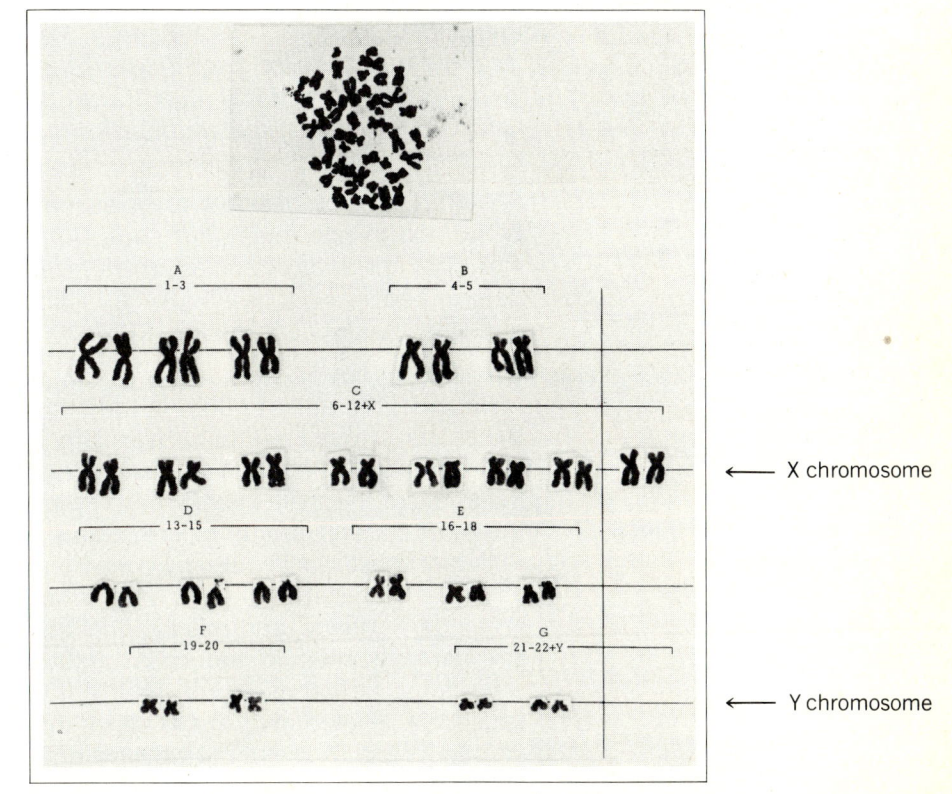

have only the X chromosome, the sperm which fertilizes an egg will determine the sex of the child conceived. The Y-bearing sperm determines that the child will be a boy, and the X-bearing sperm determines that the child will be a girl. This means that the father is the parent whose germ cells determine the sex of the children.

(It may be of interest to mention here that a common misconception held in some cultures is the idea that the mother determines the sex of the offspring. In a strongly male-oriented society, such as the China of decades ago, the worth of a wife was frequently measured in terms of the sons she was able to present her husband. And, to remain Empress, the wife of the Shah of Iran must give birth to sons, the absence of whom is regarded as a failure on her part. Poor Empress!) One active area of medical research involves the attempt to distinguish X-bearing sperm from Y-bearing sperm. There have been reports of the effects of an alkaline or an acidic uterine environment favoring one or the other, and of the differential migration of X- or Y-bearing sperm in an electrical field. However, to date, no practical differentiation method has been developed that would allow parents their choice of son or daughter.

Should it ever be possible to do so, one great benefit would be to help in controlling our population explosion. It would eliminate the large families which result when couples are determined to produce a boy or girl following a number of offspring of the opposite sex.

We still do not know how the Y chromosome determines maleness. We do know that in the human embryo, there are two sets of structures that appear early in the development of every embryo. One of the sets eventually differentiates into the ducts of the male reproductive system, and the other into the oviducts and uterus of the female reproductive system. Which of these ducts actually continues to develop while the other set degenerates depends on the development of the embryonic gonads, and this is apparently determined by the sex chromosomes in the cells of the embryo. Therefore there must be genes which the Y chromosome carries that cause the development of the testes (or represses the development of the ovaries), but we have not been able to identify individual genes and their role in sex determination.

GENETICS OF THE Y CHROMOSOME

Except for its capacity to determine maleness in higher animals, the Y chromosome has been thought to be genetically inactive. This is supported mainly by two lines of evidence. The only way we have of determining that there exists a gene which controls a particular trait is when that gene changes in some way so as to cause a change in phenotype from the normal, or what is commonly known as a *mutation*. Thus far, only one condition studied—"hairy ears" (Figure 3.3) in men of a family in India—shows signs of being determined by a Y-linked (sometimes called *holandric*) gene, and the evidence is not conclusive.

A second line of evidence that the Y chromosome is genetically "inert" stems from the apparent absence of genes that are allelic to, or that interact with, genes on the X chromosome. In other words, every gene on the X chromosome of a male is expressed, whether it is dominant or recessive. Males are therefore neither homozygous nor heterozygous for X-linked genes: they are *hemizygous*.

Recently, however, evidence has been found to indicate that the short arm of the Y chromosome may contain genes allelic to those on the short arm of the X chromosome, i.e., that the short arm of the Y may be homologous to the short arm of the X. This evidence comes from cytological studies indicating that the short arms of the X and Y chromosomes occasionally synapse during meiosis.*

In addition, Ferguson-Smith in 1965 published studies of patients suffering from a condition known as gonadal dysgenesis, or Turner's syndrome. Such patients, as we shall discuss in detail in Chapter 11, usually possess only one X chromosome in their karyotypes, and no Y chromosome. The 11 individuals whom Ferguson-Smith studied, however, appeared to have a normal X chromosome, but a severely abnormal Y chromosome missing most of the short arm. The fact that these patients manifested very similar abnormalities to those lacking an X chromosome suggests that there may be homology between the short arm of the Y chromosome and part of the X.

These constitute about the only pieces of evidence indicating that the Y and X chromosomes are in part homologous. It must be reiterated, however, that none of the genes known to

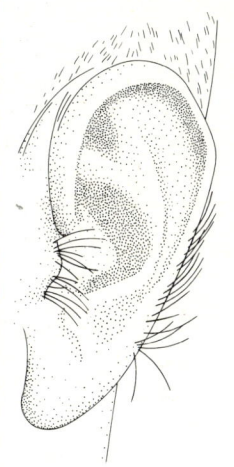

FIGURE 3.3
Hairy ears, a trait which may be determined by a Y-linked gene.

* See V. McKusick, *On the X Chromosome in Man*, American Institute of Biological Sciences, 1964, Fig. 27.

be on the X chromosome have been found on the Y chromosome, and all X-linked genes in males are expressed.

X-LINKED TRAITS

In women, on the other hand, there are genes identified as being on the X chromosome which we shall refer to as *X-linked,* and which were referred to in the past as *sex-linked.* Since there is a true pair of homologous sex chromosomes in human females, the Mendelian law of dominance and recessiveness applies to traits determined by such genes in women in the same manner as to traits determined by genes on the autosomes (or what we call *autosomal traits*).

For this reason, X-linked traits are expressed much more frequently in men than in women. When a man carries a recessive mutation on his X chromosome, such as a gene for color blindness, he will express the trait; that is, he will be color-blind. A woman can carry the same recessive gene on one of her X chromosomes, but because she is likely to have a normal dominant allele on her other X chromosome, she will be phenotypically normal (Figure 3.4).

Since the sex chromosomes segregate during meiosis in the same manner as autosomes, we can solve problems involving X-linked inheritance in the same way as we did for inheritance of autosomal characteristics in the garden pea plants. For example, consider a woman who is heterozygous for the recessive gene determining color blindness. Her genotype is *Cc,* and her phenotype is normal. Suppose that her husband also has normal vision. Since we know that he has only one X chromosome, his normal phenotype indicates that his genotype must be *CY*. The *Y* symbol indicates that there is no second allele for color blindness. (Occasionally geneticists designate the genotypes involving X-linked traits by $X^C X^c$ for the woman in this example, and $X^C Y$ for the male. For simplicity, we shall omit the *X* in our designation of X-linked genes, but you may come across the alternative designation in other books.) What incidence of color blindness can the couple in our example expect in their offspring?

The Punnett square that depicts this situation is shown in the margin. As you can see, none of the daughters would be af-

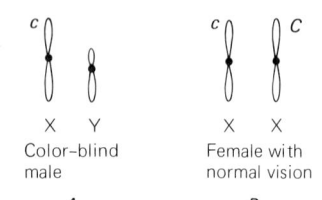

FIGURE 3.4
Sex chromosomes in (*A*) a human male and (*B*) a human female, illustrating cytological reasons for differences in the frequency of expression of X-linked recessive genes.

	C	Y
C	CC	CY
c	Cc	cY

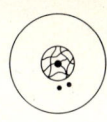

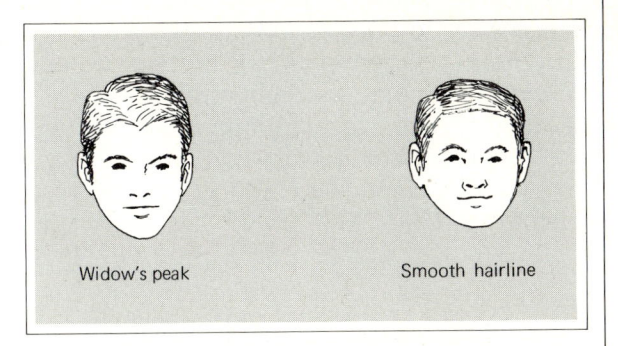

FIGURE 3.5
A widow's-peak hairline, inherited as an autosomal dominant trait, and a smooth hairline, determined by the recessive allele to the gene for the widow's peak.

fected, although half of them would be heterozygous. (The heterozygotes are also known as *carriers*.) Half the sons could be expected to be color-blind and the other half could be perfectly normal. It is possible, of course, for females to be afflicted by color blindness or any other recessive X-linked trait, if they are homozygous for the recessive gene. For this to occur, however, the mother would have to be at least a carrier, and the father must be affected. As most X-linked mutations are in fact recessive, and fairly rare, this is an unlikely situation.

We can also study combinations of autosomal and X-linked traits, and predict phenotypic and genotypic ratios on the basis of Mendel's law of random assortment, just as we analyzed combinations of traits in Mendel's experiments. Suppose that the woman in the above example is also heterozygous for the dominant autosomal gene determining a widow's-peak hairline, while the man is homozygous recessive for a smooth hairline. (See Figure 3.5 for illustration of this simple Mendelian trait in man.) The genotypes of the man and woman in this example would be, respectively, *CYww* and *CcWw*. What phenotypic ratios would be expected among their offspring? The Punnett square for this cross is shown below.

	CW	Cw	cW	cw
Cw	CCWw	CCww	CcWw	Ccww
Yw	CYWw	CYww	cYWw	cYww

The answer to the question is that all the daughters would have normal vision, and half of them would have a smooth hairline, while the other half would have a widow's peak. As for the sons, the expected ratio would be 1 with normal vision and widow's peak:1 with normal vision and smooth hairline:1 color-blind with widow's peak:1 color-blind with smooth hairline.

Note that since the man is homozygous for the *w* gene, we can reduce the size of the Punnett square by including only one of each of the two types of gametes he can produce, rather than including each twice to make a "square" Punnett square. This simplification does not change the ratios of the offspring, as it is analogous to reducing $\frac{2}{4}$ to $\frac{1}{2}$.

Another well-known X-linked trait in man is hemophilia, the so-called bleeder's disease, in which the clotting mechanism in the affected person's blood is defective. Although hemophilia and color blindness are determined by genes on the X chromosome, the traits obviously are not related to sex determination. In fact, all the characteristics that have been found to be determined by X-linked genes are unrelated to the determination of sexual characteristics.

That there must be genes on the X chromosome that influence the development of female characteristics, however, can be seen in individuals known to have three sex chromosomes, XXY, in their cells. Such persons, who are otherwise phenotypically male because of the presence of the Y chromosome, suffer from Klinefelter's syndrome (see Figure 11.7), which includes among other conditions a tendency toward breast development, perhaps from hormonal imbalance. We shall discuss abnormalities associated with aberrant sex chromosome numbers in greater detail in a later chapter. However, just as in the case of the Y chromosome, we have not yet identified individual genes on the X chromosome which are important in the normal development of the female.

SEX-LIMITED TRAITS

There do exist some traits found predominantly in men and not in women, such as baldness, that may superficially resemble Y linkage in their transmission pattern. These traits, however, are actually determined by autosomal genes, but their expression is influenced by sex hormones. The gene for baldness, for example, behaves as an autosomal dominant in males but an autosomal recessive in females. Such traits are said to be *sex-limited*.

SEX DETERMINATION IN OTHER ORGANISMS

Not all sexually reproducing organisms have the same system of sex determination as man. In wasps, for example, the state of haploidy not only is viable but also results in maleness when an egg develops *parthenogenetically,* that is, without being fertilized. Eggs that are fertilized, and are diploid, develop into females. In birds and reptiles, males have a pair of like sex chromosomes (*ZZ*), and females have unlike sex chromosomes (*ZW*). In many other higher animals, however, the determination of sex is the same as in humans; that is, the presence of the Y chromosome results in maleness.

WHY ONLY TWO SEXES?

There is another very interesting aspect of sex determination that most people who are not biologists perhaps have never even contemplated, and one about which we can only wonder. It is that throughout the natural world, enormous variety in every biological process can be found in the myriad species of living organisms that have arisen through evolutionary change, save one: all sexually reproducing species, whether plant or animal, have evolved only two different sexes.

The only known exceptions to this statement are viruses, some unicellular forms of life such as algae, and simple multicellular organisms which do not show any detectable gross physical differences between cells that unite in a simple form of sexual reproduction (Figure 3.6). All other forms of life further up the evolutionary ladder show distinct structural differences that would separate male from female. The principles underlying evolutionary change will be discussed later, but for now just consider that there seems to be no a priori reason why more than two sexes in higher organisms could not have evolved. Yet it has never happened.

The Chromosomal Determination of Sex

FIGURE 3.6
Sexual reproduction in some simple forms of life: (A) a unicellular alga, *Chlamydomonas*, and (B) a filamentous multicellular alga, *Spirogyra*. In both species, physically similar cells undergo fusion in a form of fertilization. The diploid zygotes undergo meiosis to give rise to haploid cells. In lower forms of life such as these, the free-living stage is haploid. The diploid condition is a short phase of the life cycle.

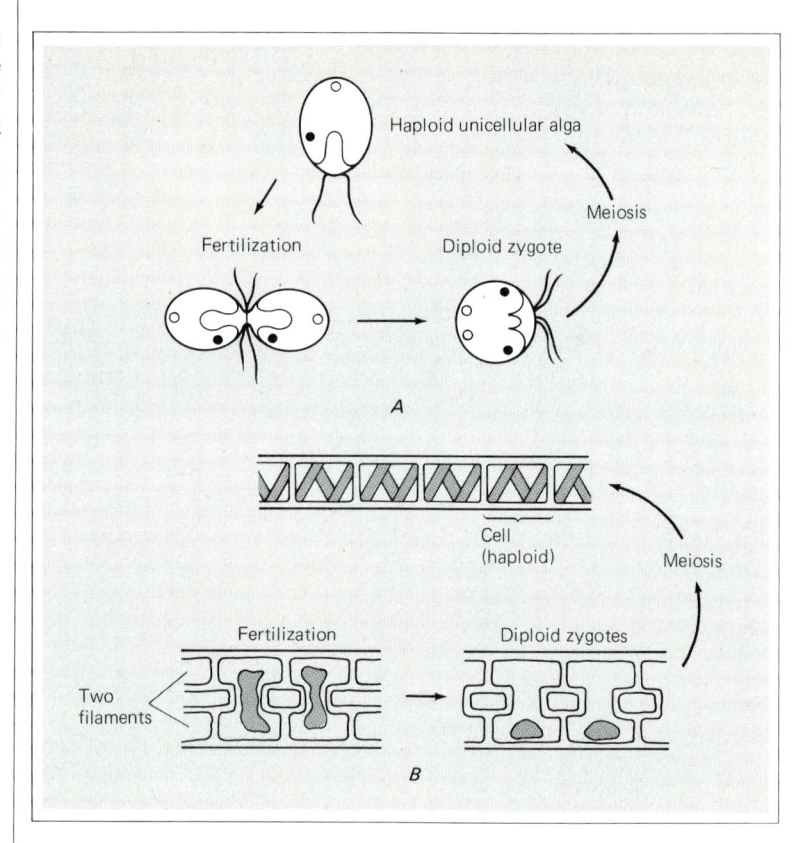

PROBLEMS

1. A man known to be a victim of hemophilia, a blood disease caused by an X-linked recessive gene *h*, marries a normal woman whose father was also known to be a "bleeder." (a) What proportion of their sons may be expected to be bleeders? (b) What proportion of all their children may be expected to be bleeders?

	h	Y
H	Hh	HY
h	hh	hY

   SOLUTION *The genotype of the man must be* hY. *His wife, having had to inherit her father's X chromosome carrying the h gene, must be* Hh. *The Punnett square of this cross is shown in the margin.*

   ANSWER (a) *Half the sons will be bleeders.*
   (b) *Half the children will be bleeders.*

2. If the man in problem 1 were normal, how would this change your answers to the two questions?

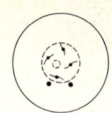

3. What phenotypes would be expected in the progeny of a cross between a color-blind woman and a normal man? (Color blindness is determined by an X-linked recessive gene.)

4. Suppose that a man is color-blind (c = color-blind) and has blond hair (m = blond hair). His wife, whose father was blond and color-blind, has normal vision (C = normal vision) and dark hair (M = dark hair). (a) What proportion of their children will have the same phenotype as the man? (b) What proportion of their children will have the genotype *MmCc*?

SOLUTION Problem 4b asks for a specific genotype. The man must be *cYmm*; the woman is *CcMm*. A *cY* × *Cc* cross would result in the Punnett square shown in the margin. One-fourth of the offspring will be *Cc*; half will be *Mm*.

	c	Y
C	CC	CY
c	Cc	cY

ANSWER (a) See Appendix A.
(b) $\frac{1}{4} \times \frac{1}{2} = \frac{1}{8}$.

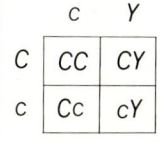

REFERENCES

A technical paper:
Ferguson-Smith, M. A., 1965: "Karyotype-Phenotype Correlations in Gonadal Dysgenesis and Their Bearing on the Pathogenesis of Malformations," *Journal of Medical Genetics*, **2**:142–155.

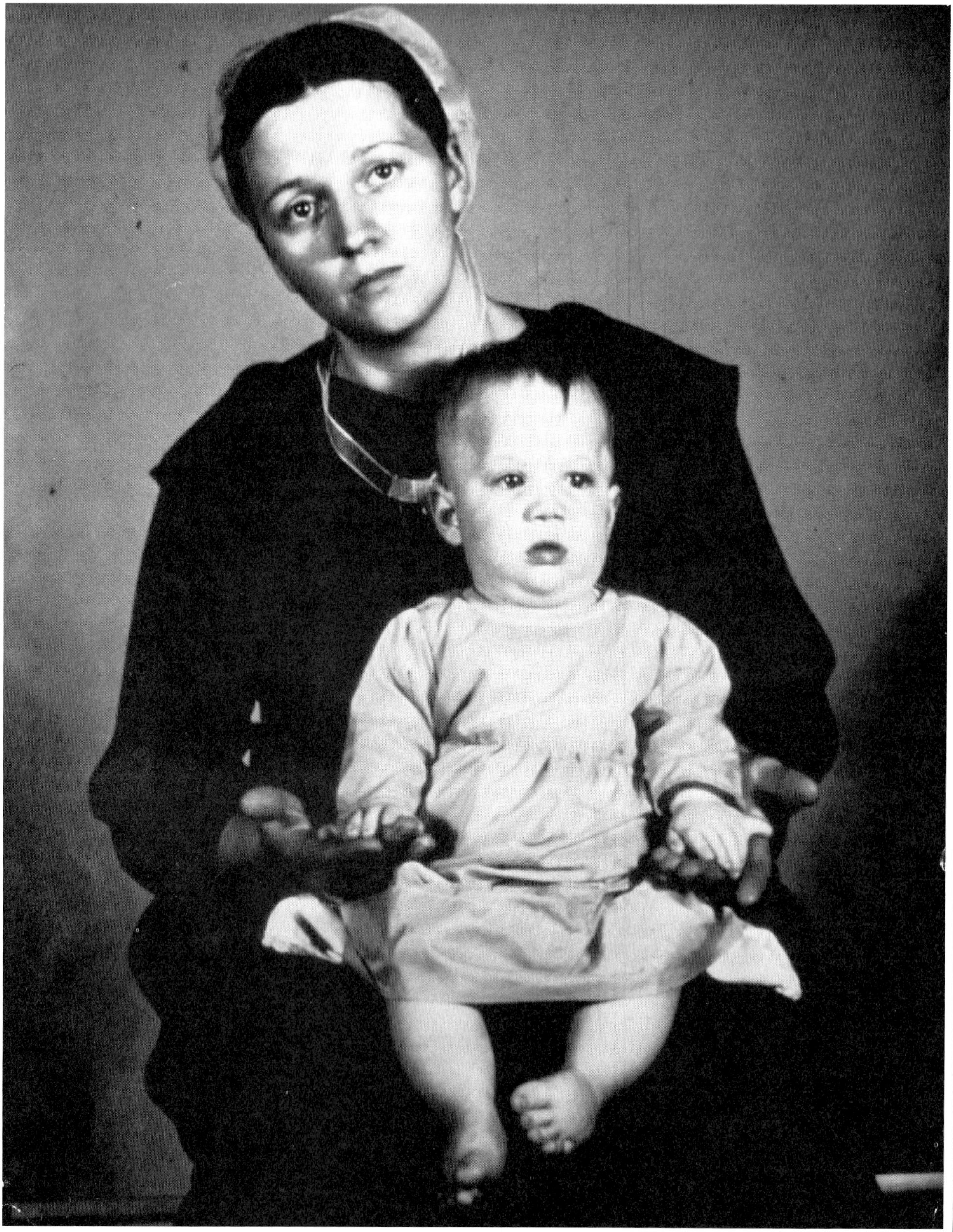

Chapter 4

Mendelian Heredity in Man

The term *laws* is assigned to Mendelian principles because they have been found to apply universally to all living organisms including man. Man is, after all, a biological organism, and it is therefore not surprising that we have been able to use inherited conditions in man as illustrations for some of the basic genetic principles in the preceding chapters.

Characteristics which appear to be determined by single pairs of alleles, such as the ones we have already discussed, are referred to as simple Mendelian traits. There are a number of such normal traits in man, as shown in Figure 4.1.

You may have noted that most of the traits described in Figure 4.1 are rather trivial. This is because, unfortunately for geneticists, some of the normal human characteristics which we consider most interesting, such as intelligence and temperament, are not simple traits, and cannot yet be assigned

SOME MENDELIAN TRAITS IN HUMANS

DIFFICULTIES IN THE ANALYSIS OF HUMAN TRAITS

Amish child with six-fingered dwarfism (the Ellis–van Creveld syndrome). [*From V. McKusick*, Human Genetics, *2d ed., Prentice-Hall, Englewood Cliffs, N.J., 1969, p. 139.*]

FIGURE 4.1
Some simple Mendelian traits found in man.

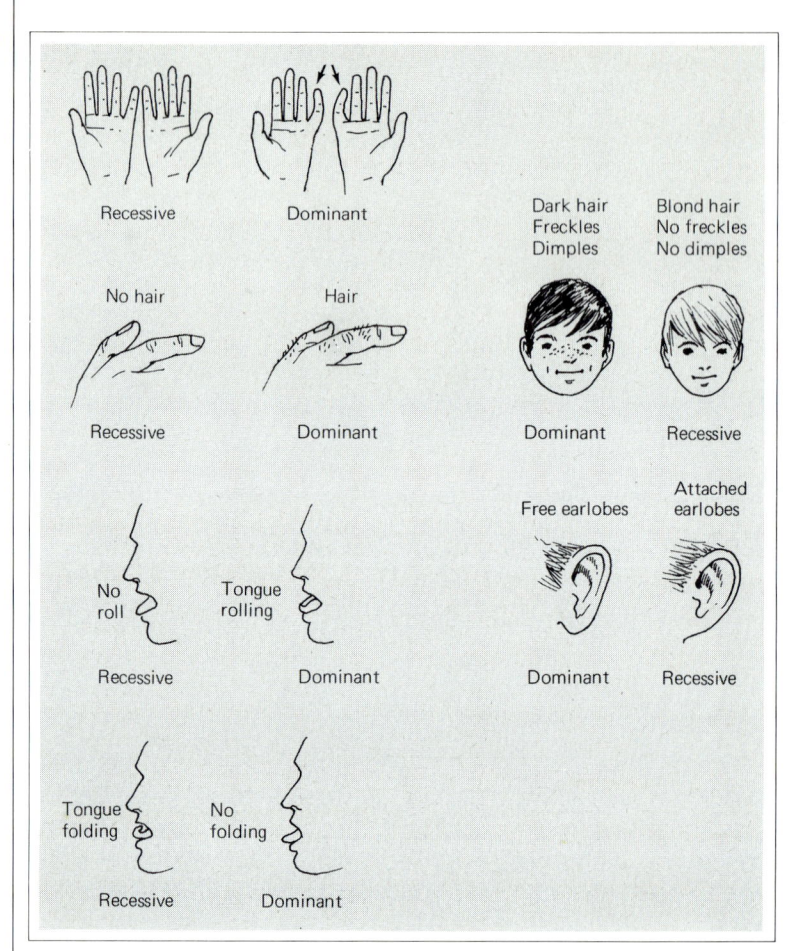

any specific gene or numbers of genes. In short, the genetic basis for such complex traits is not clear. In Chapter 5 we shall describe some of the known genetic phenomena which render genetic analysis difficult.

In addition, although man as a biological organism is subject to the laws of nature, our unique establishment of morals and codes of ethics does cause us to treat our fellow humans differently from other living organisms, at least in the analysis of the inheritance of characteristics! We cannot follow all the criteria which Mendel used in his experiments with the garden pea. For example, we cannot control and design experimental crosses that would facilitate our analysis of a particular trait. We also have problems in establishing patterns of inheritance in a family in which a condition that may be of interest appears. Recall that in order to establish any pattern, large numbers of progeny and at least three generations are needed.

Although human beings have in the past reproduced to the point of overpopulating our fragile planet, the relatively small numbers of offspring per generation in our families is a hindrance from the standpoint of genetic analysis. In addition, as geneticists themselves are obviously human beings with limited life spans, it is somewhat impractical to wait for three generations or more to appear (Figure 4.2)!

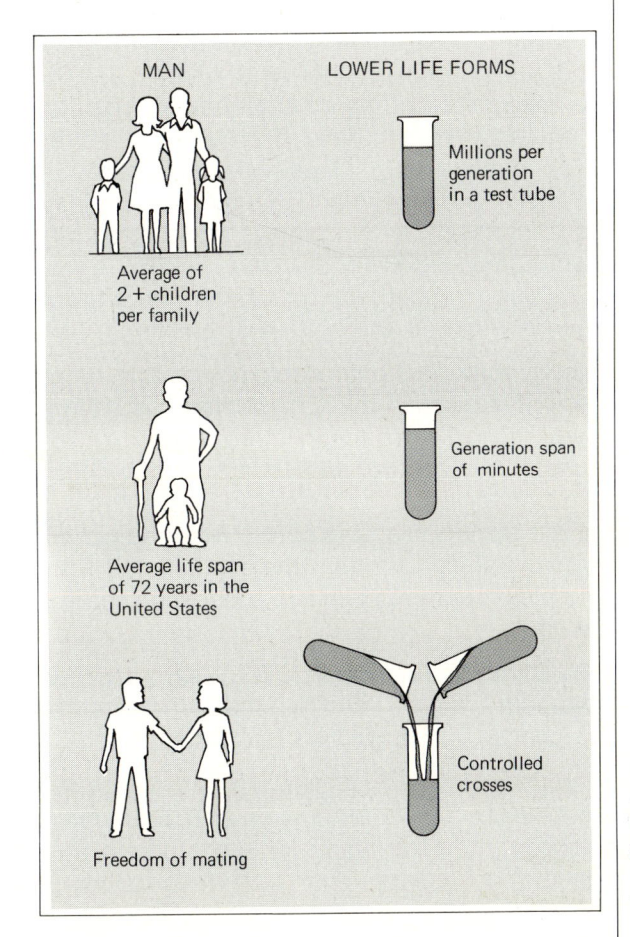

FIGURE 4.2
A schematic illustration of some of the disadvantages facing geneticists studying heredity in man, as compared with simpler forms of life such as bacteria.

66
Mendelian Heredity in Man

There is nothing we can—or would—do about making experimental crosses, but there *is* something we can do about establishing patterns of transmission of traits. Since we cannot go forward, that is, wait for future generations, we go backward: we gather information about all existing members of the family under study and as much information as possible about previous generations, and draw up what is referred to as a *pedigree chart,* a family tree of sorts. In some families, very extensive pedigrees have been compiled.

For the geneticist, however, information about the phenotype, in addition to names of the individuals, is essential if a pedigree is to be useful. The incidence of a particular condition in the pedigree will often indicate whether the gene involved is autosomal or sex-linked, dominant or recessive.

THE PEDIGREE CHART

The pedigree is the tool most widely used for the study and representation of the inheritance of human traits, and certain standard symbols have been established by geneticists in their publications. Traditionally, females are depicted by circles or by the symbol ♀; males are depicted by squares or by the symbol ♂. A marriage is indicated by a horizontal bar connecting a circle and square, and the symbols for offspring are shown suspended from a line drawn perpendicular to the marriage bar, as in Figure 4.3.

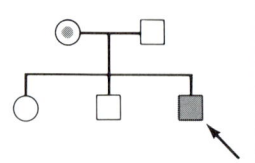

FIGURE 4.3
A simple pedigree chart of a family illustrating some of the traditional symbols used by geneticists. Explanation is in the text.

Family members expressing the trait under study are usually indicated by solid black symbols, as is one of the sons in the pedigree in Figure 4.3. Heterozygotes or carriers of the gene customarily are designated by a black dot in the middle of the symbol, as is the mother in Figure 4.3, or by coloring half of the symbol black: ◑ or ◨. If an individual is deceased, a cross is placed through the symbol, and if the phenotype of the individual is uncertain, a question mark is placed in the symbol.

Twins are indicated by ⌒ if identical, and ⋀ if nonidentical. If there are many offspring in a family, numbers are placed in the symbols to conserve space and simplify the pedigree chart. For example, ⑤ ④ would indicate that there were five normal girls and four normal boys in this *sibship* (the term used for the brothers and sisters of one generation). Occasionally, an arrow pointing at a particular affected individual, as in Figure 4.3, is added to indicate that this is the person who brought the trait to the geneticist's attention. Such a person is called the *proband,* or propositus (if male; proposita if female).

Sometimes you see pedigrees that are headed by a single person, thus: ○ □. This does not mean that the parent reproduced parthenogenetically, but only that the mate is normal and believed to be of no consequence to the analysis of the pattern of transmission.

These are the usual symbols used in making up pedigrees. Occasionally, one finds variations in the charts drawn by different geneticists, but they are usually explained by an accompanying key of some sort.

AUTOSOMAL DOMINANCE Let us turn our attention now to pedigrees which illustrate different kinds of inheritance. For the most part, we shall be using pedigrees drawn up in the study of inherited diseases rather than normal traits, simply because there are more well-documented pedigrees of diseases and abnormal conditions; however, an autosomal dominant gene, regardless of the trait it determines, normal or abnormal, will show the pattern of transmission characteristic of all autosomal dominant genes.

In Figure 4.4, a sample pedigree of a family in which an autosomal dominant trait was transmitted to many in the

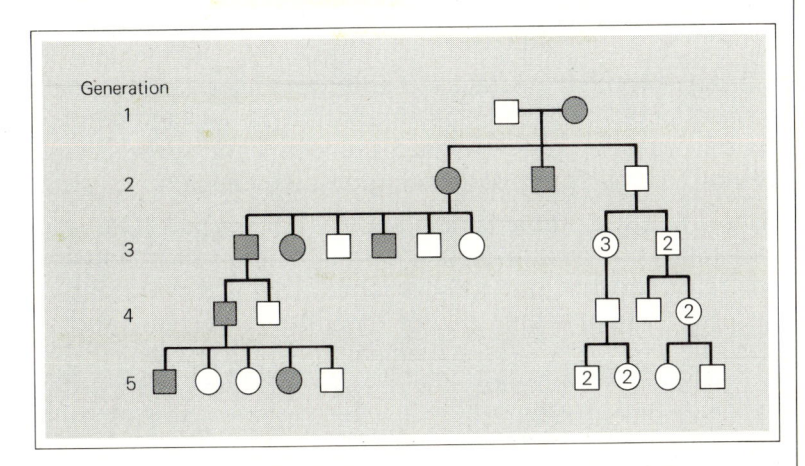

THE ANALYSIS OF
PEDIGREE CHARTS

FIGURE 4.4
Sample pedigree illustrating the pattern of an autosomal dominant trait.

kindred (a term sometimes used in place of *family*), certain characteristics of the pattern of transmission would lead a geneticist to conclude that the trait is determined by an autosomal gene. First, equal numbers of women and men are affected. Second, the affected man in the third generation passed the trait to his sons. As we discussed earlier, the Y chromosome is the only sex chromosome which a man transmits to his sons. Thus, the father could not have passed the condition to his sons if it were determined by an X-linked gene, i.e., one that would be located on the man's X chromosome.

Along the same lines, the affected male in the fourth generation produced an unaffected daughter in the fifth generation. This again would be impossible short of a mutation if the gene were an X-linked dominant, since he would have passed the gene to all his daughters, who would all be affected. An X-linked recessive would not be expressed in any daughters, since we must assume from the pedigree that the man's wife must have been normal, and she would have contributed a dominant normal allele to mask any recessive gene on the man's X chromosome.

This last point holds true for an autosomal gene as well as an X-linked one. If the gene were an autosomal recessive, there is no way that the girls in the fifth generation can be affected if the mother is homozygous normal. Therefore, if the gene is neither X-linked nor recessive, it must be autosomal dominant.

Additional evidence that the gene is dominant lies in the fact that there is no skipping of generations in the expression of the trait, because the trait is expressed whenever the gene is present. Another characteristic of dominant traits is that every affected individual has an affected parent. In order for an offspring to have inherited the dominant gene from a parent, that parent must therefore also be affected.

A third factor characterizing dominant traits is that when one generation does not express the trait, the trait is lost and will not reappear in future generations. This is so, of course, because the unaffected people must be homozygous recessive and would not have a dominant allele to transmit. The right side of the pedigree for the second, third, fourth, and fifth generations illustrates this principle.

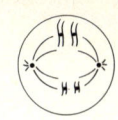

AUTOSOMAL RECESSIVENESS Figure 4.5 shows a pedigree of a family in which an inherited condition was found in a number of individuals in the fourth generation. The presence of affected daughters whose fathers and mothers are normal again indicates that the gene is not sex-linked but autosomal. It could not be an X-linked dominant because the parents are normal, and for the gene to be X-linked recessive would require that the paternal X chromosome carry the gene, which would only be possible if the father were affected.

In addition, the rarity of the condition, and the fact that two different sibships of affected individuals were involved in which the parents are normal, argues strongly against the possibility that this may be an autosomal dominant trait. Because heterozygotes are always phenotypically determined by their dominant genes, the resulting masking of recessive genes leads to "skipping of generations" in the expression of the condition under study. This is one characteristic of the pattern of transmission that distinguishes recessive genes.

Furthermore, the pedigree shows that the affected son and his affected sister in the fourth generation are the offspring of two related individuals, specifically first cousins. It is not surprising to find consanguineous marriages (marriages of related persons) when dealing with a recessive trait.

If a recessive trait is fairly rare, as are most abnormal conditions, the chances of two people being heterozygous for it are much greater if they are related than if they are not, since the two related people may have inherited the same gene from a common ancestor. The pedigree in Figure 4.5 is somewhat unusual from this point of view because the father of the

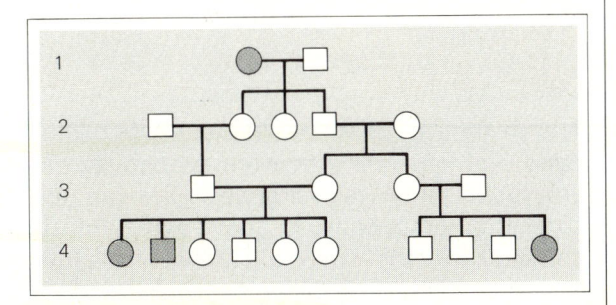

FIGURE 4.5
Sample pedigree illustrating the pattern of inheritance of an autosomal recessive trait.

GENETIC DISADVANTAGES OF MARRIAGE BETWEEN RELATED INDIVIDUALS

other sibship in the fourth generation was apparently unrelated to his wife, but by chance was a carrier for the same recessive gene.

To illustrate why it is more likely for two related persons to be heterozygous for a rare gene, let us use recessive albinism as an example. The frequency of heterozygotes for the condition in the general population is around $\frac{1}{140}$. The marriage of two heterozygotes by chance would be the same as two independent events occurring simultaneously, so using the law of probability discussed in an earlier chapter, the probability of two unrelated heterozygous people marrying is $\frac{1}{140} \times \frac{1}{140} = \frac{1}{19,600}$. The chance that they will have a homozygous recessive child would be $\frac{1}{19,600} \times \frac{1}{4} = \frac{1}{78,400}$ which you will agree is mighty unlikely!

On the other hand, what is the probability that two first cousins would produce an affected child, if a common ancestor is known to be heterozygous for the recessive gene for albinism? Figure 4.6 is a simplified pedigree of the marriage of first cousins. The woman in the parental generation is assumed to be heterozygous for a recessive gene. The probability that her offspring each inherited the recessive allele is $\frac{1}{2}$ (in other words, they could have inherited one of two alleles from the heterozygous mother, either the dominant or the recessive gene). The probability that the individuals in the third generation received the recessive allele is $\frac{1}{2} \times \frac{1}{2}$ or $\frac{1}{4}$. In this pedigree, the chance that both of the first cousins who marry are heterozygotes is, therefore, $\frac{1}{4} \times \frac{1}{4} = \frac{1}{16}$, that they would produce an albino child, $\frac{1}{4} \times \frac{1}{4} \times \frac{1}{4} = \frac{1}{64}$. This is clearly more probable than the chance of two unrelated individuals producing an affected child, which we previously calculated to be $\frac{1}{78,400}$.

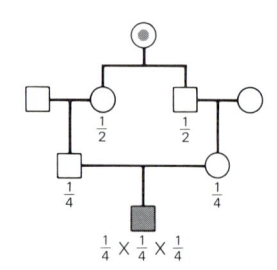

FIGURE 4.6
Pedigree illustrating the increased probability of homozygosity in offspring of consanguineous marriages, in this case, of first cousins.

THE COEFFICIENT OF CONSANGUINITY

The degree of closeness of the relationship between a husband and wife will affect the probability of their being heterozygous for the same gene. One yardstick used to measure this probability is the *coefficient of consanguinity*, defined by Victor McKusick, one of our leading human geneticists, as the "probability that any single locus is homozygous by descent from a common ancestor."* Table 4.1 is

*V. McKusick, *Human Genetics*, 2d ed., Prentice-Hall, Inc., Englewood Cliffs, N.J., 1969, p. 146.

TABLE 4.1
Coefficients of consanguinity (F) for various parental relationships

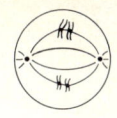

PARENTAL RELATIONSHIP	F
Father-daughter	$\frac{1}{4}$
Offspring of identical twins	$\frac{1}{8}$
Uncle-niece	$\frac{1}{8}$
Double first cousins	$\frac{1}{8}$
First cousins	$\frac{1}{16}$
Half first cousins	$\frac{1}{32}$
First cousins once removed	$\frac{1}{32}$
Second cousins	$\frac{1}{64}$
Second cousins once removed	$\frac{1}{128}$
Third cousins	$\frac{1}{256}$

Source: V. McKusick, "Human Genetics," 2d ed., Prentice-Hall, Inc., Englewood Cliffs, N.J., 1969, p. 146.

a compilation of the coefficients of consanguinity for various parental relationships.

Because of our moral codes, there are very few cases in which very close relationships such as father-daughter exist on record. It is interesting to speculate why most cultures frown on incest and even marriages between cousins, and have done so since long before genetic studies established a scientific reason for avoiding these relationships. It may very well be that the restrictions against consanguineous marriages stem from observation over the centuries that related marriages tend to result in abnormal offspring more frequently than unrelated marriages. However, this is merely conjecture, and the prohibitions may be due to cultural rather than practical considerations.

A number of studies compiling data from related and unrelated marriages show that there is almost always an increase in the incidence of mortality or disease in offspring of related individuals (see Table 4.2).

TABLE 4.2
Data showing increased frequency in mortality among offspring of consanguineous marriages in Hiroshima, Japan

RELATIONSHIP OF PARENTS	DEATHS BETWEEN 1 AND 8 YEARS*		
	n	d	p
First cousins	326	15	0.0460
1½ cousins	101	3	0.0297
Second cousins	139	3	0.0218
Unrelated	544	8	0.0147

* n = number of births, d = number dead, p = percentage of dead.

Source: W. J. Schull, "Empirical Risks in Consanguineous Marriages: Sex Ratio, Malformation and Viability," *The American Journal of Human Genetics*, **10**:294–343, 1958.

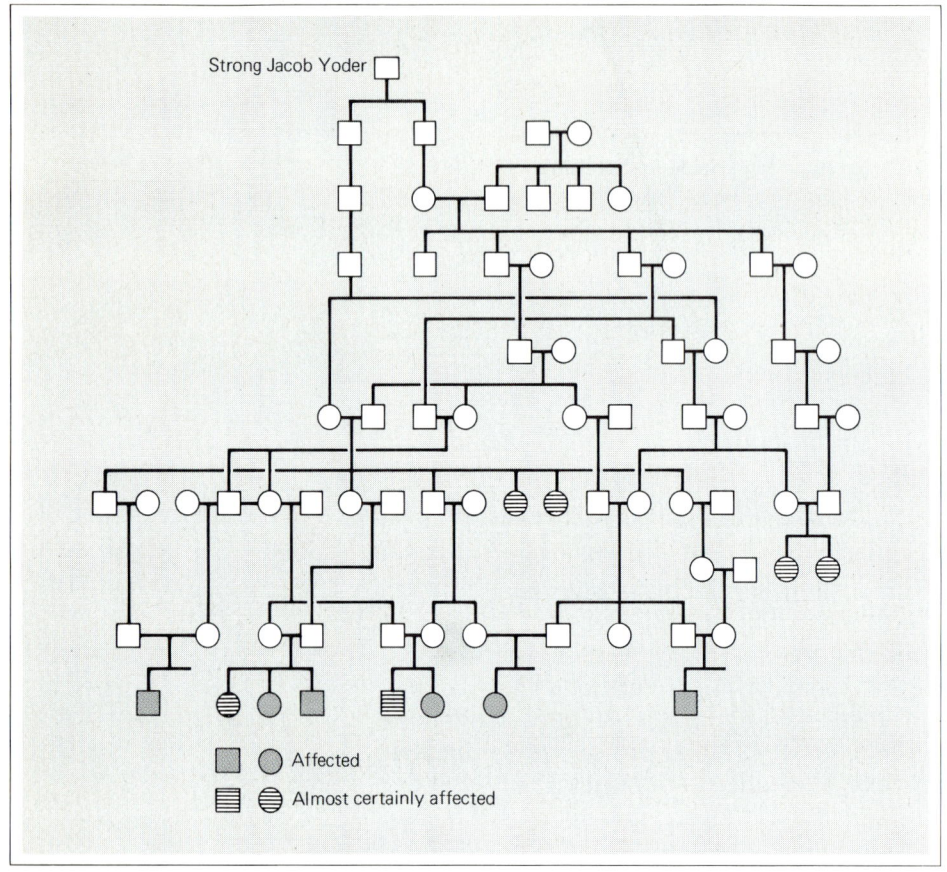

FIGURE 4.7
Pedigree showing many instances of consanguineous marriages in an Amish family with a high frequency of a rare anemia (pyruvatekinase-deficient hemolytic anemia), inherited as an autosomal recessive trait. [After V. McKusick et al., "The Distribution of Certain Genes in the Old Order Amish," Cold Spring Harbor Symposium of Quantitative Biology, **29**:104, 1964.]

Communities such as the Amish, who are prohibited from free interaction with society in general because of religious restrictions, show a larger number of genetic defects in their young than the general population. In a study by McKusick and his co-workers, four genetic disorders that are quite rare in the general population were found in relatively high frequency in the Amish populations living in Ohio, Pennsylvania, and Indiana. They are Ellis–van Creveld syndrome (a type of dwarfism, see introductory illustration), pyruvate-kinase deficient hemolytic anemia, hemophilia B, and limb-girdle muscular dystrophy.

Fortunately for human geneticists, the Amish keep very accurate family, or genealogic, records. Figure 4.7 is a detailed pedigree of the anemia mentioned above, which was found in the Amish community in Mifflin County, Pennsylvania. As you can see, the pedigree shows a good number of consanguineous marriages, and all the affected sibships can be traced to one man, Strong Jacob Yoder, who immigrated from Europe in 1742. Presumably, either he or his wife carried the gene for the anemia.

It is, of course, possible for two related persons to have a large number of perfectly normal progeny, since we are still dealing with probability, not certainty, when discussing the coefficient of consanguinity. One historical illustration of this was the ancient Egyptian practice of having pharaohs engage in brother-sister matings. This was done without any recorded incidence of inherited abnormalities in the offspring (Figure 4.8).

INBREEDING VS. HYBRID VIGOR

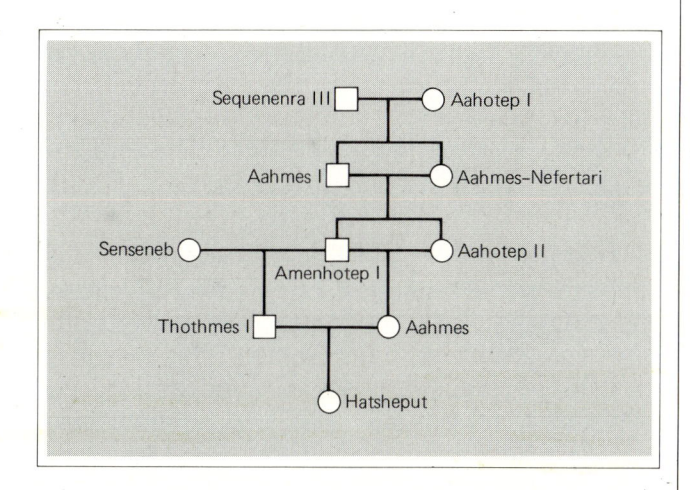

The increase in homozygosity as a result of related matings is used to advantage in domestic animals and in the breeding of plants, and in this way geneticists have made considerable contributions toward the establishment of economically advantageous stocks of plants and animals. In fact, the

FIGURE 4.8
Marriages between sibs and half sibs in the Eighteenth Dynasty of Egypt, ca. 1580–1350 B.C. [*After J. A. F. Roberts,* An Introduction to Medical Genetics, *Oxford University Press, London, 1940.*]

1970 Nobel Peace Prize was awarded to Dr. Norman Borlaug for developing strains of disease-resistant wheat that have significantly reduced food shortages in many areas of the world.

The controlled breeding of related individuals possessing certain desirable phenotypes in order to create strains which produce genetically homogeneous offspring through the resulting homozygosity at a number of genetic loci is known as *inbreeding,* and the strains thus produced are called *inbred strains.*

It might be pointed out here that much research depends on the existence of inbred strains, which offer the advantage of uniformity of traits. The effects of experimental procedures on such animals and plants may be compared to other individuals of the same strains which serve as controls.

In addition, spontaneous mutations may be discerned in inbred strains which are of interest to geneticists because they frequently result in conditions similar to diseases in man, such as muscular dystrophy and diabetes mellitus in mice. Research on such animals may one day elucidate the nature of these diseases in man. Figure 4.9 shows representative mice of different inbred strains under study at the Jackson Memorial Laboratory in Bar Harbor, Maine.

It is also possible to control the breeding of plants and animals so that there is homozygosity for only some selected loci and not for the entire genotype. In fact, all breeders of animals and plants know that they must control their inbred lines carefully, for in many cases, too much homozygosity leads to a weakening and possible dying out of the strain due to the expression of deleterious recessive genes.

It follows that if this is so, then increased heterozygosity should lead to stronger, though heterogeneous, strains, and this is true. The phenomenon of greater viability from the opposite of inbreeding, which is called *out-crossing,* is well known to geneticists, and is frequently referred to as *hybrid vigor,* or *heterosis.* Presumably this results from the masking of recessive genes by dominant alleles, which implies that dominant genes tend to be more favorable than recessive ones. Why this may be so will be clearer when we discuss natural selection and evolution later.

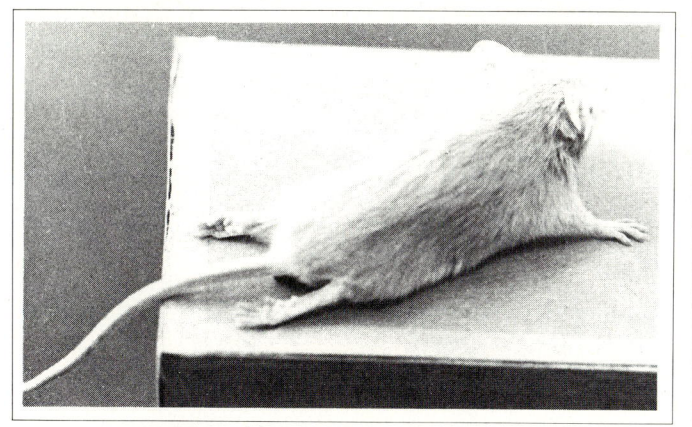

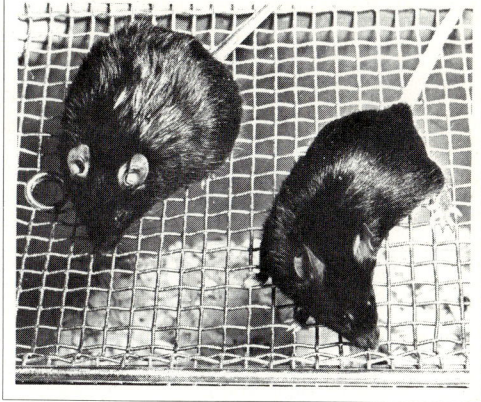

FIGURE 4.9
Photographs of some mutant mice. (*A*) A mouse suffering from muscular dystrophy. (*B*) Normal mouse on the *right;* diabetic mouse on the *left*. (*C*) Hairless mouse. (*D*) Obese mouse. (*E*) Mouse with loop tail. All of these conditions are inherited. It is hoped that studies of those that resemble human diseases will lead to an understanding of and possible cures for the conditions. [*Courtesy of the Jackson Memorial Laboratory.*]

One can put these principles to use in everyday life, for example, in the choice of a dog or cat as a family pet. Dog lovers know that some of the nicest, healthiest, most intelligent dogs are mongrels. They tend to be more even-tempered than many purebred strains that have been inbred for many generations; however, when two mongrels breed, there is no way of knowing what the offspring will be! The advantage of purebred strains, of course, is that the offspring can be guaranteed to be of certain size, shape, etc., barring unusual events such as mutation.

Figure 4.10 includes photographs of an experimental cross between a purebred basenji male and a female cocker spaniel. The F_1 generation do not resemble either parental strain, although the coloring is that of the mother and the hair texture that of the father. The F_2 generation correspondingly exhibits a wide variation in phenotypes. On the other hand, backcrosses of the F_1 to either parental strain result in offspring strongly resembling the parental strain.

FIGURE 4.10
These photographs illustrate the effects of hybridization between two breeds of dogs. (*A*) In the parental generation, a purebred Basenji male (left) was mated to a cocker spaniel female (right). The F_1 consisted of a male and female which did not really resemble either parent, although the coloring was that of the cocker's and the hair texture that of the Basenji's. (*B*) (*facing page*) When the F_1's were backcrossed to the parents, the offspring bore strong resemblance to the parental types.

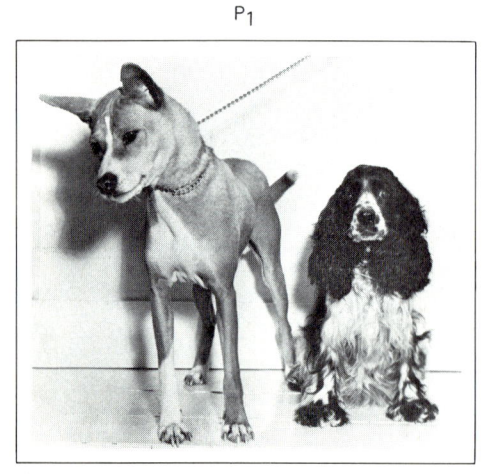

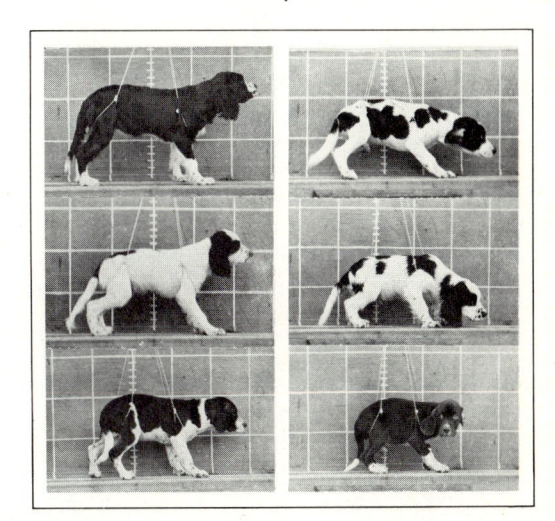

Backcross of F_1 ♀ to Basenji ♂

Backcross of F_1 ♂ to Cocker ♀

B

78
Mendelian Heredity in Man

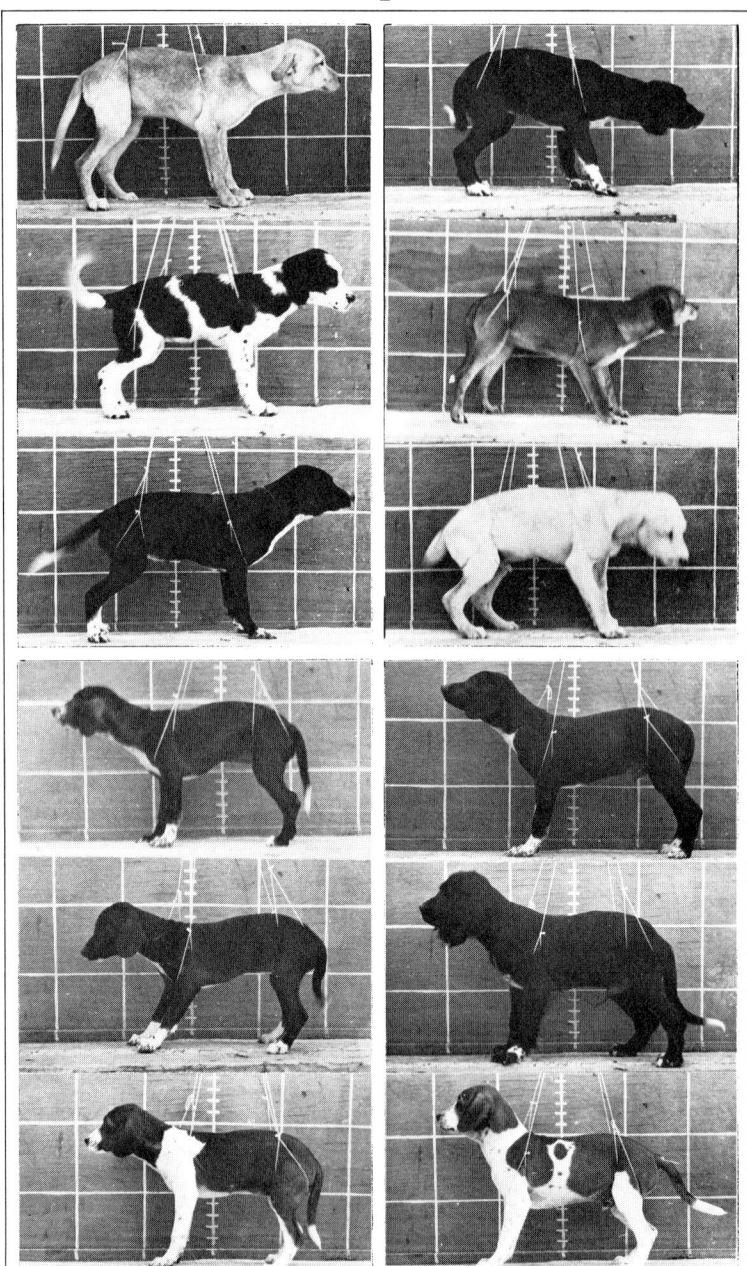

F_2

FIGURE 4.10 (cont.)
(C) In contrast to the offspring resulting from the backcrosses, the F_2 generation which resulted when the F_1 hybrids were crossed to each other shows a wide variety of coloration and body structure. This is due to the fact that hybrids, when crossed with each other, produce more varieties of genotypes in the offspring than when hybrids are backcrossed to homozygotes. Thus, using one pair of genes illustrate this: $Aa \times Aa \to AA, 2Aa, aa$ (3 genotypes); $Aa \times AA \to AA, Aa$ and $Aa \times aa \to Aa, aa$ (only 2 genotypes). (D) A summary of the breeding experiments described above. [*Courtesy of Dr. J. Paul Scott, Bowling Green State U., Bowling Green, Ohio.*]

C

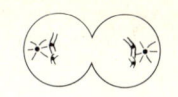

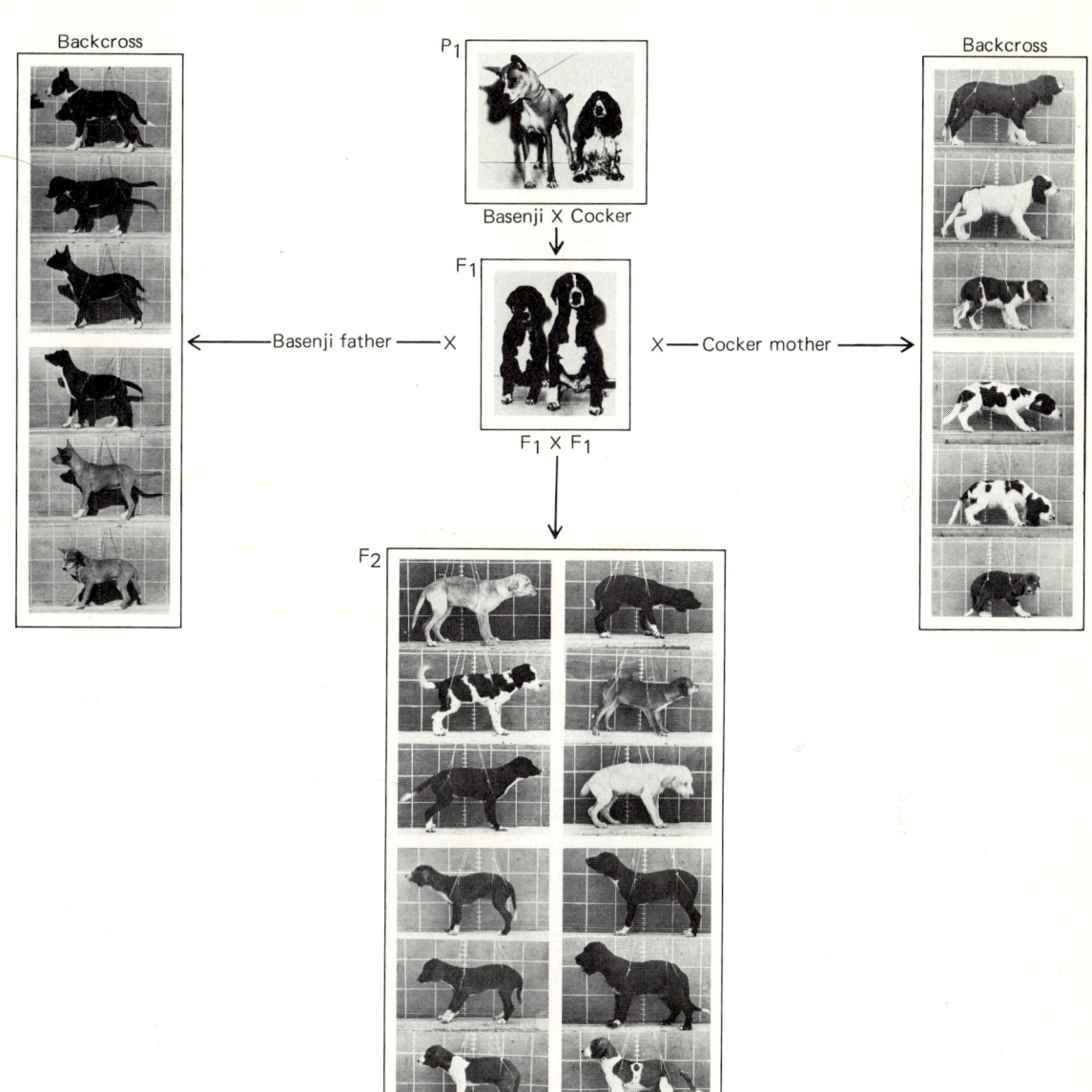

Backcross

P₁
Basenji X Cocker

F₁
F₁ X F₁

Backcross

← Basenji father — X

X — Cocker mother →

F₂

D

The genetic basis for these results can be explained in the following manner. If we assume the two parents to be homozygous for different alleles determining a particular trait, the F_1 would be hybrids. A cross of $F_1 \times F_1$ to produce an F_2 generation yields three different genotypes (just as in our classical Mendelian monohybrid cross, Tt × Tt →TT, Tt, tt). On the other hand, a backcross of the F_1 to a homozygous parent yields only two genotypes: the hybrid, and the homozygote like the parent, for example, Tt × tt → Tt, tt. There would therefore be more of a tendency for the offspring of a backcross to resemble the parental strain than for the F_2 to resemble the parental strain.

ANALYSIS OF PEDIGREE CHARTS X-LINKED RECESSIVENESS

To return to our discussion of the analysis of human traits, Figure 4.11 is a pedigree illustrating a classic example of X-linked inheritance, that of hemophilia in the royal families of Europe. That the condition is X-linked is immediately indicated by the fact that only males are affected.

Dominance of the trait can be dismissed because all the affected males had normal parents. The absence of any affected females strongly suggests that the gene is not autosomal recessive. The pattern fits beautifully what we would expect of an X-linked recessive, in which carrier females are unaffected but transmit their gene to half their sons, who are then affected. From the pedigree, it appears that the original carrier was Queen Victoria, who was conceived by a germ cell carrying a mutation for hemophilia. Of course, there is no way of determining whether the mutation originated in her paternal or maternal gamete.

Incidentally, since we were speaking of consanguineous marriages earlier, it might interest you to note that Queen Elizabeth II and Prince Philip of England are third cousins whose great-grandparents were siblings, Edward VII and Alice of Hesse, respectively. Fortunately for the British royal house, the gene for hemophilia was not transmitted to Edward, and although Alice was a carrier, her daughter, who was Philip's grandmother, apparently did not inherit the gene. Elizabeth, Philip, Margaret, and their children can therefore rest secure in the knowledge that they cannot transmit the disease to future generations!

FIGURE 4.11

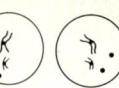

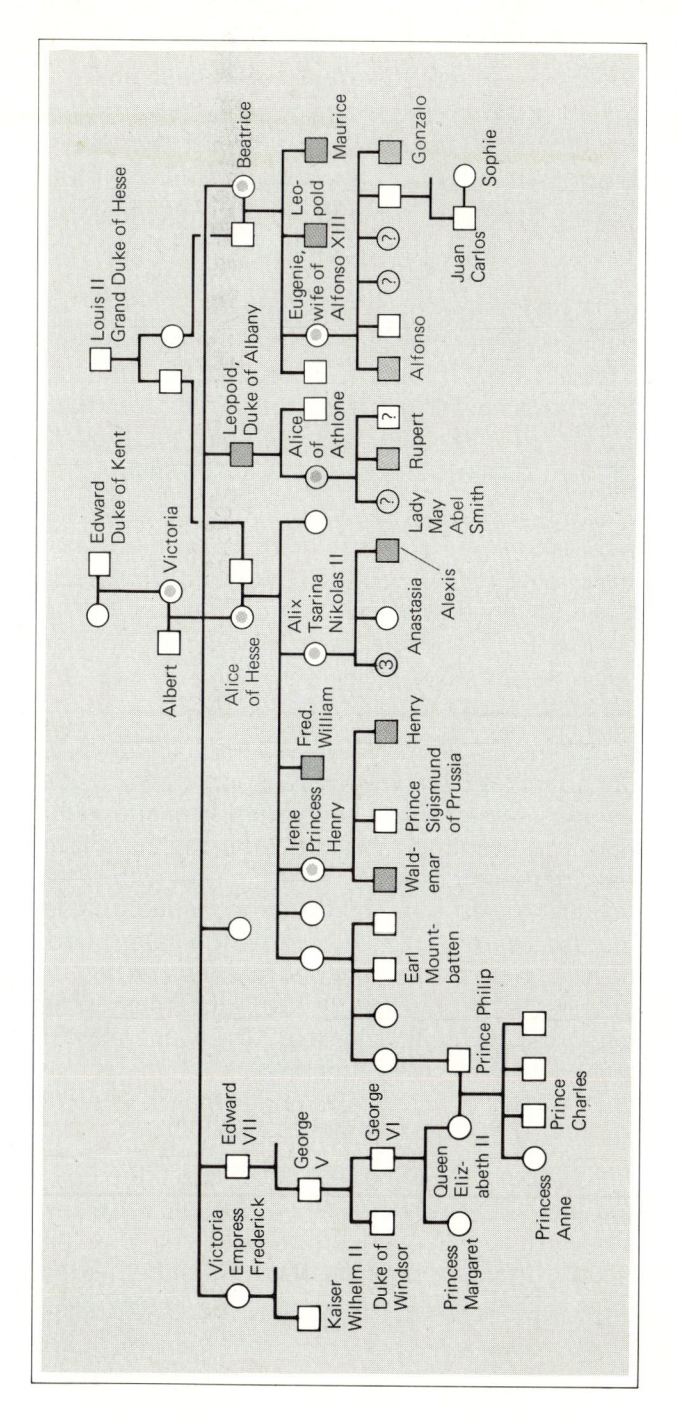

Pedigree chart of the royal families of Europe, and the transmission of hemophilia, a sex-linked recessive trait. [*After V. McKusick, Human Genetics, 2d ed., 1969, p. 56. Adapted by permission of Prentice-Hall, Inc., Englewood Cliffs, N.J.*]

GENETIC COUNSELING

The importance of pedigree charts is not limited to determining the genetic nature of a condition solely for pedagogical purposes. The practical application of the analysis of pedigrees is in the area of genetic counseling.

When an unusual condition appears in a family, especially if it is a condition detrimental to the individual, one of the first questions that must be answered is whether it is inherited, for if it is, there would then be the possibility of other offspring inheriting the same condition. From our preceding discussion, you may have gained the impression that this is a relatively easy thing to do: simply draw up a pedigree chart and by using the process of elimination as we did, analyze the pattern of incidence and predict the probability that the conditions may be expected to occur again.

In some cases this is possible, and if the pattern clearly indicates the type of gene determining the disease, the geneticist would then advise the parents of the probability of further incidence and let them decide whether to have more children. This decision is always an agonizing one, and becomes proportionally more difficult with the degree of seriousness of the trait under study.

Huntington's chorea is an example. This is an autosomal dominant trait, and involves a progressive deterioration of the nervous system which causes physical and mental degeneration and eventually death. What makes this inherited disease difficult for the counselor to deal with is that it does not usually express itself until the individual reaches his reproductive years. Thus, a man whose father had the disease must live with the knowledge that he has a 50 percent chance of having the disease, and must wrestle with the dilemma of whether or not to have children. For if he should prove to have inherited the dominant gene, then his children would have to live under the same burden that must be a most difficult one to bear.

But suppose the abnormal condition is the one incidence in the family? In a situation like this, it verges on the impossible to make any genetic analysis. For one thing, without a pattern, there is no way to determine whether the condition is in fact inherited at all. Geneticists are very aware that conditions which *appear* to be the same as one known to be inherited may in fact be due to some trauma suffered by the developing fetus which resulted in the abnormality. We refer to such noninherited conditions which mimic an inherited one as *phenocopy*.

Figure 4.12 illustrates this by showing two individuals with similar deformities, the absence of normal extremities; however, the individual in Figure 4.12A is homozygous for a rare recessive gene, phocomelia, and the individual in Figure 4.12B is a victim of an environmental factor encountered during gestation: he is one of the thousands of "thalidomide babies" born a decade ago.

You may recall that at that time in many parts of Europe babies were born with deformed or absent limbs. The cause was eventually traced to a new sleeping pill given pregnant mothers during the first trimester of pregnancy, when metabolic changes frequently cause sleeplessness. Although it had been tested without effect on pregnant mice, the drug, which in other ways was superior to barbiturates (it left no hangover in the morning, and does not cause death by overdose), in some way interfered with normal limb development in humans. The "thalidomide baby" syndrome could be considered a phenocopy of phocomelia.

Because of phenocopies, which are many and well-documented in experimental animals, it is impossible to determine if a spontaneous occurrence of an abnormal condition in a family is inherited. Perhaps by chance the parents are heterozygous for a particular recessive gene. But the abnormality could also be due to a new mutation (which would have to be a dominant one, since it is highly unlikely the same recessive mutation would occur in both parental gametes) or to an environmental factor, in which case the probability of recurrence in the family, genetically speaking, is zero.

About the only way available to determine whether the condition was inherited is to study karyotypes of the parents and the affected child for abnormal chromosome content. This is now routine procedure in many hospitals. However, given

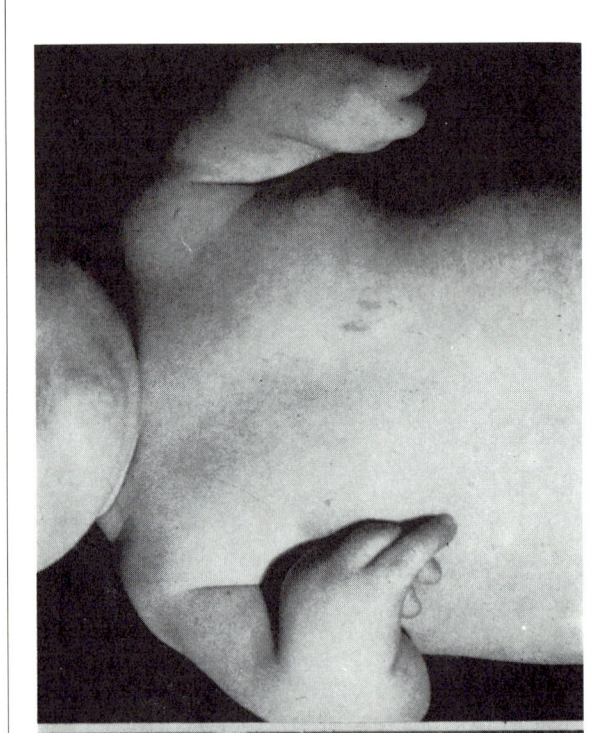

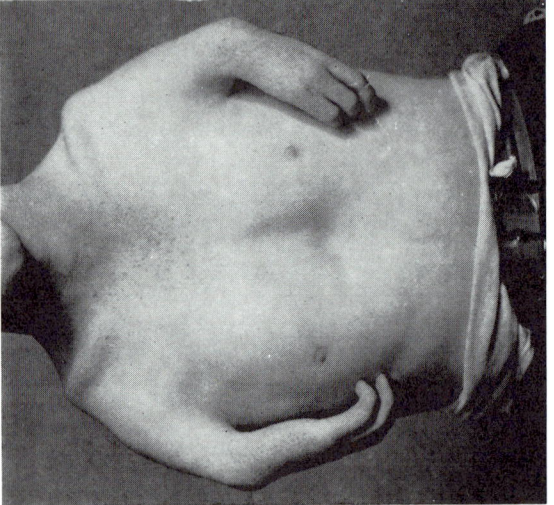

84
Mendelian Heredity in Man

FIGURE 4.12
A case of phenocopy: (A) phocomelia and (B) thalidomide syndrome. [From R. E. Cooke (ed.), The Biologic Basis of Pediatric Practice. Copyright © 1968 by McGraw-Hill, Inc. Used with permission of McGraw-Hill Book Company.]

no visible evidence of genetic abnormality from such studies, no conclusions can or should be reached by the counselor.

It is important to bear in mind the difficulties inherent in the genetic analysis of human traits due to our long life span and small sibships as we discuss the genetic implications of environmental pollution in a later chapter.

PROBLEMS

1. (a) Analyze the pedigrees in the accompanying figure for mode of inheritance. (b) Assuming the condition is determined by the gene r, give the genotype of each of the individuals marked a to g in the pedigree.

    SOLUTION *The trait is very likely recessive since the parents of three affected males in generation F_4 and F_5 are normal and related. Although the predominance of affected males may indicate that this is X-linked, close observation will show it cannot be and is therefore autosomal. The reason is that the normal males a and c are most likely the sources of the mutation inherited by the fourth and fifth generations, since their wives are unrelated and presumably homozygous normal. Therefore, the gene must be autosomal, for if it were X-linked, then a and c would have to express the mutant phenotype.*

    ANSWER (a) *Autosomal recessive.*
    (b) *See Appendix A.*

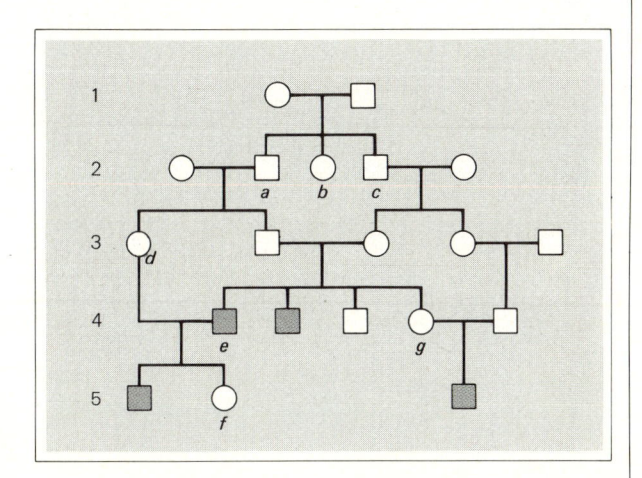

PROBLEM 1

Mendelian Heredity in Man

PROBLEM 2

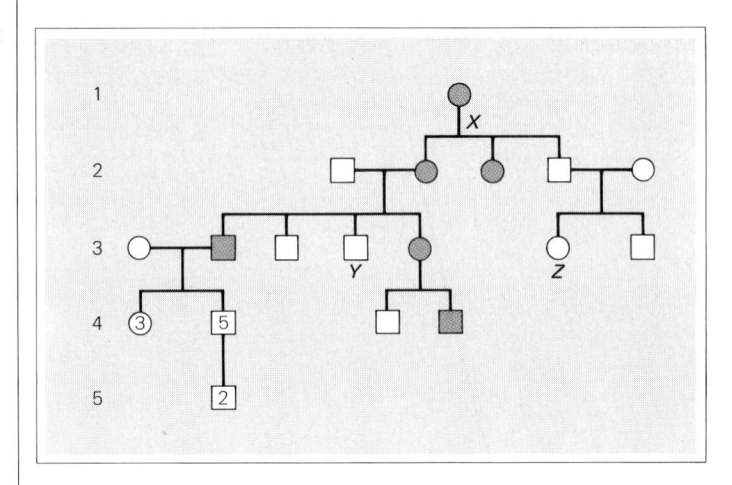

2 Give the genotypes of individuals X, Y, and Z in the accompanying figure.

3 Give the genotypes of individuals marked A to E in the accompanying figure.

PROBLEM 3

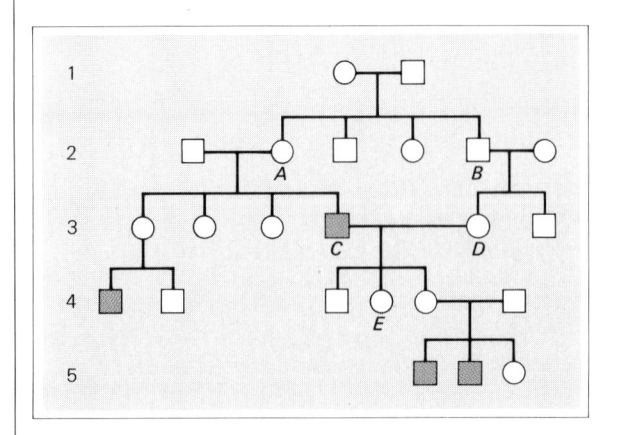

REFERENCES

The following volumes deal primarily with human genetics. The first two mentioned are more for students of biology and genetics. Thompson and Thompson's book is written at the level of the medical student or practicing physician. Roderick's volume is perhaps the most easily understood for the nonscientist, since it is written on a more elementary basis than the others.

McKusick, V. A., 1969: *Human Genetics,* 2d ed., Prentice-Hall, Inc., Englewood Cliffs, NJ.

Roderick, G. W., 1968: *Man and Heredity,* St. Martin's Press, Inc., New York.

Scheinfeld, A., 1971: *Heredity in Humans,* J. B. Lippincott Company, Philadelphia.

Stern, C., 1973: *Principles of Human Genetics,* 3d ed., W. H. Freeman and Company, San Francisco.

Work cited:

McKusick, V. A., J. A. Hosteter, J. A. Egeland, and R. Eldridge, 1964: "The Distribution of Certain Genes in the Old Order Amish," *Cold Spring Harbor Symposium of Quantitative Biology,* **29**:99–114.

Schull, W. J., 1958: "Empirical Risks in Consanguineous Marriages: Sex Ratio, Malformation and Viability," *American Journal of Human Genetics,* **10**:294–343.

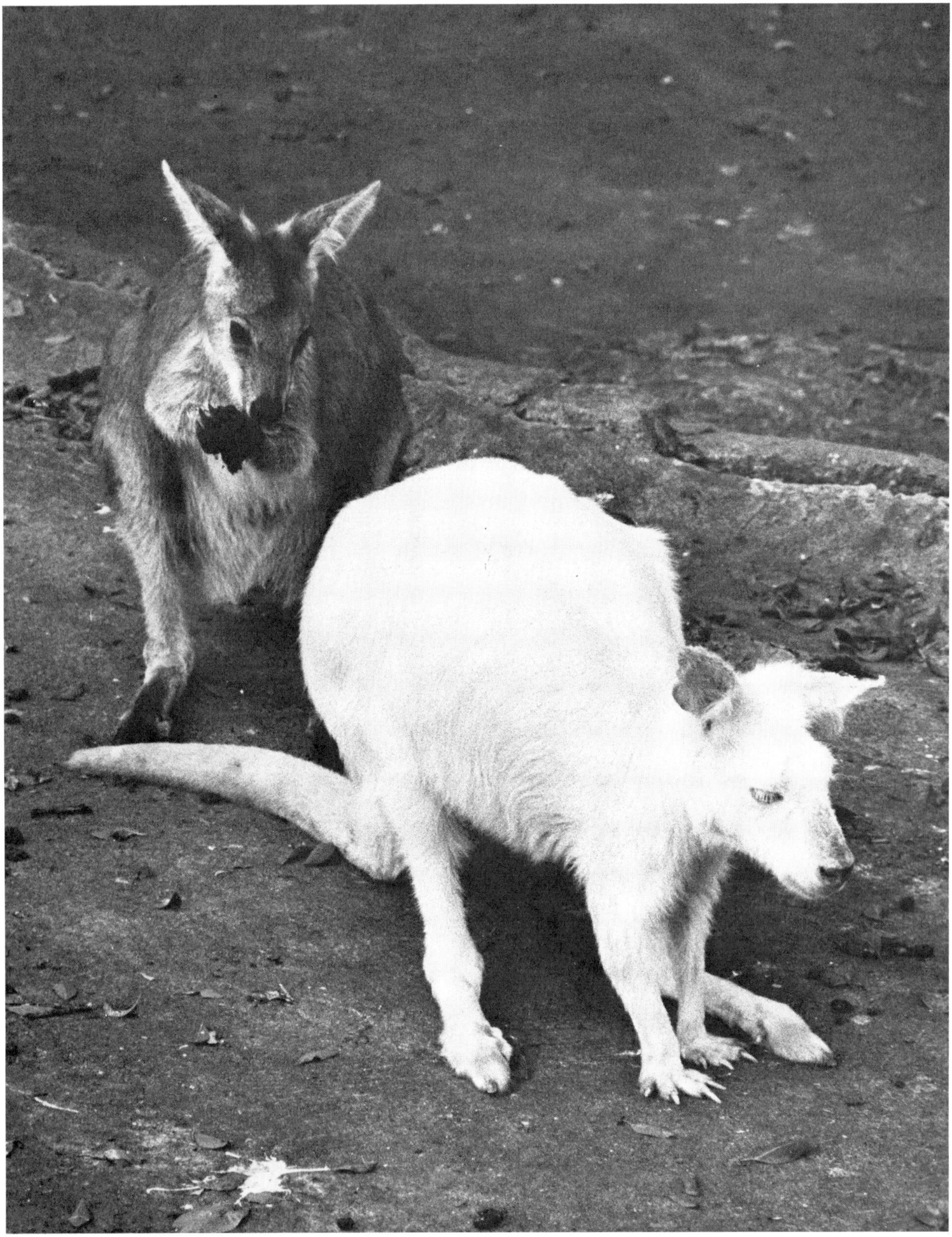

Chapter 5

Beyond Mendelian Genetics

In the first chapter, we mentioned that luck was one of the factors that allowed Mendel to succeed in formulating genetic laws where others had failed. One aspect of this luck was that by chance he chose simple traits to study. Thus far, our own discussion has been limited to the same kinds of traits, ones that are clearly determined by sets of alleles that are dominant or recessive, autosomal or sex-linked. Were all traits in living organisms similarly inherited, there would be very little unknown about even the complex organisms.

Unfortunately, the majority of characteristics cannot be analyzed in a simple Mendelian fashion. In this chapter we shall discuss a number of genetic phenomena which obscure the genetic nature of many interesting traits in complex organisms (and which would have thrown Mendel completely off the track had he encountered them in his studies!).

Thus far, we have used only examples of complete dominance and recessiveness in which heterozygotes are phenotypically indistinguishable from homozygous dominant in-

INCOMPLETE DOMINANCE AND CODOMINANCE

An albino kangaroo. The effect of the albinism gene masking all other pigmentation genes is a phenomenon known as epistasis. [*Courtesy of Taronga Zoo, Sydney © (Tom McHugh—Photo Researchers) 1972.*]

dividuals. A large number of alleles found in plants and animals do not follow this system, however. For example, in four-o'clock flowers, if a true-breeding plant with white flowers is crossed with a true-breeding plant with red flowers, the offspring (which are heterozygotes) produce pink flowers. And if these F_1 are then crossed, they produce a ratio of F_2 plants as shown in Figure 5.1. This is known as *incomplete dominance*.

In humans, one normal trait determined by incomplete dominance is the type of hair. Curly hair is incompletely domi-

FIGURE 5.1
Flower color inheritance in four-o'clock plants as determined by incomplete dominance.

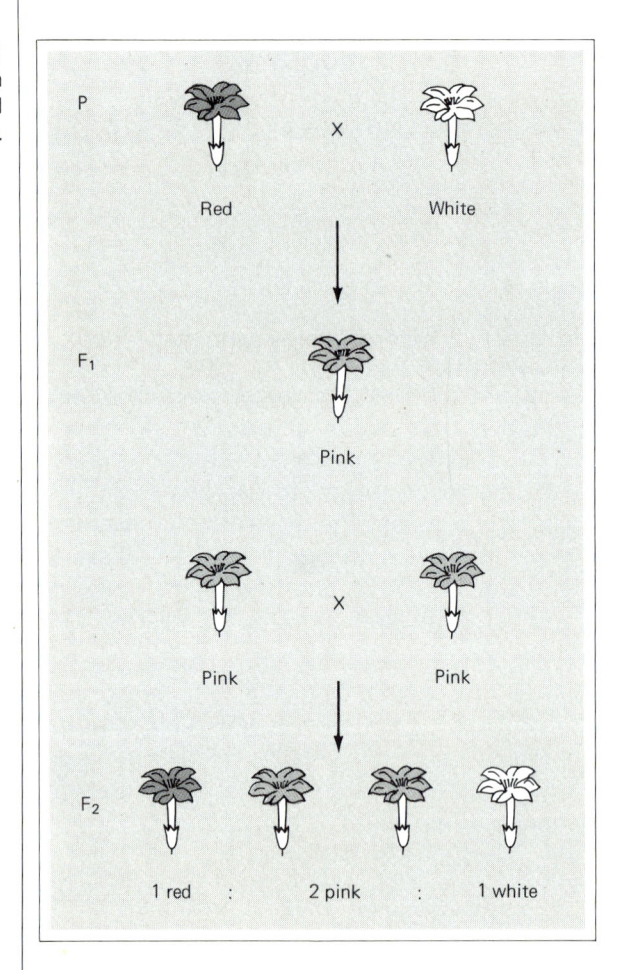

nant over straight hair, which is determined by the homozygous recessive condition. Heterozygotes have wavy hair. Singing voice is also believed to be determined by incomplete dominance. The alleles are designated y^1 and y^2, bass and soprano voices have the genotype y^1y^1; tenor and alto are homozygous for y^2, y^2y^2. The heterozygotes, y^1y^2, are baritone and mezzo-soprano.

Another type of dominance we have already mentioned is seen in the inheritance of blood groups in man (see page 14). You may recall that this is known as *codominance*, in which heterozygotes for two codominant alleles express both traits determined by the alleles. There are a number of different blood groups in humans, but the one best known to the general public is probably the ABO group.

In the ABO blood group, the two codominant alleles involved determine chemical substances on the surface of red blood cells. In other words, I^A determines that the red blood cells of the individuals carrying this gene will have a specific type of substance on the blood cells that categorizes them chemically as being type A. I^B determines a different substance on the red blood cell that results in type B blood. Codominance is seen in the I^AI^B heterozygotes who have *both* A and B substances, and consequently they are said to have type AB blood. Individuals who are homozygous recessive, or blood type O, have no comparable substances on their red blood cells. Table 5.1 shows a summary of the possible genotypes and resulting phenotypes found in the ABO system. We shall go into the details of blood-type reactions in a later chapter.

Although incomplete dominance and codominance are variations of dominance and recessiveness, the genes none-

TABLE 5.1
Determination of blood type by codominant alleles

BLOOD TYPE	BLOOD-TYPE SUBSTANCE	GENOTYPES
A	A	I^AI^A, I^Ai
B	B	I^BI^B, I^Bi
AB	A,B	I^AI^B
O	—	ii

	$S/^B$	$s/^B$
$s/^A$	$Ss/^A/^B$	$ss/^A/^B$
si	$Ss/^Bi$	$ss/^Bi$

1 wavy hair, type AB
1 straight hair, type AB
1 wavy hair, type B
1 straight hair, type B

theless conform to expected patterns of transmission, according to the laws of segregation and random assortment. For example, what offspring could be expected from a marriage of two people whose genotypes were $Ss/^B/^B$ and $ss/^Ai$, assuming S = curly hair, Ss = wavy hair, and ss = straight hair? The Punnett square for this problem and the phenotypic ratio which would be the answer to our question are shown in the margin.

POLYGENIC INHERITANCE

In 1909, a Swedish geneticist named Nilsson-Ehle analyzed the inheritance of kernel color in wheat and contributed an extremely important concept to genetics. In crossing true-breeding red-kernel plants with true-breeding white-kernel plants, the F_1 obtained were medium red, indicating incomplete dominance. However, in using different strains of the parental plants, he found that F_1 crosses derived from the different strains gave different phenotypic ratios among the F_2. In some cases, there was the expected ratio of 3 reds (1 red, 2 medium reds) to 1 white, indicating that a single pair of alleles was involved. Others, however, showed ratios approximating 15:1, or even 63:1.

Nilsson-Ehle realized that these figures could be explained in the following way: the 15:1 ratio could reflect the presence of two independently assorting pairs of genes, which we have thus far associated with the classical 9:3:3:1 ratio in the offspring of dihybrid crosses. In the case of the 15:1 ratio, however, all plants with at least one dominant gene—which would include the 9, 3, 3 groups—would be red, and the only white offspring would be homozygous recessive for both pairs of genes. (See Table 5.2 for this analysis.) In the same manner, the 63:1 ratio can be ascribed to independent assortment of three pairs of genes. The parental genotypes here would be: $R^1R^1R^2R^2R^3R^3 \times r^1r^1r^2r^2r^3r^3$.

The great significance of Nilsson-Ehle's results in that they showed that traits are not always determined by single gene pairs, and that there are traits determined by a number of different gene pairs which interact to produce a particular phenotype. In the case of kernel color in wheat, the interaction is an additive one: the more dominant genes a plant

P $R^1R^1R^2R^2 \times r^1r^1r^2r^2$
       Red ↓ white

F_1 $R^1r^1R^2r^2$
       Medium red

F_2

	R^1R^2	R^1r^2	r^1R^2	r^1r^2
R^1R^2	$R^1R^1R^2R^2$ Dark red	$R^1R^1R^2r^2$ Medium dark	$R^1r^1R^2R^2$ Medium dark	$R^1r^1R^2r^2$ Medium red
R^1r^2	$R^1R^1R^2r^2$ Medium dark	$R^1R^1r^2r^2$ Medium red	$R^1r^1R^2r^2$ Medium red	$R^1r^1r^2r^2$ Light red
r^1R^2	$R^1r^1R^2R^2$ Medium dark	$R^1r^1R^2r^2$ Medium red	$r^1r^1R^2R^2$ Medium red	$r^1r^1R^2r^2$ Light red
r^1r^2	$R^1r^1R^2r^2$ Medium red	$R^1r^1r^2r^2$ Light red	$r^1r^1R^2r^2$ Light red	$r^1r^1r^2r^2$ White

1 dark red + 4 medium dark + 6 medium red + 4 light red = 15 red to 1 white

TABLE 5.2
Nilsson-Ehle's analyses of inheritance of kernel color in wheat, illustrating polygenic inheritance

inherits, the darker the color. This kind of situation is referred to as *polygenic inheritance,* and usually involves the inheritance of *quantitative traits.*

Note that the larger the number of gene pairs involved in determining a single trait, the more difficult it becomes to distinguish between the phenotypes. For example, the difference in kernel color between $R^1r^1R^2r^2R^3r^3$ and $R^1r^1R^2r^2r^3r^3$ would be much subtler than between $R^1r^1R^2r^2$ and $R^1r^1r^2r^2$.

It is thought that some of the more complex human traits, such as intelligence, which we have never been able to study genetically, may be polygenic, resulting from a large number of different gene pairs. This leads to such small differences between phenotypes that a continuous type of variation exists. Other examples of human traits thought to be polygenic are pigmentation, blood pressure, and fingerprint ridge count.

The interaction between different pairs of genes not only can result in enhancement of effect, as in the case of kernel color

EPISTASIS

in wheat, but also can work in the opposite manner: one pair of genes can completely mask the expression of another pair of genes. This masking effect is referred to as *epistasis*.

For example, black coat color in mice is determined by a dominant gene *B*. Its recessive allele *b* results in brown coat color in the homozygous condition. A different pair of genes determines whether any pigment will be present at all. *CC* or *Cc* animals have pigmentation; the homozygous recessive animals are albino, regardless of what alleles they possess for coat color. Thus, *BBcc* mice are albino, whereas *bbCc* animals are pigmented and have brown coat color. In this situation, we consider the *cc* genotype *epistatic* to the *B* alleles.

Note that this is not the same as dominance, which is a term limited to the interaction between alleles. Epistasis describes the interaction between different pairs of nonalleles.

Albinism also exists in humans, and is similar to the albinism described in mice. A person who is homozygous recessive for an albinism gene will not have pigmentation anywhere in his body regardless of what other genes he may have for hair color, etc. This results in white hair and in pink eyes due to lack of pigment in the iris of the eyes. (The pink color results from the red color of blood which normally is hidden by pigmentation in the iris.) Figure 5.2 illustrates albinism in humans.

MODIFIERS

There are a number of unidentified genes that influence the expression of a trait. We refer to them simply as *modifier genes* because they are capable of modifying the phenotype in ways we do not understand at present.

For example, in humans there is a gene-determined ability to taste a particular chemical, phenylthiocarbamide, called PTC for short. This ability is apparently determined by a single pair of genes, and tasting is a dominant trait; those who are homozygous recessive cannot taste PTC at all. However, modifiers apparently exist which result in differences in degree of ability to taste, ranging from a slight sensation to a very strong reaction. The different degrees of expression of a trait are called *expressivity*. Two individuals may therefore be genetically identical for PTC taster genes and yet be phenotypically different due to the mysterious modifiers.

One important concept brought out by studies of modifying genes is that the phenotype of an individual is the result of the *interaction of all his genes*, not just isolated genes. It is because we must take into account the whole of the geno-

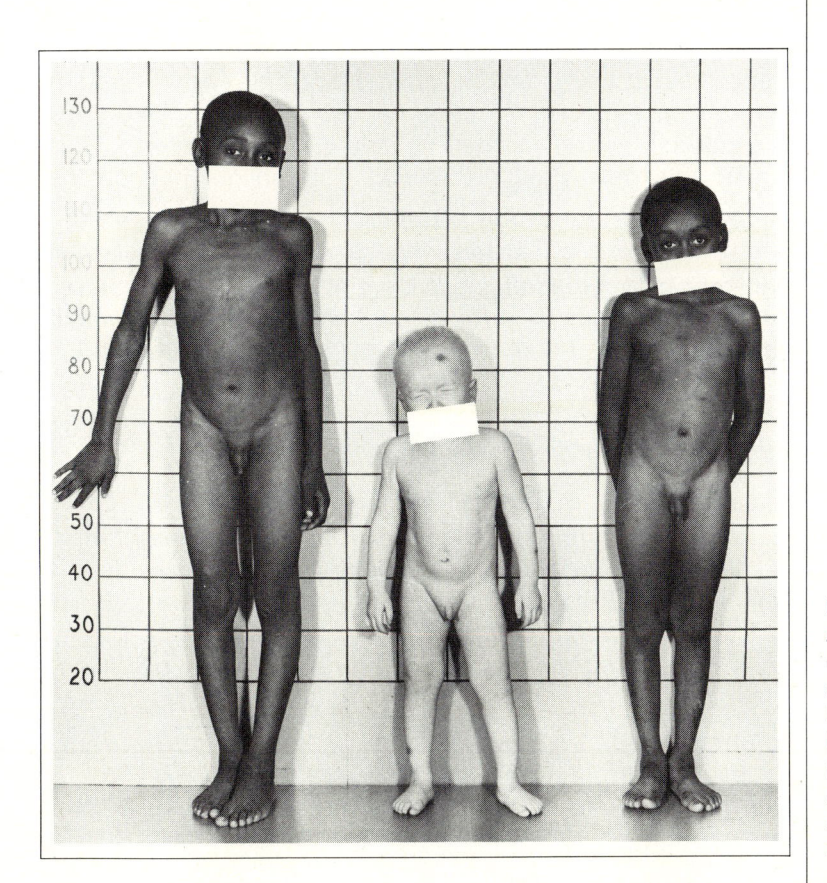

FIGURE 5.2
Albinism in humans. The albino child is squinting because the lack of pigmentation in his eyes causes him to be especially sensitive to light, in contrast to his normally pigmented siblings. [*Courtesy of Dr. Victor McKusick, Johns Hopkins University School of Medicine.*]

Beyond Mendelian Genetics

LINKAGE

One of the most important genetic phenomena that we have not discussed in detail, though we alluded to it in our discussion of sex linkage, is the fact that chromosomes contain far more than a single locus. That there are many genes on a chromosome was theorized as early as 1903 in the Sutton-Boveri hypothesis on the "qualitative" difference between chromosomes. The actual proof came less than a decade later in T. H. Morgan's studies on the fruit fly, *Drosophila melanogaster*.

If you think about it, there must be a number of different genes on a chromosome, because as complex as we humans are, we possess only 23 pairs of chromosomes in our cells. It would be highly unlikely that only 23 pairs of alleles could determine the myriad number of traits in our phenotypes.

All genes located on the same chromosome are said to be *linked* and form a *linkage group*. Now think about this for a moment before reading on: If Mendel had by chance chosen to study two traits determined by linked genes, which of his laws would have been most difficult for him to ascertain? If you thought random assortment, you would be absolutely right.

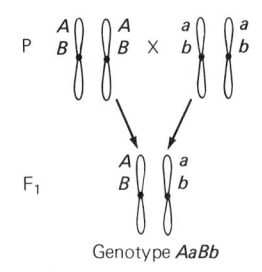

FIGURE 5.3
A cross involving two linked genes.

LINKED GENES CANNOT ASSORT INDEPENDENTLY

Let us illustrate this by following meiosis in the germ cell of an individual who is the offspring of individuals with the genotypes $AABB \times aabb$, assuming that genes A and B are linked. The chromosomes of the homozygous parents and their dihybrid offspring are diagrammed in Figure 5.3.

Notice that in contrast to Figure 2.17, which shows meiosis in a dihybrid with two pairs of randomly assorting genes, the F_1 dihybrid for two linked genes can produce only two kinds of gametes following meiosis, as shown in Figure 5.4. A cross of two such dihybrids would then yield the results shown in the margin.

	AB	ab
AB	AABB	AaBb
ab	AaBb	aabb

Phenotypic ratio: 3:1
(assuming complete dominance)

FIGURE 5.4
Abbreviated meiosis in the germ cells of a dihybrid for two linked genes.

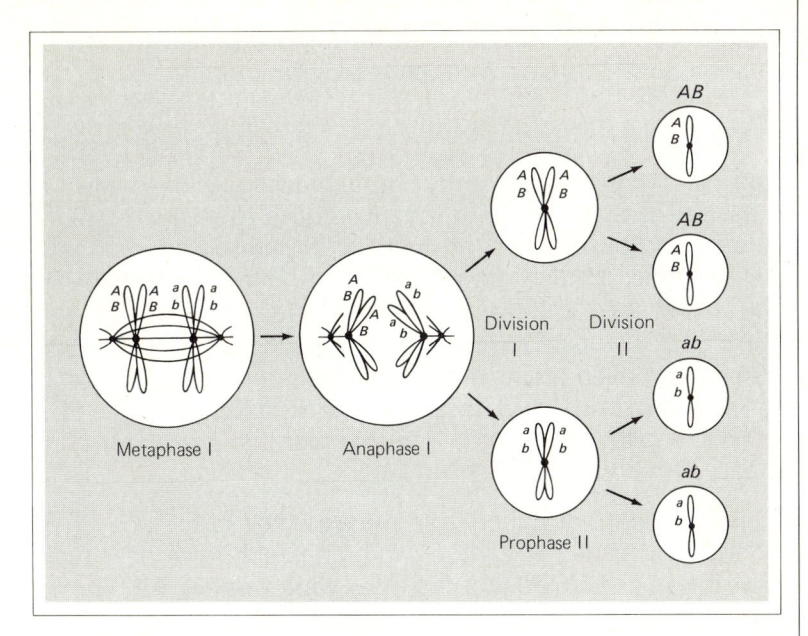

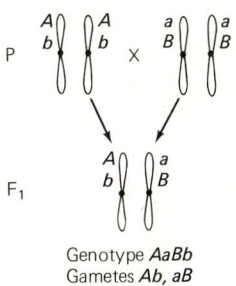

Assuming complete dominance and recessiveness, the phenotypic ratio here is 3 dominant to 1 recessive, certainly different from the 9:3:3:1 ratio obtained in random assortment in a cross of two dihybrids.

When the dominant genes are linked on the same chromosome as in our dihybrid in Figure 5.3 and the recessive alleles are in the homolog, the genes are said to be in the *cis arrangement,* sometimes called the *coupling phase.*

Given different parental combinations, however, there could be a mixture of dominant and recessive genes on the same chromosome. For example, if two parental types of the genotypes *AAbb* × *aaBB* are crossed, their offspring would still be *AaBb* but the genes would be in different combinations on the chromosomes from the cis position we discuss above (see Figure 5.5). This arrangement of genes is called the *trans position,* or *repulsion phase.* Notice that the phenotype of the dihybrids in cis and trans, as diagrammed in the Punnett squares shown in the margin, would be the same; how-

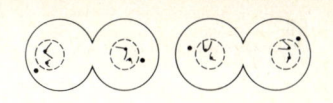

Genotype *AaBb*
Gametes *Ab, aB*

FIGURE 5.5
Chromosomes of a dihybrid (the F_1) for two linked genes in the trans position. Following meiosis, gametes of such an individual will contain two combinations of genes, *Ab* and *aB.*

CIS-TRANS POSITIONS

Trans

	Ab	aB
Ab	AAbb	AaBb
aB	AaBb	aaBB

1:2:1 phenotypic ratio

Cis

	AB	ab
AB	AABB	AaBb
ab	AaBb	aabb

3:1 phenotypic ratio

ever, if each dihybrid were crossed to its own kind, the F_2 generation would have different phenotypic ratios.

CROSSING-OVER

You can imagine how confusing it must have been for the classical geneticists who by chance chose to study the transmission of linked genes. Around 1905 Bateson and Punnett, for example, studied flower color and pollen-grain shape in the sweet pea; both were known to be simple Mendelian traits. Let us assume *A–* = purple flowers, *aa* = red flowers; *B–* = long pollen grains and *bb* = round pollen grains.* In a cross of dihybrids, they obtained approximately the ratio of 9 purple long to 1 red long to 1 purple round to 3 red round, instead of the classical 9:3:3:1.

* Remember that when a dash is used in depicting a particular genotype, such as *A–B–*, it means the second allele is of no consequence in determining phenotype.

This significant deviation from the expected ratios was never adequately explained by the two pioneers, but we now know that it was due to the fact that they were working with linked genes, probably in the cis arrangement. But if you look at Figure 5.4 or the dihybrid crosses above, there should only be two or three phenotypic classes in the F_2, depending on whether the dominant genes are in coupling or repulsion. Where did the four phenotypic classes of progeny in the Bateson and Punnett experiment come from?

The explanation came later when it was found that chromosomes in meiosis often form crosslike configurations when in tetrad formation in late prophase I. Such formations are known as *chiasmata* (singular, *chiasma*). It appears that there is actually an exchange of genetic material between the homologous chromatids at this time, an act which is called *crossing-over*. Figure 5.6 diagrams crossing-over in cells of a dihybrid with genes linked in the cis position, similar to the material studied by Bateson and Punnett in the sweet pea.

This reciprocal exchange, which occurs infrequently during gametogenesis, would result in a small percentage of gametes that show a new combination of genes, namely, *aB* and *ab*. The gametes containing the new combination of traits are known as *recombinant types*; the others, the ones expected, are the *parental types*.

But if there are four types of gametes from crossing-over, how do we distinguish this from independent assortment?

By the *frequency* of recombinant gametes. In true independent assortment of unlinked genes, the four gamete types are produced in equal numbers. In crossing-over, the recombinant gametes are usually in the minority, as seen in the data obtained by Bateson and Punnett.

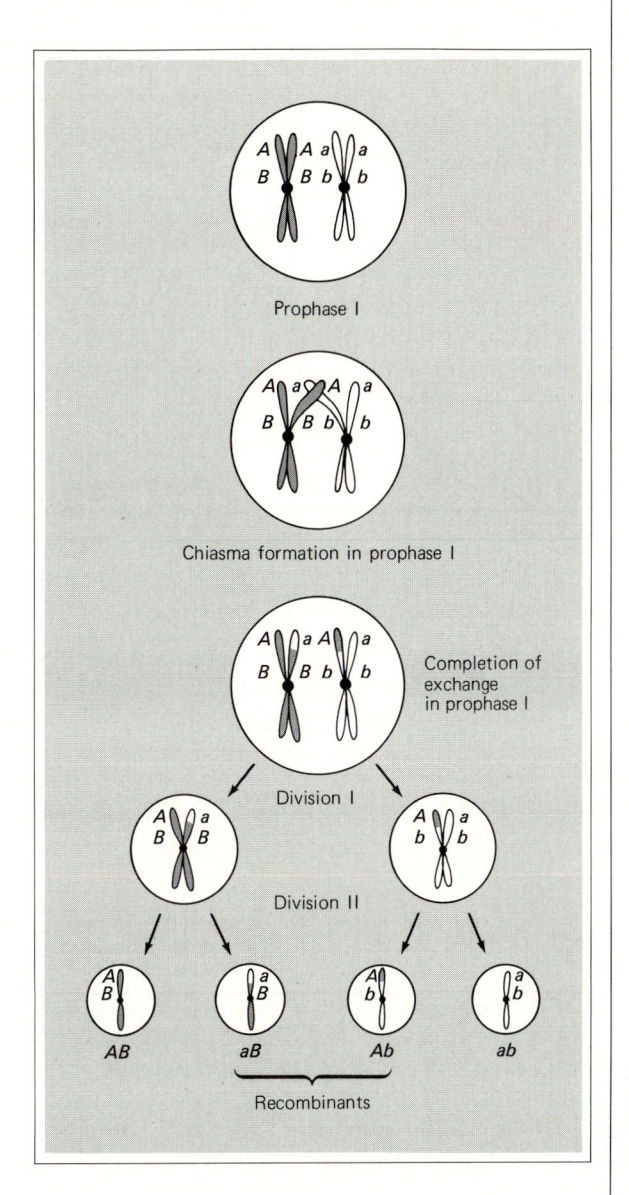

FIGURE 5.6
Chiasma formation and crossing over. For clarity, one of the homologs has been colored black.

MAPPING CHROMOSOMES BY CROSSING-OVER DATA

In fact, geneticists have been able to use the frequency of crossovers between genes to indicate the relative distances between them. We can do this because the farther apart two genes are, the greater the likelihood of crossing-over between them; the closer they are, the less likely the chances that an exchange will occur. See Figure 5.7 for an illustration of this point.

By calculating percentage of recombinant types among the progeny in studies of linked genes, one can assign what are called *map units* between them to indicate their relative positions on a chromosome.*

*For illustrations of how geneticists calculate map units between genes and the usefulness of knowing the distance between genes, see Appendix C.

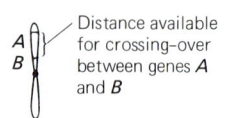

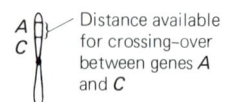

Linked genes may seem to assort independently only if they are approximately 50 units apart. In such an instance the crossover gametes (50 percent) would equal parental types (50 percent) and the result would indicate absence of linkage. This would be discerned only if the two genes were later found to be linked to other genes on the same chromosome less than 50 units distant.

With the discovery of linkage and crossing-over, the classical geneticists were able to compile data and formulate extensive chromosome maps indicating where many of the mutations they found, and therefore the normal alleles of these mutations, were located on the chromosomes of their object of experimental study, the fruit fly *Drosophila melanogaster*.

Figure 5.8 gives a partial representation of the genes located on the four pairs of chromosomes found in *Drosophila*. Notice that a number of genes are involved in the normal formation of a particular structure such as the eye or the wing. Some 40-plus loci have been found that determine either eye color or shape, for example, indicating that in this organism there are at least that many genes determining normal eye phenotype!

FIGURE 5.7
The greater the distance between genes, the more likely there will be crossover exchanges between them. If we assume that A is twice as far from B as it is from C, we can expect approximately twice as many recombinant events between A and B as between A and C.

A, B: Distance available for crossing-over between genes A and B
A, C: Distance available for crossing-over between genes A and C

MAPPING CHROMOSOMES IN MAMMALS

In mammals, the most extensively studied linkage groups are those found in inbred mice. The mouse has 20 pairs of chromosomes, and 19 linkage groups have been discovered, including, of course, the genes on the sex chromosomes. To date about 16 of these linkage groups have been assigned to specific chromosomes.

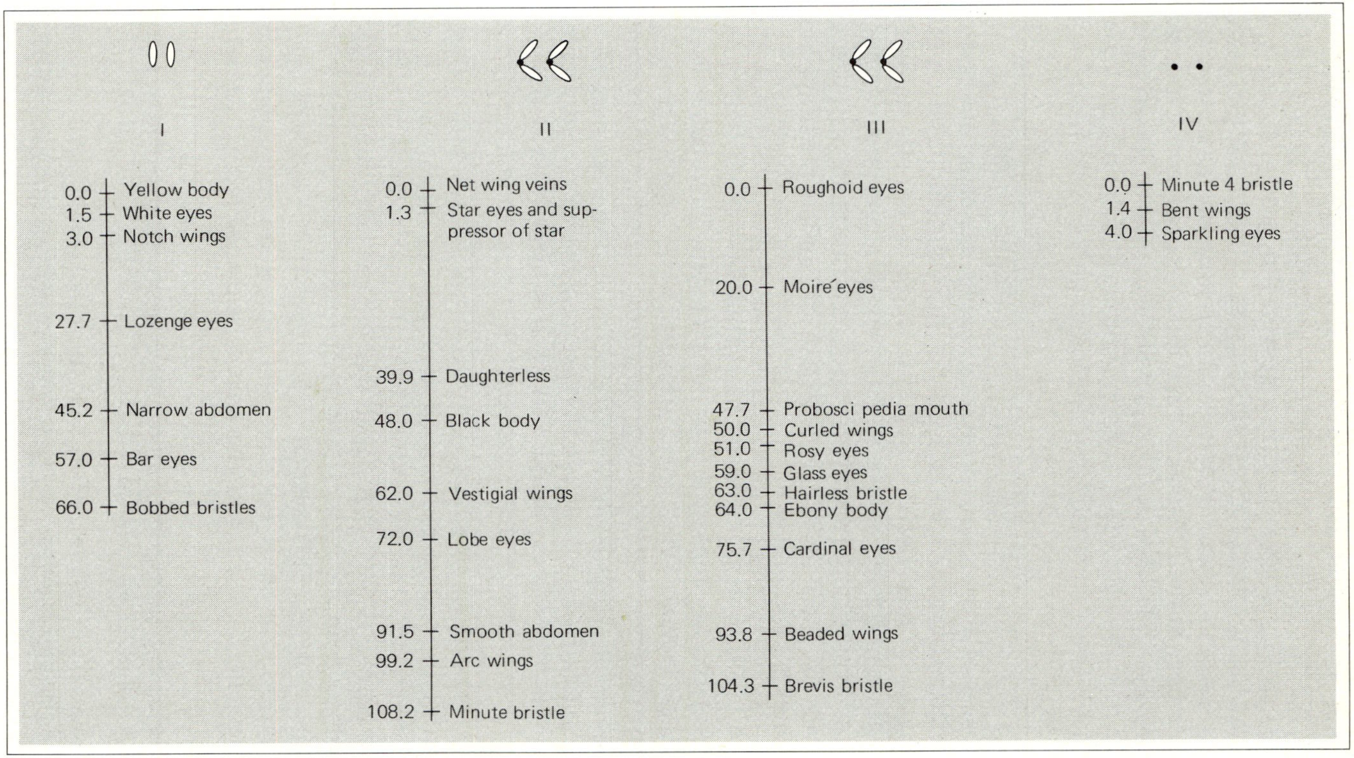

I	II	III	IV
0.0 — Yellow body	0.0 — Net wing veins	0.0 — Roughoid eyes	0.0 — Minute 4 bristle
1.5 — White eyes	1.3 — Star eyes and suppressor of star		1.4 — Bent wings
3.0 — Notch wings			4.0 — Sparkling eyes
27.7 — Lozenge eyes		20.0 — Moiré eyes	
45.2 — Narrow abdomen	39.9 — Daughterless	47.7 — Probosci pedia mouth	
57.0 — Bar eyes	48.0 — Black body	50.0 — Curled wings	
66.0 — Bobbed bristles	62.0 — Vestigial wings	51.0 — Rosy eyes	
	72.0 — Lobe eyes	59.0 — Glass eyes	
		63.0 — Hairless bristle	
		64.0 — Ebony body	
	91.5 — Smooth abdomen	75.7 — Cardinal eyes	
	99.2 — Arc wings	93.8 — Beaded wings	
	108.2 — Minute bristle	104.3 — Brevis bristle	

FIGURE 5.8
A partial representation of genes that have been mapped on the four chromosome pairs in the fruit fly *Drosophila*. The actual shape of the chromosomes is given above the chromosome "maps." [After E. W. Sinnott, L. C. Dunn, and T. Dobzhansky, Principles of Genetics, 5th ed. Copyright © 1958 by McGraw-Hill, Inc. Used with permission of McGraw-Hill Book Company, New York, 1958.]

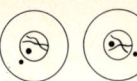

It is not difficult to understand why evidence for linkage and crossing-over is hard to come by in studies on human genes. To identify linkage by crossing-over, the geneticist needs to be able to make the desired crosses between certain genotypes, and since recombinants are usually a fairly small minority of the progeny, a large number of offspring are necessary. X linkage is obviously one example of linked genes in humans. However, although dozens of X-linked conditions have been analyzed, few autosomal genes have been localized.

One fairly recent technique, first developed in 1960, that may prove of great help in localizing human genes on autosomes involves the occasional fusion of diploid cells in tissue culture. It has been found that cells of entirely different species, such as mouse and human cells, will fuse. As such cells containing two different sets of chromosomes undergo mitosis, there is a constant loss of human chromosomes from the fusion cell.

In 1968, Migeon and Miller found that when cells from a mutant mouse lacking a particular enzyme, thymidinekinase, fused with normal human cells (which can produce the missing enzyme), the human chromosome containing the normal gene for determining the production of the enzyme was retained. By staining the chromosomes, they deduced that the gene determining thymidinekinase is located on one of the chromosomes in the E group. It has now been identified as chromosome 17.

With the aid of techniques such as cell fusion and more precise identification of individual chromosomes and chromosomal defects, it is hoped that increasing numbers of specific genes will be localized on specific chromosomes in the near future. This avenue of approach to the study of linked genes in man is far more plausible than the reliance on progeny data which, as we have already pointed out, requires greater numbers than exist in human families.

Table 5.3 lists genes which have been assigned to specific chromosomes in man.

In addition, evidence has been found for about a dozen cases of linkage between at least two loci without assignment to a

TABLE 5.3
Genes assigned to chromosomes in man

GENE OR LOCUS	CHROMOSOME
Duffy blood group locus	1
Nuclear cataract	1
Nucleoside phosphorylase*	14
Haptoglobin	16
HL-A	16
Phosphoglucomutase	16
Thymidine kinase	17

* F. Ricciuti, and F. H. Ruddle, "Assignment of Nucleoside Phosphorylase to D-14 and Localization of X-linked Loci in Man by Somatic Cell Genetics," *Nature*, **241**:180–182, 1973.
Source: V. McKusick, *Mendelian Inheritance in Man*, Johns Hopkins Press, 1971, p. xli.

specific chromosome. One is the linkage between alleles for the ABO blood groups and a dominant mutation known as the nail-patella syndrome, so named because of the abnormal nails and knees of the affected people. It is believed that the two loci are about seven units apart.

Although we will not go into detail on them in this book, it is important for you to know that there are factors in chromosomes which apparently interfere with crossing-over, in some cases suppressing it, in other cases increasing the frequency of recombinants beyond what is expected. What these factors are remains unclear. For example, in *Drosophila* males, there is no crossing-over at all between any of the chromosome pairs. In addition, we know that there are cases of more than one crossover event at the same time between two homologs, and sometimes exchanges between three or four of the chromatids in a tetrad. These factors all serve to further complicate an already complicated situation!

One of the genetic advantages of crossing-over must be pointed out here, and that is the fact that this phenomenon increases variability in the gametes. As we have indicated in our discussion on heterosis earlier, and as you will see when we discuss evolution later, the greater the amount of variability, the better it is for the individual and the species.

LETHAL GENES

There are other phenomena which are known to result in offspring ratios other than what is expected from Mendel's laws. We need not go into all exceptions here, but for example, there exist in various species *lethal genes*, which, in the homozygous condition, cause death in early embryonic stages so that the affected individuals are resorbed and never develop to term. This would result in the absence of one expected class of offspring. In a sense, one could regard lethal genes as being epistatic to all genes that would have determined normal development.

Lethal genes and the other genetic phenomena we have discussed in this chapter fortunately did not exist in Mendel's peas, at least in the traits he studied. Perhaps you now realize just how fortunate Mendel was in his choice of material. All seven traits showed complete dominance. Further, and this is where his luck is almost incredible, he always obtained independent assortment, indicating that each gene pair was located on a different pair of chromosomes. Do you know how many chromosome pairs there are in the garden pea? Seven!

PROBLEMS

1. A normally pigmented man whose father is albino, a condition determined by an autosomal recessive gene *a*, marries a woman known to be carrying the gene for albinism. In addition, they are both heterozygous for the dominant gene *M* for dark hair. (*a*) What proportion of their children will be dark-haired? (*b*) What proportion of their children will be blond?

    SOLUTION (*a*) *The cross is of dihybrids, AaMm × AaMm. The proportion of offspring that have pigmentation is $\frac{3}{4}$ (that is, three-fourths of the progeny of Aa × Aa have at least one dominant gene for pigmentation). The proportion who have a dominant M gene is also $\frac{3}{4}$.*

    ANSWER $\frac{3}{4} \times \frac{3}{4} = \frac{9}{16}$. *If you make a Punnett square for this cross, you will see that actually 12 individuals have at least one dominant M gene but 3 of these are also homozygous recessive for albinism and will therefore produce no pigmentation.*

2. It has been recorded that two albinos, each known to be homozygous for a recessive gene, married and produced all normally pigmented offspring. What would be a geneticist's explanation for this (excluding infidelity, of course!)?

3. A wild mouse was captured which had a peculiarly short tail. When mated to normal-tailed mice, the offspring were found to

be in a ratio approximating 1 short-tail to 1 normal tail. (a) What type of gene determines short-tailness in mice? (b) When the short-tailed animals were mated, the ratio obtained was 2 short-tailed to 1 normal-tailed. Can you explain the deviation from an expected 1:2:1 ratio?

4 A woman is known to be heterozygous for hemophilia and color blindness, both X-linked genes. Her father was color-blind, with normal blood picture. If she marries a normal man, what phenotypic ratios can she expect (assuming no crossing-over occurs in the woman's gametes): (a) in her sons? (b) in her daughters?

SOLUTION *The woman's paternal X chromosome contains the gene for color blindness and a normal gene for blood clotting (part A of the accompanying figure). As she is known to be heterozygous for hemophilia as well, her maternal X chromosome must be as shown in part B of the figure. Therefore, she can transmit either chromosome to her sons with equal probability.*

ANSWER *Half the boys will be color-blind, and half will be hemophiliacs.*

5 (a) Give the genotypes of individuals A to F in the accompanying pedigree. (b) How can you account for individual F? (Hint: Draw the sex chromosomes of each individual to facilitate your analysis.)

Problems 6 to 8 are for the benefit of those who have read Appendix C:

6 Assume that two linked genes, r and s, are 20 units apart. What genotypes of offspring, and in what proportions, may be expected from a cross of RrSs × rrss if the dihybrid genes are in cis position?

SOLUTION *The chromosomes of the individuals involved are as shown in the accompanying figure. Since there is 20 percent crossing-over, the parental gametes from the dihybrid, which are RS and rs, will be produced 80 percent of the time, so that 40 percent of all gametes will be RS and 40 percent rs. The recombinant gametes will be Rs and rS, and will together form 20 percent of all gametes. The Punnett square for this cross will be the following:*

ANSWER

	0.4RS	0.4rs	0.1Rs	0.1rS
rs	0.4RrSs	0.4rrss	0.1Rrss	0.1rrSs

PROBLEM 4

PROBLEM 5

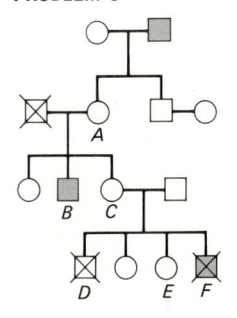

■ ● Hemophilia
⊠ ⊠ Color blindness

PROBLEM 6

7. Assume that the dihybrid genes of problem 6 are in trans position. What genotypic ratios may be expected in offspring of *RrSs* × *rrss*?

8. Female fruit flies dihybrid for two linked genes, *G* and *H*, with the genes in the cis position, are mated to male fruit flies homozygous recessive for the two genes. The offspring obtained are as follows:

G–H–	310
gghh	290
G–hh	195
ggH–	205
	1000

What is the distance between the *G* and *H* loci?

SOLUTION The parental-type offspring are G–H– and gghh, and total 60 percent (600 out of 1000). The others must be presumed to be recombinant types, and total 400 out of 1000.

ANSWER The genes are 40 crossover units apart.

9. (a) If you were not informed that *G* and *H* are linked, what aspect of the data in problem 8 would indicate linkage? (b) If there were no linkage, out of 1000 progeny in a *GgHh* × *gghh* cross, what genotypes and in what proportions would be expected?

REFERENCES

Work cited:

Bateson, W., and R. C. Punnett, 1905–1908: "Experimental Studies in the Physiology of Heredity," *Reports to the Evolution Committee of the Royal Society,* nos. 2, 3, 4. Also reprinted in L. Peters (ed.), *Classic Papers in Genetics,* Prentice-Hall, Inc., Englewood Cliffs, N.J., 1959.

Nilsson-Ehle, H., 1909: "Kreuzungsuntersuchungen und Hafer und weizen," Lunds Universitet. Arsskrift, Avd., ser. 2, pp. 1–122.

A general review article on cell hybridization:

Marx, J. L., 1973: "Somatic Cell Hybrids: Impact on Mammalian Genetics," *Science,* **179**: 785–787.

More technical detailed discussions on cell hybridization can be found in:

Ephrussi, B., 1972: *Hybridization of Somatic Cells,* Princeton University Press, Princeton, N.J.

Migeon, B. R., and C. Miller, 1968: "Human-Mouse Somatic Cell Hybrids with Single Human Chromosome (Group E): Link with Thymidine Kinase Activity," *Science,* **162:** 1005–1006.

Ricciuti, F., and F. H. Ruddle, 1973: "Assignment of Nucleoside Phosphorylase to D-14 and Localization of X-linked Loci in Man by Somatic Cell Genetics," *Nature,* **241:** 180–182.

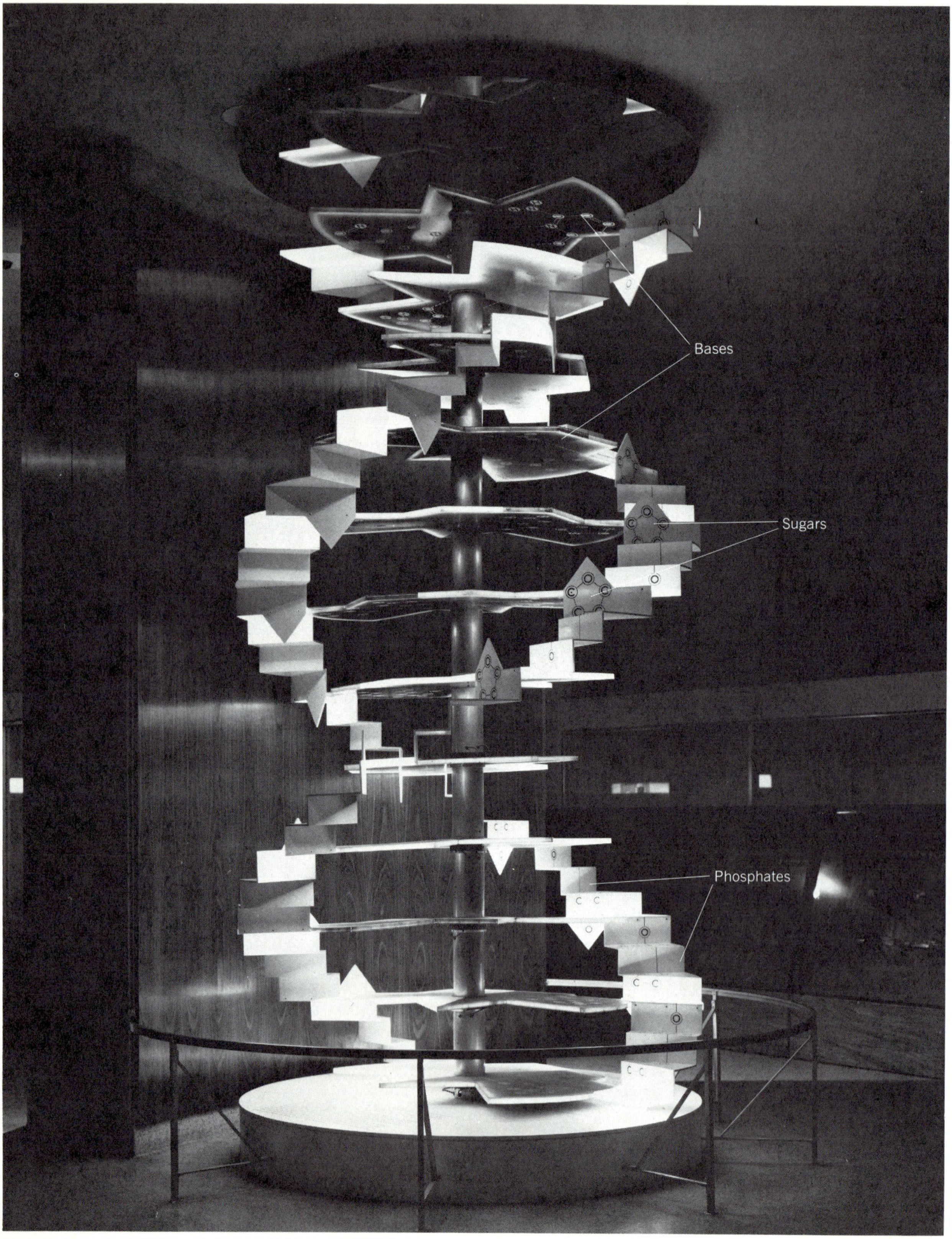
Bases
Sugars
Phosphates

Chapter 6

What Is a Gene?

In their studies on linkage and crossing-over, it appeared to the classical geneticists that the exchange of genetic material occurred only *between* genetic loci. This implied a certain indivisibility of the gene. The concept of "beads on a string" thus arose to describe the localization of genetic material determining a particular trait. The individual beads were the genes, and the string on which they were held together was the rest of the chromosome.

Although the rapidly expanding numbers of geneticists were contributing vast amounts of information on the transmission of inherited traits, it was still obvious that there was a major gap in genetic knowledge, namely, what happens between the gene and the phenotype. What was not understood at all was *how* the gene actually determines phenotype, how it actually controls the development of characteristics in a living organism. Basically, this ignorance stemmed from the fact that no one knew what a gene was.

Although it had been established beyond doubt that genetic information was located on the chromosomes, still, *where* in

PROTEINS AND NUCLEIC ACIDS IN CHROMOSOMES

A model of DNA. [*Courtesy of the American Museum of Natural History.*]

the chromosome was not known. This was a crucial question because chromosomes are composed of two different substances, proteins and nucleic acids. These are two important but separate classes of *organic compounds,* substances so named because they are found in living organisms and contain the element carbon. It was necessary to decide which of these compounds—proteins or nucleic acids—is actually the gene.

THE NATURE AND FUNCTION OF PROTEINS

First let us briefly and simply discuss the nature of these two vital substances, proteins and nucleic acids, since we shall be devoting a good deal of attention to them in this and ensuing chapters. Indeed all research activity in the newest, most exciting and productive branch of genetics, molecular genetics, has centered on the relationship between these two organic compounds.

Proteins are very large, complex molecules found in all our cells. They play an essential role in the structure of our cells: all cellular membranes are composed of special kinds of proteins. Therefore, in order for growth and specialization of cells to be possible, the cells must be able to manufacture new proteins to form the structural elements of new and growing cells.

There is also a class of protein molecules that is essential for all metabolic reactions occurring in a cell. Metabolic reactions in a cell must occur at extremely fast rates in order to supply the living cell with enough products of these reactions to carry on life processes. A class of proteins known as *enzymes* serves to catalyze biochemical reactions so that the cell may continue to function. A cell is capable of thousands of biochemical reactions, and each of these reactions is regulated by a specific enzyme. Since these enzymes are eventually broken down, a cell must be capable of replacing them.

The structure of a protein molecule is usually fairly complex. Figure 6.1 is a schematic diagram of a generalized protein molecule.* Basically, proteins are made up of a string of subunits known as *amino acids,* of which 20 different kinds have been identified in the natural world. Such a string of amino acids is called a *polypeptide* because the kind of

* In this and other illustrations of the chemistry of various molecules, rather than burden the reader with the true chemical formulas, we shall use generalized shapes to depict various important subunits.

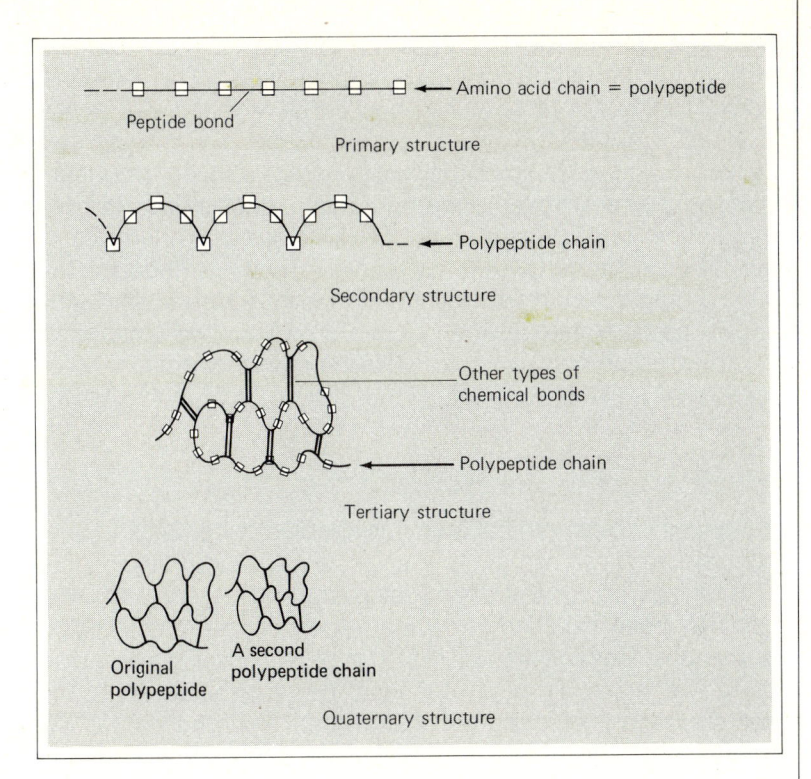

FIGURE 6.1
Diagrammatic illustrations depicting the basic structure of a protein and the various levels of structure and complexity.

chemical bond that keeps the amino acids together is known as a *peptide bond*. The number and sequence of amino acids are the protein's *primary structure,* which distinguishes one protein from another.

Hydrogen bonds between some of the amino acids confer a helical shape to the molecule that is known as its *secondary structure*. In addition, there are bonds formed between different parts of a polypeptide that cause the molecule to undergo further changes in shape. This dimension is known as the *tertiary structure*. Finally, more than one polypeptide chain can form associations which confer upon the protein its *quaternary structure*. It is not until the protein has achieved the various levels of structure, depending on its complexity, that it becomes functional. Considering the numbers and variety of proteins known to be synthesized by a cell, a mechanism that is capable of such flexibility is one to be held in real wonder.

What Is a Gene?

NUCLEIC ACID: DNA

All true chromosomes found in complex organisms are made of proteins, but all also contain nucleic acids. Like proteins, nucleic acids are large molecules, containing three different kinds of subunits: a *sugar, phosphoric acid,* and *nitrogenous bases*. Figure 6.2 shows the relationship of these three substances. When linked together, these substances form a *nucleotide,* and a chain of nucleotides such as the ones in the chromosomes is called a *polynucleotide*. Because the kind of sugar found in the nucleic acid of chromosomes is known as deoxyribose, the nucleic acid in chromosomes of complex organisms is called *deoxyribonucleic acid,* or *DNA* for short.

IDENTIFICATION OF THE GENETIC SUBSTANCE

Advances in biochemical analysis enabled geneticists to concentrate on these two major components of the chromosome in their search for the chemical basis of the gene. Anyone who has been exposed to topical reports in the mass media in the past decade knows that the search eventually led to DNA, which has probably received more publicity than any one chemical in the history of science! Nonetheless we will briefly trace the historical development of events that ultimately led to the recognition of the chemical nature of the gene, since such an examination will reveal the important role of microorganisms in modern genetics experimentation.

FIGURE 6.2
Schematic diagram of the subunits which form a nucleic acid.

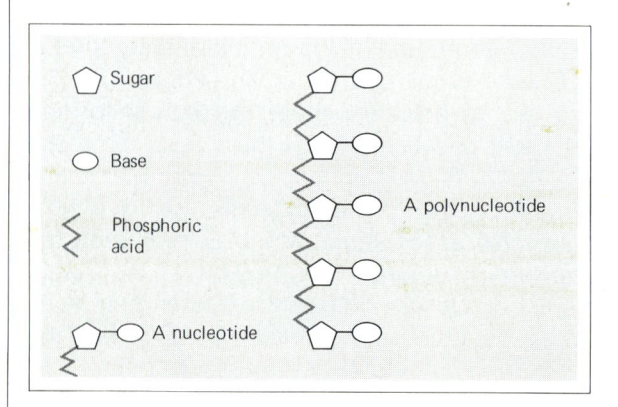

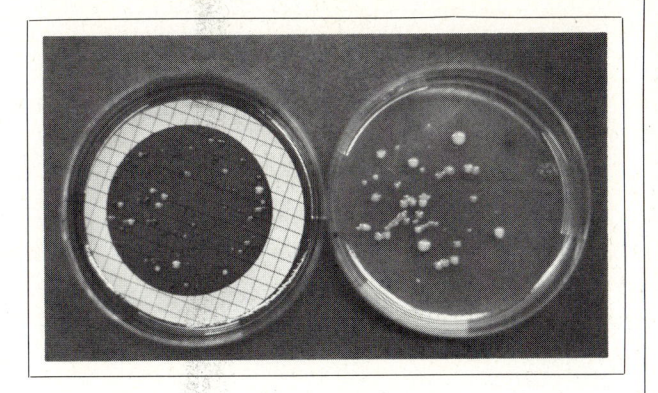

FIGURE 6.3
Photographs of bacterial cultures grown on agar plates. Different strains of bacteria form colonies which differ in size and other characteristics such as smoothness of the surface of the colonies. [*Courtesy of Dr. Leah Koditschek, Montclair State College.*]

One of the first clues to the question of the chemistry of the gene came from the work of F. Griffith, who in 1928 published some baffling results obtained in studies he had carried out on a type of bacterium that causes pneumonia in mice, *Diplococcus pneumoniae.* The disease-producing organisms are normally covered by a shell of material which causes a group of such cells (called a *colony*), grown on plates containing agar and other nutrients, to appear smooth and shiny. But in the course of his studies, Griffith found a mutation among the cultures. The mutant bacteria did not have the shell around them, and when grown on agar, they formed colonies that were small, rough, and dull in appearance. (See Figure 6.3 for examples of bacterial cultures.)

Griffith proceeded to test his mutants (Figure 6.4). He injected some of the mutants (rough type) into one group of experimental animals; he then killed some of the virulent smooth type and injected them into another group. Into a third group he injected a combination of live rough mutants and heat-killed smooth bacteria. The results observed with the first two experimental groups were not surprising: neither group succumbed to the disease. Apparently the virulence had been lost together with the smooth shell from the mutant bacteria. What was perplexing was the result with the third group, which had received a mixture: disease and death. Furthermore, autopsy of the animals and culture of bacteria found in them isolated living smooth bacteria! When he published his work in 1928, Griffith offered no conclusions

SOME HISTORICAL MILESTONES

FIGURE 6.4
The experiment by F. Griffith which led to the eventual discovery that the genetic substance was DNA.

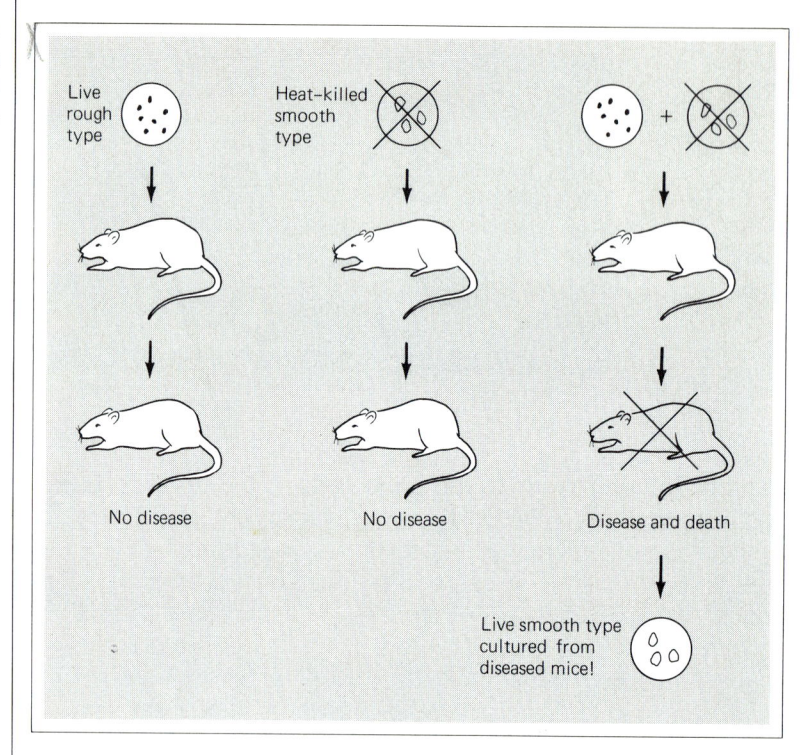

about these results, but the two logical possibilities were that either the dead virulent bacteria had caused the transformation of the mutant bacteria into virulent bacteria, or that somehow the mixture had caused the dead virulent bacteria to come back to life.

It was not for another decade and a half that the first alternative was proved to be the case by three men, D. T. Avery, C. M. MacLeod, and M. McCarty, who jointly tested different parts of the killed bacteria and finally isolated the substance which could cause the transformation of rough type into smooth type with no other components of the cell present. What was this substance? DNA!

Apparently, in the mixture of heat-killed virulent bacteria and live mutants used by Griffith, some of the DNA from the smooth type which was not destroyed by the heat treatment had entered the mutants and contributed its genes for the production of the shell, transforming the mutants into the virulent type.

The identification of DNA as the transforming substance was strong evidence that the genetic material was made of nucleic acid, for the change was heritable, indicating that it was indeed a genetic change. This fact was the reason live mutant bacteria were recovered in Griffith's experiment: the transformed mutant cells gave rise to offspring which had inherited the newly acquired trait of being virulent.

To this day we use the term *transformation* to describe the genetic changes caused in a cell by the introduction of genetic material from another cell. This was also one of the first indications to biologists that simple one-celled organisms such as the bacteria, which are normally haploid, do in fact undergo exchanges of genetic material in simplified versions of sexual reproduction (Figure 6.5).

A few years later, further evidence for the role of nucleic acids was made available by the studies of A. D. Hershey and M. Chase on a system that is even simpler than the bacterium: the virus. Specifically, they were studying reproduction of a *bacteriophage,* viruses which parasitize bacteria. One of the unique properties of all viruses is that they must

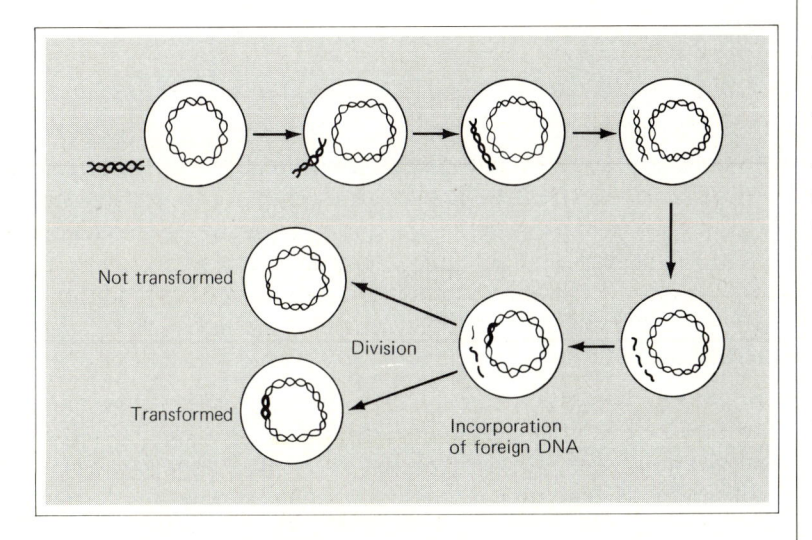

FIGURE 6.5

Transformation in bacteria. Only a portion of one of the strands of the transforming DNA is incorporated into the host cell, so that with the first division, one of the daughter cells is the same as the parental cell, while the second daughter cell contains a foreign piece of DNA and is transformed.

infect a cell in order to reproduce. They cannot reproduce on their own, but they are able, in effect, to "take command" of the cell's genetic machinery and direct the cell to produce viral proteins and nucleic acids to form new viral particles.

In their studies, Hershey and Chase determined that when a viral particle infects a bacterial cell, the protein coat of the virus remains outside the cell, only the contents of the virus managing to enter the cell's interior. They showed that this material is nucleic acid. The viruses in their experiments contained DNA. Figure 6.6 shows some drawings of common viruses, and the manner in which a bacteriophage infects a host cell.

THE VALUE OF MICROORGANISMS IN GENETIC STUDIES

In view of what we have already discussed about the need for large numbers of progeny and short life spans for genetic analysis, it should not be surprising that geneticists seized upon microorganisms for study. For here was the opportunity to work with a life span of minutes, and organisms so small that a test tube contains progeny by the millions! The fact that they have only a single strand of genetic material was an additional advantage, considering the complexities of diploidy that stymie analysis of higher organisms.

THE SIGNIFICANCE OF DNA

Once the importance of nucleic acids was established by microbial geneticists, a frenzy of effort ensued all over the world to characterize the nature of DNA, for only if scientists understood the molecular nature of DNA could they begin to understand how the gene determines the function or structures of a cell. The significance of the quest was not lost on anyone, for what was being sought was the basis to all life. Since all matter is made of atoms and molecules, only if we can reduce cellular processes to the level of molecules can we truly understand the manner in which living organisms function.

The knowledge which has come from this area of genetics, commonly called *molecular genetics*, has been so fundamental to our concepts of life, and the implications so profound and far-reaching, that it is absolutely essential for

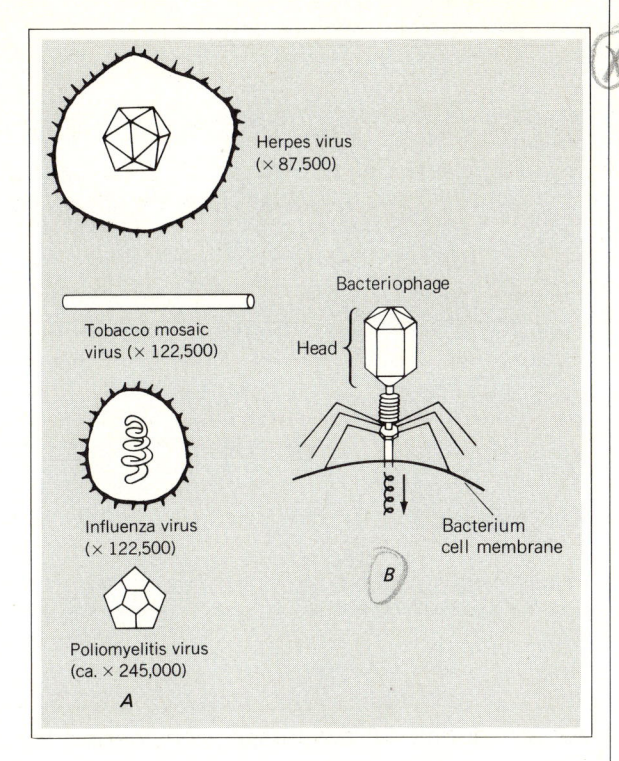

FIGURE 6.6
(A) Some common viruses are drawn as they would appear under high magnification. (B) A bacteriophage infecting a host bacterial cell. DNA from the inside of the head of the virus is injected (arrow) into the host cell, leaving the protein "shell" of the virus outside. [*After R. W. Horne, "The Structure of Virus." Copyright © Jan. 1963 by Scientific American, Inc. All rights reserved. p. 128.*]

all nonscientists to have some basic understanding of the molecular nature of the gene and gene action. We shall discuss the implications of molecular genetics for man and his society in a later chapter.

THE STRUCTURE OF DNA

The excitement was great, the competition keen, and in 1953, in a paper only 2½ pages long, the solution was presented by an Englishman, Francis Crick, and a young American, James Watson. Their analysis led them to formulate a model of the DNA molecule as a double-stranded helical structure, or what is now known as the Watson-Crick double helix.

THE DOUBLE HELIX

The beauty of the molecule is in its simplicity. The introductory illustration is a model of DNA, and Figure 6.7 depicts a

FIGURE 6.7
The molecular structure of the genetic material DNA (*A*) unwound and (*B*) as it exists in the double helical form.

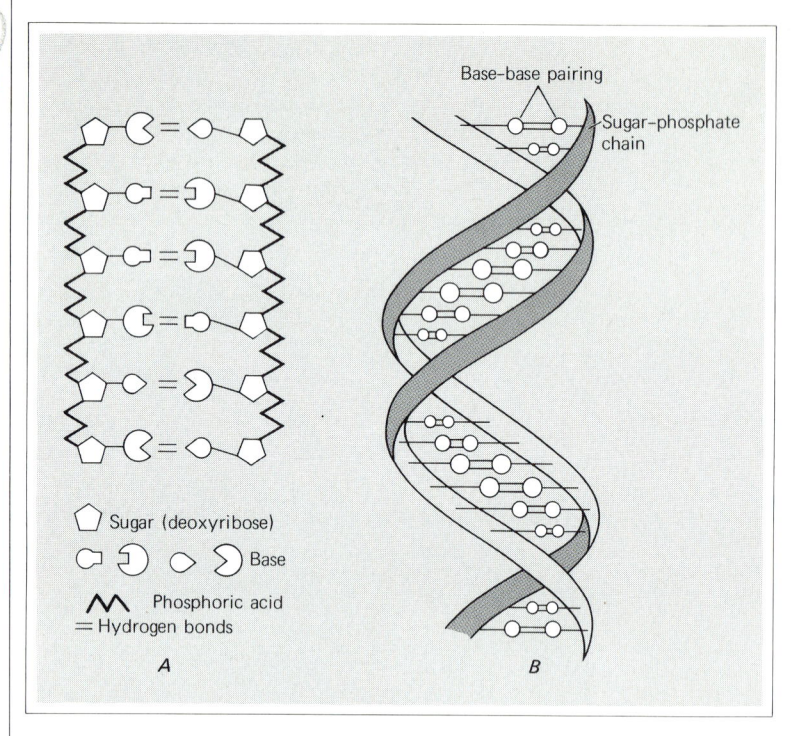

diagram of the basic structure of DNA. If we untwist the double helix, the molecule resembles a ladder, with the sugar and phosphoric acids linked in chains as the sides of the ladder. Forming the rungs of the ladder are the nitrogenous bases, one attached to each sugar molecule and bonded together in the middle of the rung by the element hydrogen.

One of the tests of the significance of a scientific discovery is not only its immediate contribution but also its heuristic value, its capacity to lead toward further discoveries. This test is passed with flying colors by the Watson-Crick double helix, as is witnessed by the number of Nobel Prize winners in the last two decades who have in some way contributed to our knowledge of the molecular nature of gene action. Many of these studies were direct offshoots of the concept of the double helix. Table 6.1 lists some of these notable accomplishments.

TABLE 6.1
Nobel Prize winners of the past two decades, and their contributions to our understanding of the gene and various aspects of gene action and protein synthesis

YEAR	NOBEL WINNERS	RESEARCH
1958	G. Beadle and E. Tatum; J. Lederberg	Biochemical genetics in fungus; bacterial recombination
1959	S. Ochoa; A. Kornberg	Artificial synthesis of nucleic acids
1962	J. Watson and F. Crick; M. Wilkens	Molecular nature of DNA
1965	F. Jacob, A. Lwoff, and J. Monod	Protein synthesis in virus
1968	R. W. Holley; H. G. Khorana; M. W. Nirenberg	Deciphering the genetic code
1969	M. Delbruck; A. Hershey; S. Luria	Sexual reproduction and genetics of bacteriophage
1972	R. Porter; G. M. Edelman (physiology and medicine);	Chemical structure of antibodies (see Chapter 9)
	C. B. Anfinsen; S. Moore and W. H. Stein (chemistry)	Chemical structure and activity of the enzyme ribonuclease

THE SIGNIFICANCE OF SPECIFIC BASE PAIRING

One question whose answer was suggested by the double helix was the manner in which gene duplication occurs so precisely in cell division, something that had been known since the nineteenth century but had never been fully understood. Cells undergo mitosis millions of times a day in complex organisms, for the most part resulting in cells exactly identical to each other. The precision involved has long fascinated biologists. We now know this precision is a function of the nitrogenous bases.

The one variable component of DNA is the nitrogenous base. The sugar is the same molecule in all DNA, and phosphoric acid is the same in all DNA. There are four different bases, however, which vary in quantity from organism to organism and from species to species. These four bases are adenine, thymine, guanine, and cytosine; for simplicity we shall refer to them as A, T, C, and G.

Even before the double helix was discovered, biochemists knew that the amount of A always equaled the amount of T in the DNA of any organism, and the amount of C always equaled the amount of G. With the clarification of the molecular nature of DNA, the reasons for this phenomenon and for the precision of duplication of genes in mitosis became apparent.

To put it very simply, the consistency of the ratio of $C + T = A + G$ is a matter of the size of the bases and the space available between the sugar molecules in the double helix. If we assume that there is room for three "units" between sugar molecules and that A and G are two units big, while C and T, being smaller molecules, are only one unit big, then it is easy to see how only certain of the bases can be linked in the space available—namely, A and G (each two units) with C and T (each one unit) (Figure 6.8). A and G together would form four units, too big for the space, and C and T together would form only two units, not big enough to fill the space between two sugars. In addition, certain details of structure, which we need not go into, allow *only* linkage of A with T, and C with G. This is known as the *specificity of base pairing*.

Experiments indicate that when a cell is preparing to undergo division, the double helix within each chromosome unwinds in a progressive manner from one end. (An enzyme has been found which causes the hydrogen bonds holding the bases together to break, resulting in single polynucleotide strands with the bases exposed.) From the nucleoplasm, new molecules of the bases are bonded to the exposed bases, but because of the specificity of base pairing, the original partners of the exposed bases are exactly replaced by new *but chemically identical molecules* (identical to themselves), as illustrated in Figure 6.9. An enzyme, DNA polymerase, then causes the joining of the nucleotides.

The two new strands of polynucleotides will be the complement of their partner strand, and each will be identical to the original strand that is not their partner. In Figure 6.9, strands 1 and 4 are the original partners. Upon duplication, new strands 2 and 3 are mirror images to their partners 1 and 4, respectively, and because of the base pairing, 1 and 3 are chemically identical. The end results are two double helices that are exactly alike: they will separate in mitosis into daughter cells which are therefore genetically identical.

In higher organisms, whose cells contain chromosome pairs rather than the single strand of DNA found in bacteria, each

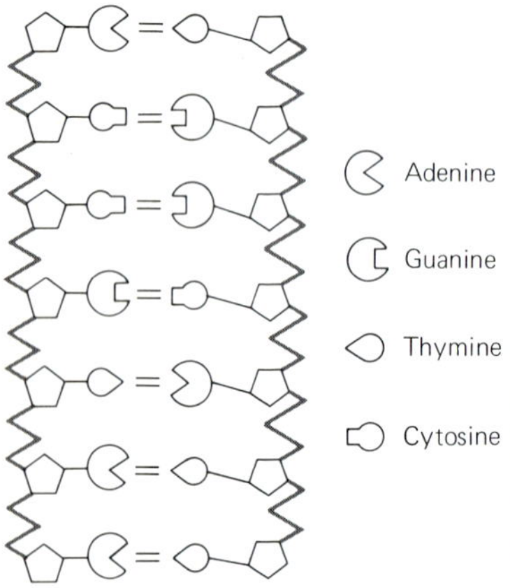

FIGURE 6.8
The specificity of base pairing found in all DNA allows the base A (adenine) to pair only with T (thymine), and C (cytosine) only with G (guanine).

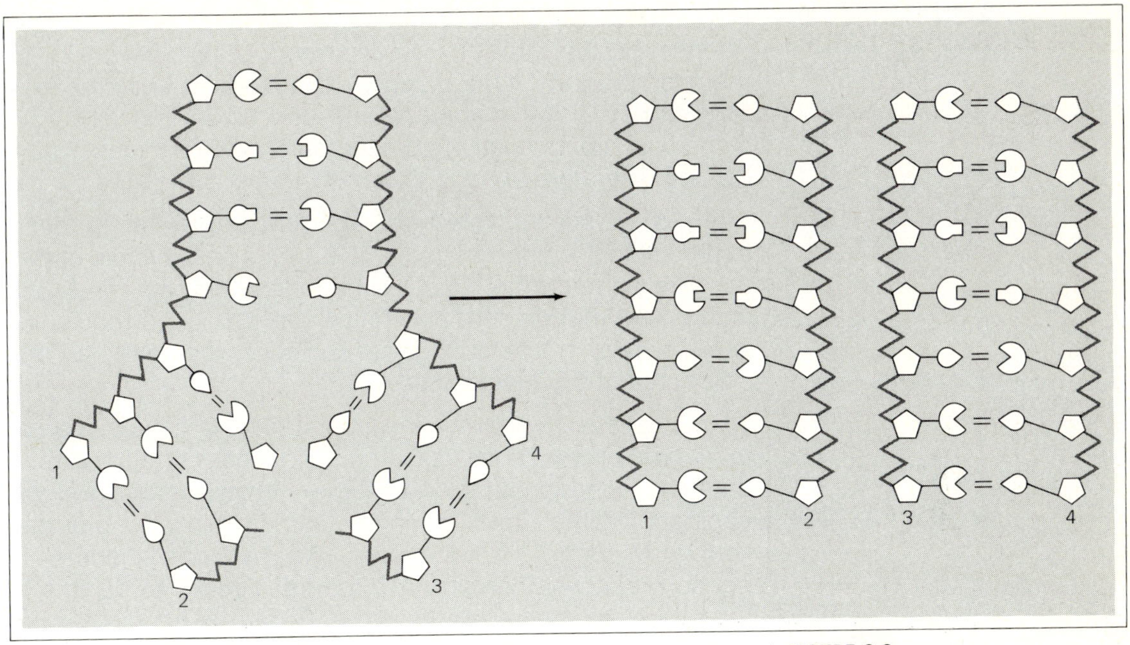

chromosome consists of a double helix of DNA. During mitosis and meiosis, one of the double helices resulting from the synthesis described above will be in each of the sister chromatids. For this reason the sister chromatids may be considered genetically identical, barring events such as crossing-over.

Just as all chemical reactions that occur in a cell are under the control of enzymes, so are all the steps involved in the duplication of DNA, which is referred to as *replication*. This type of duplication, in which the resulting double helices consist of one old and one new strand, is frequently referred to as *semiconservative replication*. We shall not go into the chemical aspects of the steps or the enzymes involved in replication, but they have been fairly well detailed, to the point where molecular biologists have been able to synthesize DNA in a test tube in the presence of a small amount of natural DNA to serve as a "primer."

FIGURE 6.9
Diagrammatic illustration of the replication of genetic material during cell division, resulting in two chemically identical double helices. Strands 1 and 4 are original strands; 2 and 3 are new strands.

THE SIGNIFICANCE OF THE MOLECULAR NATURE OF THE GENE

How far we have come in such a short time! Genetics as a science began with the "rediscovery" of Mendel's experiments at the turn of the century. Little more than half a century later, man has made DNA in a test tube—DNA, the substance that controls all life processes, that determines what a cell and therefore an organism will be and can do.

Although we shall devote a good portion of the later chapters to the subject, some reflection now on the possibilities of this accomplishment may help you to understand why many scientists feel that present and future discoveries in molecular genetics hold greater impact for society as a whole than any other area of the life sciences.

Why? Because what we can understand, *we have a chance to control.* Through the research of countless fine workers in the field, in addition to the Nobel Prize winners, we are approaching the depth of understanding that may one day lead us to control life processes. What worries scientists is not so much whether we have the intelligence to do this in the future as whether we have the wisdom to handle our potential capabilities for the benefit of man and our natural world.

REFERENCES

Work cited:

Avery, D. T., C. M. Macleod, and M. McCarty, 1944: "Studies on the Chemical Nature of the Substance Inducing Transformation of Pneumococcal Types. Induction of Transformation by a Deoxyribonucleic Acid Fraction Isolated from Pneumococcus Type III," *The Journal of Experimental Medicine,* **79**:137–158. Also reprinted in J. Peters (ed.), *Classic Papers in Genetics,* Prentice-Hall, Inc., Englewood Cliffs, N. J., 1959.

Griffith, F., 1928: "The Significance of Pneumococcal Types," *The Journal of Hygiene,* **27**:113–159.

Hershey, A. D., and M. Chase, 1952: "Independent Functions of Viral Protein and Nucleic Acid in Growth of Bacteriophage," *The Journal of General Physiology,* **36**:39–56.

Watson, J. D., and F. C. Crick, 1953: "Molecular Structure of Nucleic Acids. A Structure for Deoxyribose Nucleic Acids," *Nature,* **171**:737–738. Also reprinted in J. Peters (ed.), *Classic Papers in Genetics,* Prentice-Hall, Inc., Englewood Cliffs, N.J., 1959.

Other good sources:

The Molecular Basis of Life: An Introduction to Molecular Biology, with introductions by Robert H. Haynes and Philip C. Hanawalt, *Readings from Scientific American,* W. H. Freeman and Company, San Francisco, 1968.

Facets of Genetics, with introductions by Adrian M. Srb, Ray D. Owen, and Robert S. Edgar, *Readings from Scientific American,* W. H. Freeman and Company, San Francisco, 1968.

A self-instructional supplement at a basic level:

Pardee, G., W. A. Reynolds, and R. Reynolds, 1966: *DNA: The Key to Life,* Educational Methods, Inc., Chicago.

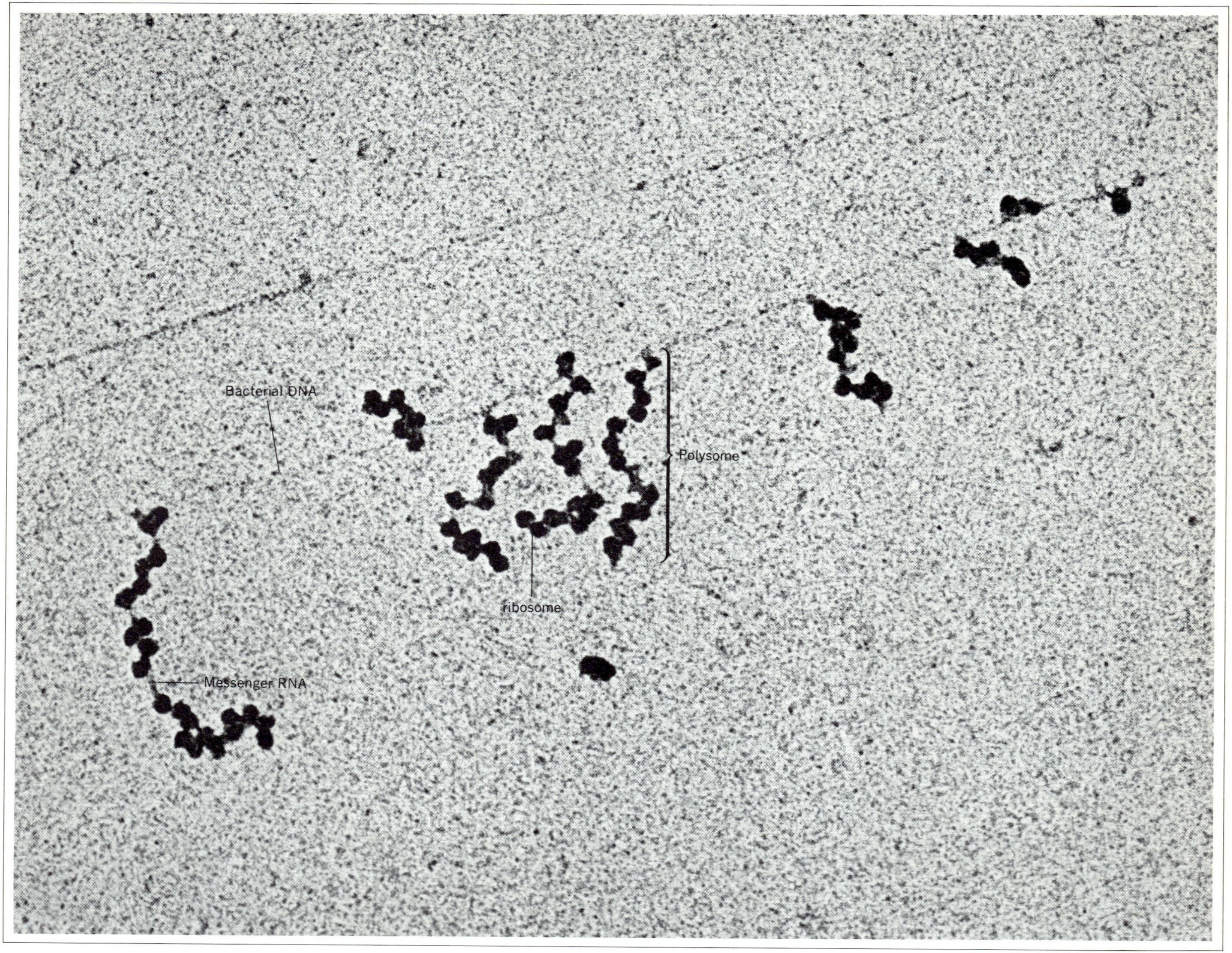
Bacterial DNA
Polysome
ribosome
Messenger RNA

Chapter 7

What Does a Gene Do?

Prior to the discovery of the double helix, some definite ideas had already been formed about the actual function of a gene. We mentioned the essential roles of the class of organic compounds known as proteins earlier. You recall that they are components of the structure of a cell, and, in the form of enzymes, determine the metabolic reactions of a cell. In other words, the structure and function of a cell—which are actually its phenotype—are determined by the proteins in the cell.

GENES AND ENZYMES

It was, therefore, a logical step to investigate the relationship between genes and proteins. The classical experimentation that clearly pointed the way for studies of gene action was performed by G. Beadle and E. Tatum in 1941 on a microorganism that has since proved to be extremely valuable for genetic studies, the bread mold *Neurospora crassa*, a fungus (Figure 7.1).

Like bacteria, *Neurospora* offered the advantages of being small, having a short generation span and large numbers of progeny, and being haploid, which reduces complexity by

Bacterial gene in the process of transcribing messenger RNA molecules. Magnification: 158,000×. [*Courtesy of Dr. O. L. Miller, Jr. and Dr. Barbara A. Hankalo, Biology Division, Oak Ridge National Laboratory.*]

eliminating dominance and recessiveness. Thus, there is no masking of the expression of genes in the bacterial and fungal cells of a haploid organism, and any mutations of genes can therefore be immediately detected.

BEADLE AND TATUM'S EXPERIMENTS ON *NEUROSPORA*

What Beadle and Tatum set out to do was to expose cultures of normal *Neurospora* (sometimes the normal phenotype is referred to as *wild type*) to ultraviolet radiation, which is known to cause inherited changes (i.e., mutations), and to analyze the effects on the cells.

Wild-type *Neurospora* can be grown in a medium containing only the bare minimum number of nutrients necessary for survival, since the fungus is able to manufacture many substances essential for growth through its metabolic reactions. Such a medium is known as *minimal medium*. Beadle and Tatum found that after exposure to radiation, some of the cultures could no longer survive on minimal medium. But if they added supplemental nutrients to the medium, the cells were able to live.

By adding individual nutrient supplements to the medium one at a time, Beadle and Tatum found that a mutation always resulted in the inability of the cell to synthesize a particular substance. In other words, a particular metabolic reaction which normally would have led to the production of an essential substance was blocked, resulting in the cell's inability to survive on minimal media.

THE "ONE-GENE-ONE-ENZYME" HYPOTHESIS

In their analysis, Beadle and Tatum felt that this result reflected a definite relationship between genes and those vital substances that control metabolic reactions, enzymes:

The development and functioning of an organism consist essentially of an integrated system of chemical reactions controlled in some manner by genes. It is entirely tenable to suppose that these genes which are themselves a part of the system, control or regulate specific reactions in the system either by acting directly as enzymes or by determining the specificities of enzymes.*

* G. W. Beadle and E. L. Tatum, "Genetic Control of Biochemical Reactions in *Neurospora*," *Proceedings of the National Academy of Science,* **27**:499–506, 1941.

Thus, each mutant obtained by irradiation was unable to synthesize a particular substance because the gene which

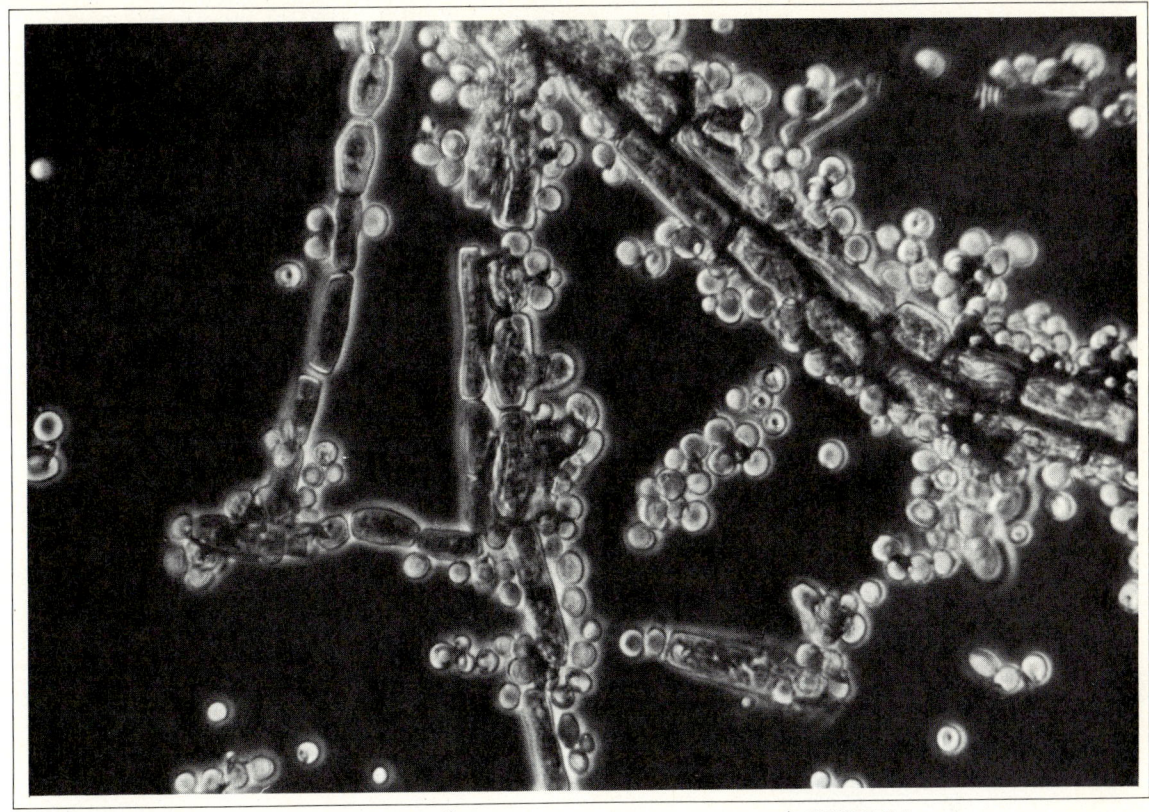

FIGURE 7.1
Photomicrograph of *Neurospora*, a fungus. [*Photo—Carroll H. Weiss, RBP.*]

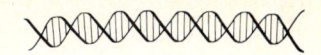

had been changed was no longer able to determine the presence of the enzyme needed for that particular reaction. From their analysis arose a hypothesis that came to be known as the "one-gene–one-enzyme" theory, and although modern genetics has led to modification of this theory, the close relationship between genes and enzymes has been confirmed as the basis of modern concepts of gene action.

In humans, the relationship between genes and enzymes was discovered many decades before Beadle and Tatum's work. In 1901 a British physician, Sir Archibald Garrod, was studying a condition called alkaptonuria. You might remember that Mendel's laws were just receiving their due attention in

INBORN ERRORS OF METABOLISM

the scientific world at the time. Garrod noted that several patients suffering from the disease were offspring of consanguineous marriages. With the help of classical geneticists, he concluded that alkaptonuria was a heritable disease which was determined by a recessive gene.

Further study showed a number of heritable diseases which seemed to result from metabolic disorders, and so in 1908 Garrod published a book called *The Inborn Errors of Metabolism,* in which he linked these biochemical defects with genes. Since that time, over 100 conditions have been shown to be due to either the deficiency or absence of specific enzymes. Figure 7.2 shows the relationship between some of the disorders and the particular metabolic reaction blocked by lack of the essential enzyme.

Fortunately, a few of these conditions can be treated after detection. For example, galactosemia is a condition in which the enzyme for the breakdown of a sugar, lactose, found in milk, is missing and could, if no steps are taken, result in severe mental retardation and other symptoms. However, if the condition is detected, the mental retardation can be avoided by simply eliminating milk from the diet of the homozygous babies. In other cases, such as albinism, such

FIGURE 7.2
The relationship between various metabolic reactions and inherited disorders (shaded). Smooth arrows indicate the direction in which one substance is metabolized to another. Jagged arrows indicate blockage of the reactions by mutated genes resulting in an absence of essential enzymes.

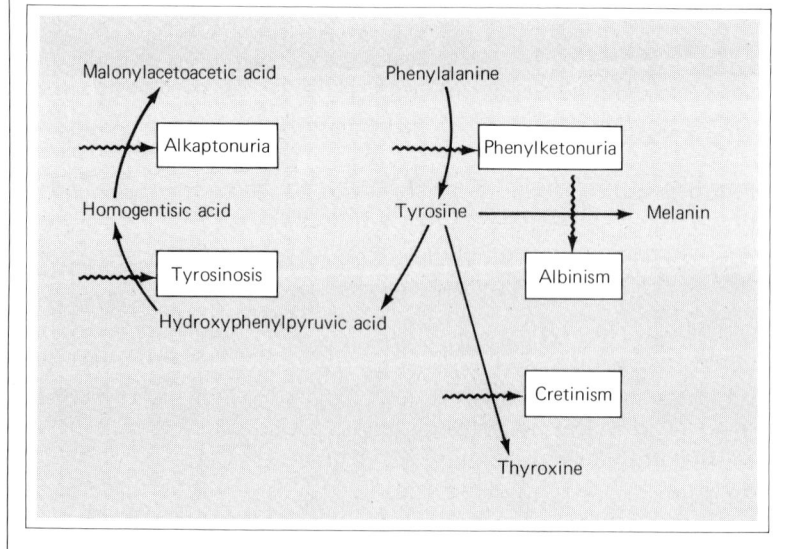

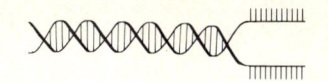

cures are not yet possible. We know what the problem is, but we do not yet have the ability to deal with it.

THE MECHANISM OF GENE-DETERMINED PROTEIN SYNTHESIS

In attempting to discern the mechanism by which genes determine enzymes and other proteins in a cell, biologists had to reflect on the fact that the synthesis done by a cell occurs in the cytoplasm of the cell, whereas the genetic information for this synthesis resides in the DNA which never leaves the nucleus, being bound up in the chromosomes. It appeared likely that there is some system of nucleocytoplasmic cooperation.

First, there has to be some channel of communication between the nucleus and the cytoplasm. Evidence of this came from different areas of biology. For example, in the 1940s, biologists used dyes that did not harm a living cell but which were able to enter the cell and stain various organelles. They found that some particles leave the nucleus by traversing the nuclear membrane to enter the cytoplasm. More recently, the high magnification of powerful electron microscopes showed that the nuclear membrane actually is full of tiny pores through which material can pass into and out of the nucleus. Figure 7.3 shows the pores in an electron micrograph, a picture taken through the electron microscope.

MESSENGER MOLECULES FOR THE GENE

The nature of the system of communication between nuclear and cytoplasmic elements was hypothesized by a team of French scientists at the Pasteur Institute, F. Jacob and J. Monod, who proposed that there must be a molecule that serves as a kind of messenger to carry the genetic message from the gene to the ribosome, the organelle that was known to be associated with protein synthesis in cells.

Subsequent experimentation showed that there was indeed such a messenger molecule, and that it was in fact a type of nucleic acid, slightly different from DNA, called *ribonucleic acid* or RNA. The major differences between DNA and RNA are that the latter has a different sugar, ribose, which accounts for its name. Also, RNA is single-stranded and has uracil (or U) instead of thymine (T) as one of its four bases.

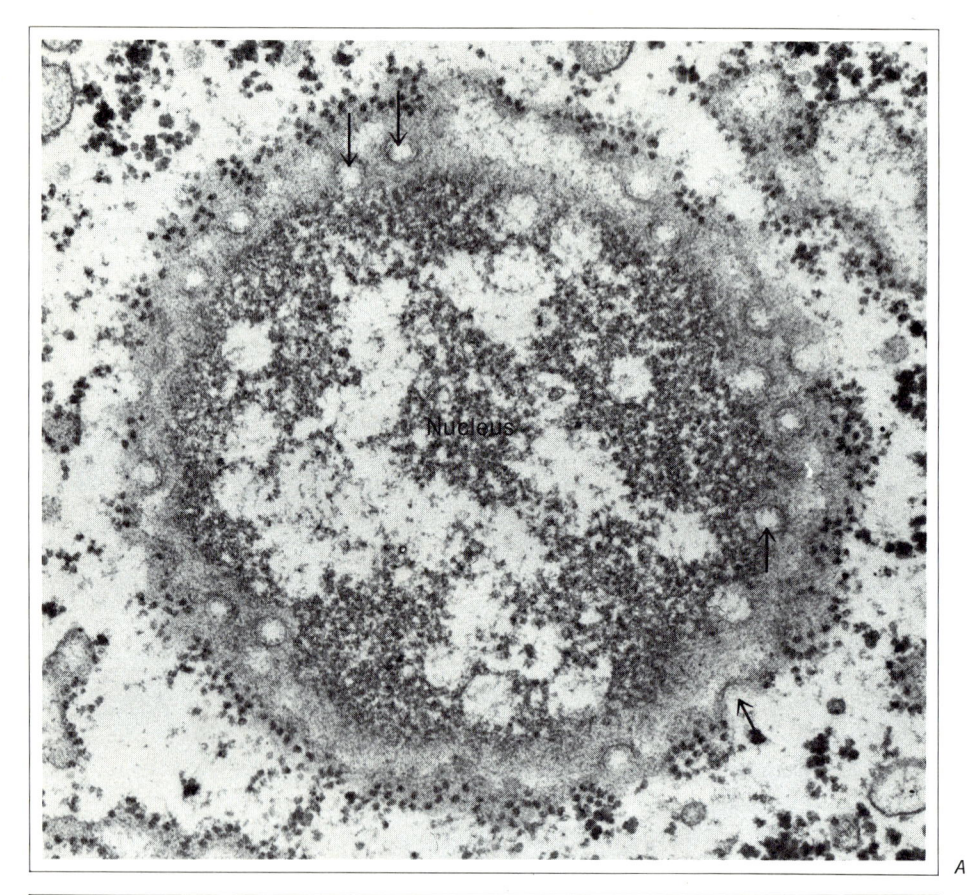

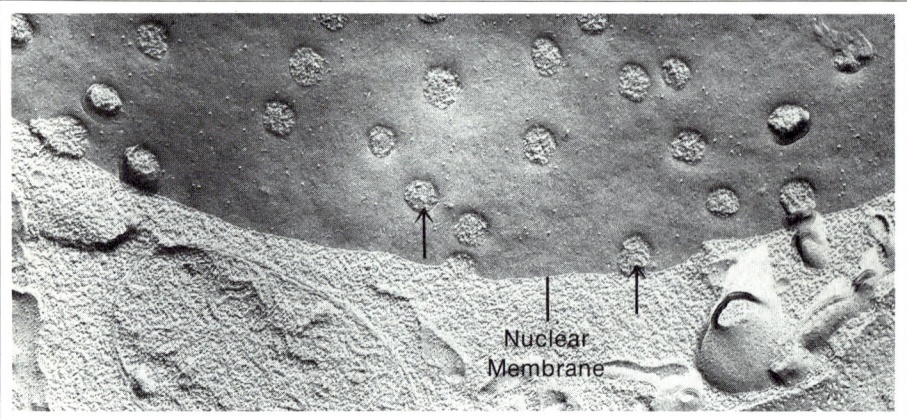

FIGURE 7.3
Electron micrographs showing the presence of pores (arrows) in the nuclear membrane through which substances can pass into and out of the nucleus and cytoplasm. (A) A micrograph of a section through the nucleus of a cell. [*Courtesy of Dr. G. E. Palade, Rockefeller University.*] (B) A surface view of the nuclear membrane. [*Courtesy of Dr. Daniel Branton, University of California, Berkeley.*]

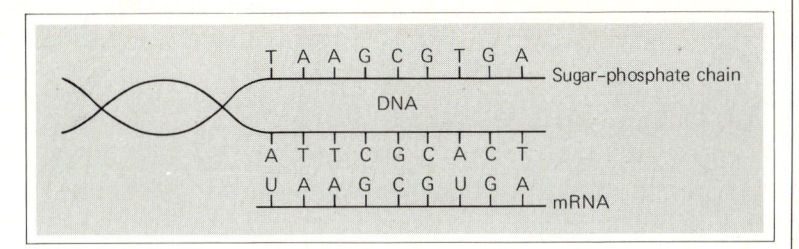

FIGURE 7.4
Schematic illustration of transcription, or the formation of a molecule of messenger RNA. Since one of the strands of DNA serves as the template, the RNA molecule will contain a sequence of bases which is the complementary sequence of the template strand, and identical to the other strand of DNA, with the exception of uracil molecules instead of thymine.

The messenger RNA, or mRNA, is formed when a particular segment of DNA—that is, a gene—becomes active. In a manner similar to the replication of DNA during cell division, an active gene serves as the template for the formation of a piece of RNA which then carries a sequence of bases that is the mirror image of one of the strands of DNA and identical to the other strand of DNA. Each sequence of three bases on the mRNA is called a *codon*. (We shall discuss later why the codon is composed of a sequence of only three bases, no more or less.) Why only one strand serves as the template for mRNA is not clearly understood. This process, illustrated in Figure 7.4, is known as *transcription*.

TRANSLATION OF THE GENETIC MESSAGE

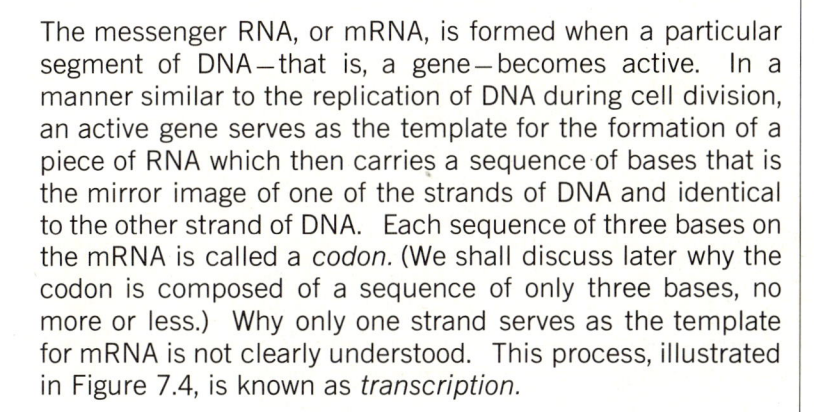

Once the mRNA has been transcribed, it then leaves the nucleus and becomes attracted to ribosome particles in the cytoplasm which tend to form clusters known as *polysomes*. Ribosomes have been found to contain RNA and protein as well, and because they are the sites of synthesis in a cell, they have sometimes been called the "factories" of the cell. Figure 7.5 shows the relationship between a mRNA strand and a polysome.

In the cytoplasm there is yet another class of DNA-determined RNA molecules which become involved in the process of synthesis when mRNA attaches to polysomes. These molecules are referred to either as soluble RNA (sRNA) or, more commonly, transfer RNA (tRNA), and are quite different in form and function from mRNA. Recent evidence has indicated that the tRNA molecule is a very interesting one, featuring parts of the tRNA which exist as double-stranded nu-

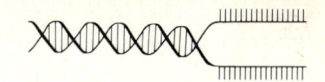

FIGURE 7.5
The relationship between a molecule of mRNA and ribosomes in the cytoplasm.

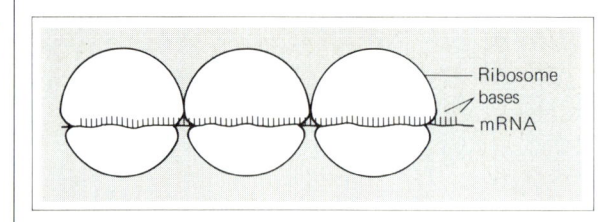

cleic acid. Figure 7.6 illustrates what was thought to be the structure of a tRNA molecule.

As you can see in the figure, tRNA has other unique features: it is joined at one end to a single molecule of amino acid, and at the apex of one of the loops found in tRNA there is a short sequence of bases (called the *anticodon*) that is the source of attraction between the tRNA and the mRNA. Every tRNA that has a sequence of bases at the loop complementary to a sequence on the mRNA forms a bond with the messenger similar to the bonding between the double strands of the DNA helix. The two other loops are believed to be the sites of attachment of the tRNA to the ribosome, and of the enzyme catalyzing the acceptance of an amino acid by the tRNA.

In 1973, S. H. Kim and co-workers at MIT published their findings of studies on a molecule of tRNA from yeast cells

FIGURE 7.6
A conventional diagram of a molecule of tRNA showing portions doubled over to form double strands. Large arrow points to exposed bases that will be attracted to complementary sequences on the mRNA. The sequence of three bases is known as the tRNA's anticodon. A molecule of amino acid is bound to one end of the tRNA.

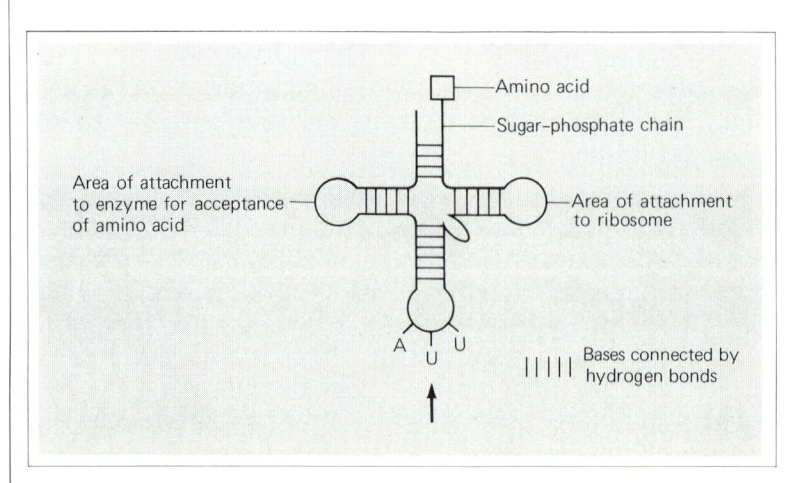

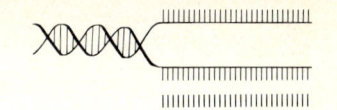

that modifies the conventional concepts of the structure of tRNA. Rather than a cloverleaf pattern, the molecule forms two main portions that are more or less at right angles to each other, with each portion doubled over by bonds holding complementary bases together. Figure 7.7 illustrates the most recent concepts of tRNA structure.

Several tRNA molecules are attracted to the long messenger, and they bring their amino acids with them. The result of this attraction between mRNA and tRNA is a progressively lengthening series of amino acid molecules lined up next to one another. Figure 7.8 illustrates this interaction, which is a phase of gene-determined protein synthesis known as *translation*.

It is not difficult to envision the establishment of peptide bonds between the amino acids and the enzymatically caused release of the amino acids from the tRNA, resulting in the formation of a chain of amino acids. What had we described earlier as a chain of amino acids? A polypeptide, the primary and secondary structure of a protein.

At this point in the discussion of gene action on the molecular level, the accomplishments of molecular geneticists in dissecting the physiology of the cell to such a high degree seem almost unbelievable. Cells, remember, are for the

COMPLETION OF THE POLYPEPTIDE

FIGURE 7.7
Representations of the structure of tRNA as studied in yeast cells. (*A*) Actual appearance of the folding of the molecule. (*B*) Diagrammatic illustration of the possible function of different parts of the molecule. [*After S. H. Kim, "Three-Dimensional Structure of Yeast Phenylalanine Transfer RNA: Folding of the Polynucleotide Chain,"* Science **179**:285–288. Copyright © 1973 by the American Association for the Advancement of Science.]

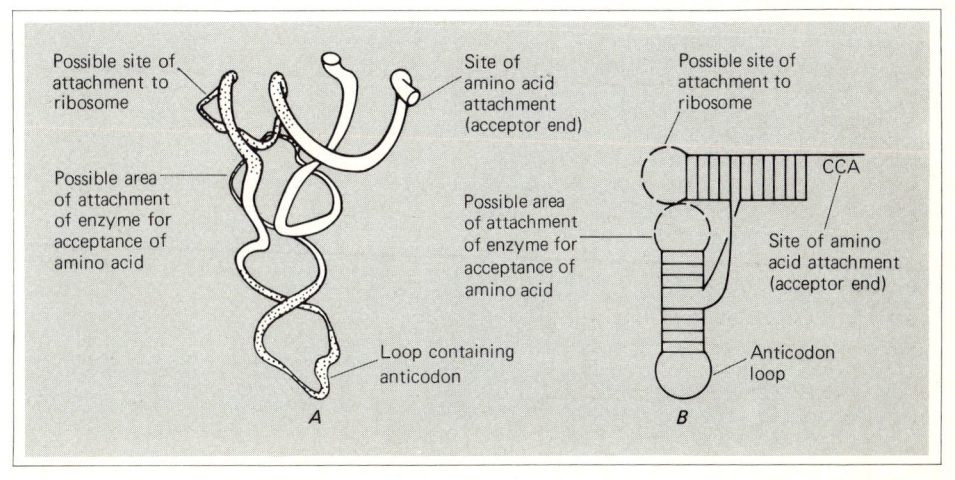

most part so small as to be invisible to the naked eye. Yet here we are discussing how the *components* of a cell are interacting with each other!

As the growing polypeptide is added to the next amino acid in the sequence, the amino acid is enzymatically released from the tRNA, which then returns to the cytoplasm to pick up another amino acid molecule, the same kind it had before. (The specificity of amino acids for each tRNA molecule will be explained later in this chapter.) After the message has been completely translated, the polypeptide chain is released, and continues to undergo change to assume secondary and tertiary configurations, the mechanisms for which are not clearly understood. The ribosomes break up

FIGURE 7.8
The interaction among mRNA, tRNA, and ribosomes in the formation of a polypeptide chain, known as the translation phase of protein synthesis. (A) General appearance of polysome as polypeptides increase in length. (B) Detailed schematic diagram of tRNA molecules bringing in and releasing amino acids as they are added to the growing polypeptide chain. The sequence of amino acids in the chain is indicated by numbers and is determined by the sequence of triplet codes in the mRNA.

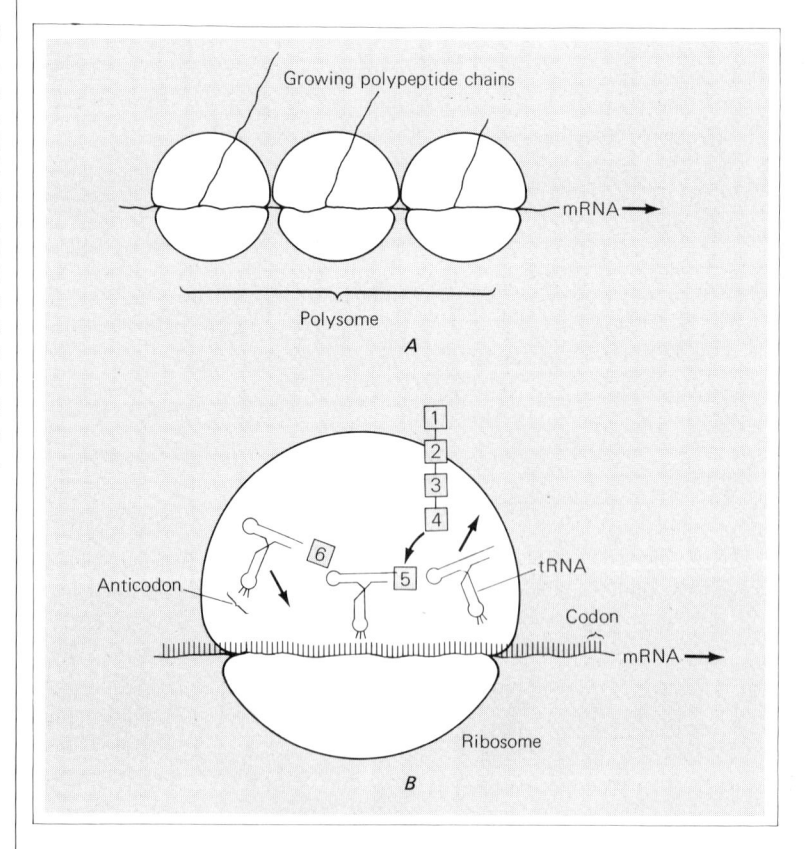

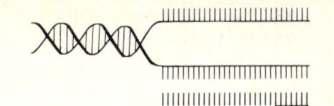

into two parts of different sizes which reassemble when polysomes are formed, preparatory to further protein synthesis.

The mRNA is a short-lived molecule, and once its message has been translated into protein it is destroyed. This may seem wasteful, since each message is for a protein the cells need, but actually it is necessary in order to maintain a balance of reactions in the cell. If a particular message is allowed to be translated continuously, it could cause the finely tuned balance of metabolic reactions to be thrown off. For one thing, the same amino acids would be needed over and over again, depleting the supply of such molecules in the cytoplasm faster than they could be replaced. If a protein is needed in large quantities by the cell, it is obtained by the continual production of more mRNA bearing the same genetic message.

In some cases, it may be detrimental to a cell or an organism to have too much of a very potent substance. Consider, for example, the hormone adrenalin, synthesized in the adrenal glands, a pair of small glands situated on top of the kidneys. In times of emergency or excitement, one of the ways the body can respond is by producing adrenalin in response to signals from the nervous system. When circulated throughout the body in the bloodstream, adrenalin causes the symptoms of excitement so familiar to all of us: rapid breathing, dilation of the pupils, spurts of energy and strength, etc. Imagine how exhausting it would be to be in this state constantly! It is not only fortunate but imperative for our survival that our cells and tissues do not continually produce all the substances they are capable of producing.

One exception to the short-lived nature of most mRNA is found in the cytoplasm of egg cells. During the maturation of an egg cell, messenger molecules containing information for the divisions of the cell after fertilization are released to the cytoplasm. These mRNA molecules are apparently protected from enzymes that destroy mRNA by being bound to protective protein "coats." As the zygote grows, the mRNA for division and synthesis of new cell structures becomes active in a manner we do not understand. That the information is stored in the cytoplasm is evidenced by experiments in which nuclei are removed from amphibian egg cells. These cells can then be pricked by a needle and will begin to divide.

After a number of divisions, however, the cell will die, presumably because the messenger molecules are used up.

The magnificent contributions of molecular biologists to our understanding of gene action, which is essentially an understanding of life itself, have thus revealed that each tiny cell, which are present by the billions in a human being, is a microcosm unto itself, with different parts communicating with each other. It is an understatement to say that the more we know about life processes, the more wonderful, in every sense of the word, is life.

HOW DOES A CELL PRODUCE THE CORRECT PROTEIN?

Following the discovery of the steps involved in protein synthesis, a concerted effort was undertaken by many laboratories to determine how the cell is able to synthesize the correct protein. It is obvious that not all cells contain the same proteins. In Chapter 2 you saw that no two cell types look alike, and we know that no two cell types function in the same way; muscle cells are able to contract because of special contractile proteins present in their highly specialized cytoplasm, for example, and no other cells have these proteins, which is why no other cells are capable of contraction.

Recall that earlier in our discussion we had said that the individuality of a protein is determined by the number and sequence of amino acids in its polypeptide chains. Therefore, in order to determine that a specific protein will be made by a cell, the genetic material, DNA, must have in its substance the information for a particular number and sequence of amino acids.

VARIABILITY IN PROTEINS AND DNA

Let us turn our attention once again to the double helix and attempt to discern where in this unique structure the messages for the myriad types of proteins that exist in living organisms may be stored.

Because of the great numbers of proteins found in nature, it was evident to molecular biologists that the answer did not lie in the sugars and phosphoric acids of DNA. These subunits of DNA are the same in all cells, and could not, therefore, account for variability in anything. What is variable in DNA is the sequence and numbers of nitrogenous bases in cells from individual to individual, although the proportional amounts of the four bases are consistent in cells from the same individual.

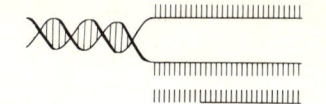

There must be a correlation, then, between the sequence of bases in DNA and the sequence of amino acids in proteins. In other words, the message for particular proteins must be encoded in some way in the sequence of nitrogenous bases. The next area of molecular genetics to receive intensive investigation was the nature of this "genetic code."

THE GENETIC CODE

There are 20 amino acids; there are 4 different bases in DNA. How many bases would form a code for a particular amino acid? Those among you with some mathematical training will recognize that the code could not be a doublet code. With 4 bases to work with, there are only 16 possible combinations of 2 bases, for example, A–T, A–C, A–G, A–A, C–T, C–C, C–G, G–C, etc. (Note that here we are discussing the *sequence* of bases on one strand of DNA, *not* the base pairs between the two partner strands of a double helix.) Whatever the code is, it has to allow for at least 20 different combinations to cover the 20 different amino acids that exist.

It was evident, then, that the code was probably a triplet code, which allows for 64 combinations. It is beyond the scope of our discussion to go into the many elegant experiments conducted; for details, turn to some of the references listed at the end of the chapter as supplemental reading. It is enough to say that scientists have established that the genetic code *is* a triplet code, and have completely solved the "meaning" of each of the 64 triplets. This has to stand as one of the great scientific achievements of all time.

Table 7.1 shows the complete unabridged "dictionary" of the genetic code. Note that the code contains U and not T in the triplets. This is due to the fact that geneticists refer to the triplets carried on the mRNA molecules as the components of the genetic code rather than the sequence in the DNA itself, because it was from work on the mRNA that the code was deciphered. Each triplet representing the information for a particular amino acid is referred to as a *codon*.

What Does a Gene Do?

TABLE 7.1
The deciphered genetic code, showing the amino acid coded by each of the codons

UUU UUC	phen	UCU UCC		UAU UAC	tyr	UGU UGC	cyst
UUA UUG	leu	UCA UCG	ser	UAA UAG	nonsense	UGA UGG	nonsense try
CUU CUC CUA CUG	leu	CCU CCC CCA CCG	prol	CAU CAC	his	CGU CGC CGA CGG	arg
				CAA CAG	gln		
AUU AUC	ileu	ACU ACC ACA ACG	thr	AAU AAC	asn	AGU AGC	ser
AUA				AAA AAG	lys	AGA AGG	arg
AUG	met						
GUU GUC GUA GUG	val	GCU GCC GCA GCG	ala	GAU GAC	asp	GGU GGC GGA GGG	gly
				GAA GAG	glu		

Abbreviated words to the right of triplets represent amino acids. Note that some amino acids such as arginine (arg) are coded by as many as six different codons. Explanation for the "nonsense" codons is in the text.

Obviously 64 possible combinations in triplet sequence are more than enough to represent 20 amino acids, and indeed, there has been found a certain "redundancy" in the genetic code, whereby more than one triplet codes for a particular amino acid. The term geneticists have given for this kind of redundancy in the code is *"degeneracy"*; it is a *degenerate code*. Further, it has been found that the first two bases in the codon are the important ones, and the third base may in some cases be any of the four bases. The triplet would still code for the same amino acid as long as the first two bases are the same and have the same sequence.

Knowledge of the genetic code has helped to clarify some of the events in protein synthesis. The tRNA molecules are attracted to the codons on the mRNA because at the apex of one of the hairpin loops the molecules have a triplet which is complementary to the codon. This triplet on the tRNA, as was stated before, is called an *anticodon,* and the amino acid that each tRNA carries is the one determined by the codon to which the anticodon of the tRNA is complementary. In other words, the tRNA molecules do not attach to any amino acid at random, but only to the one specified by its anticodon.

Figure 7.9 recapitulates the scheme of protein synthesis as it is now theorized by molecular geneticists. One can make the analogy of the sequence of triplets on a messenger molecule to the sequence of letters forming words and sentences in any language. In this manner the proteins synthesized by a cell are read from messages determined by a gene.

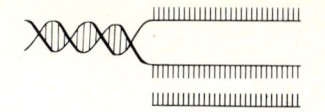

We have already stated that chromosomes contain numerous genes, and we now know that each "gene" must consist of scores of nucleotides. What causes transcription of mRNA to occur only over the length of a particular gene and not overlap into neighboring DNA?

The answers to these questions are not completely known. The initiation of gene activity, that is, of the transcription of mRNA, is probably due to "feedback" information which the nucleus obtains from substances in the cytoplasm. The nature of the chemical state of the nucleoplasm, which must reflect the state of the cytoplasm from which it receives nutrients and replacement substances, must be the signal which triggers transcription of certain mRNA to replace or synthesize new proteins needed by the cell.

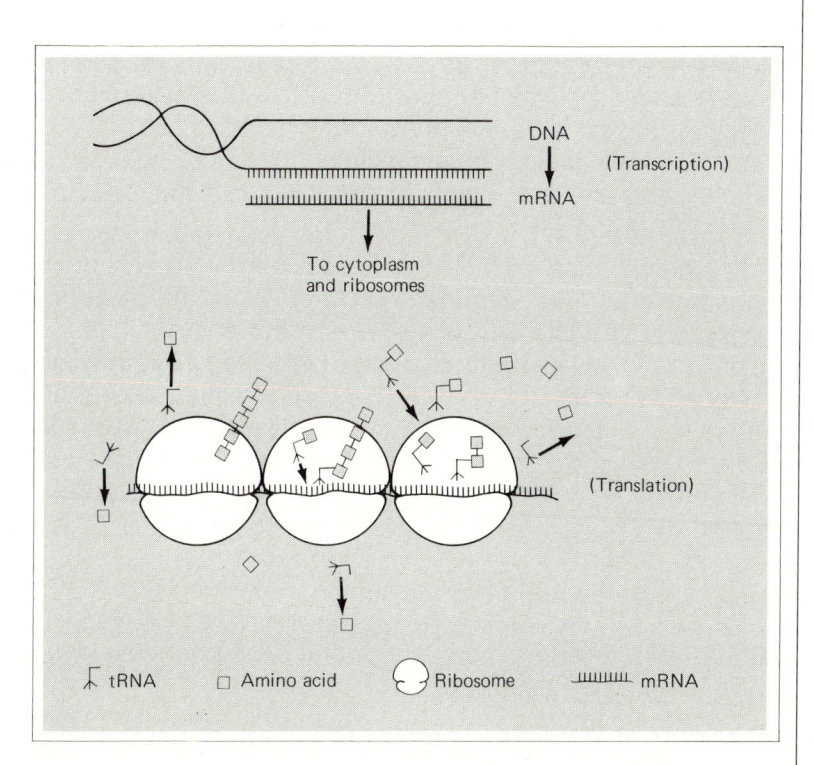

TERMINATION OF THE MESSAGE

FIGURE 7.9
A summary of present concepts of how a gene determines a specific protein synthesized by the cell.

Experiments have shown that only one of the strands of the double helix serves as the template for messenger molecules. It appears that the message being formed ends at codons which do not code for any amino acids. If you will refer to Table 7.1 again, you will see that there are three such codons which do not code for amino acids. It is thought that the positions of these *nonsense codons* serve to terminate the message. There is some evidence that they also cause the release of the polypeptide chains from the tRNA molecules.

UNIVERSALITY OF THE CODE

The deciphering of the genetic code has served to underscore the relationship between all living beings from one-celled organisms to man, for it has been found that the code is a *universal* code. That is to say, what codes for a particular amino acid in bacteria also codes for the same amino acid in complex organisms.

One of the decisive experiments done to prove the universality of the code came from the laboratory of G. von Ehrenstein and F. Lipmann in 1961. They took tRNA molecules from the intestinal bacterium *E. coli* and mRNA and ribosomes from red blood cells of rabbits. Because techniques have been developed to isolate these parts of the cells without damaging their biological capabilities, they were able to incubate the three components together in a test tube.

The result was the production of hemoglobin, a very important protein found only in red blood cells, and which, of course, is not found in bacteria (Figure 7.10). The fact that bacterial tRNA molecules picked up the correct rabbit amino acids and inserted them into the correct positions in the polypeptide chain for hemoglobin determined by rabbit mRNA was clear evidence that the code is indeed universal. If it were not, the bacterial anticodons could not have been attracted to rabbit codons to produce a highly specialized protein.

SOME PROBLEMS OF GENE ACTION IN COMPLEX ORGANISMS

As was stated earlier, many of the major discoveries dealing with molecular aspects of gene action have come from studies on microorganisms. After the discovery of the mechanism for gene-controlled protein synthesis, biologists found that the same basic mechanism also exists in complex organisms since they have isolated the major components of the pathway from mammalian cells. This had led some scientists to enthusiastically claim that what can be done in microbial systems should be possible in human systems—that "what is true for *E. coli* is true for elephants."

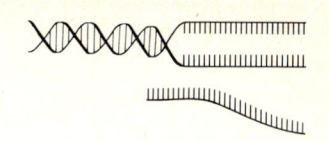

For this reason, much publicity was given to recent developments such as the artificial synthesis of a functional gene and the isolation of another one which determines the production of the enzyme that can break down a sugar lactose in *E. coli*. The implications of these achievements for the future are great, and we will discuss them at length later in the book.

However, always bear in mind that despite the universality of the DNA-protein pathway and the code, there are many problems in extrapolating results obtained in experimentation on one-celled organisms to multicellular organisms with their different levels of organization (see Figure 7.11). Remember that the genetic material in bacteria and viruses are naked strands of DNA, and there is only one strand per cell. In man, on the other hand, DNA is bound up with proteins in the chromosomes, which probably necessitates additional facets of gene action that are not found in bacteria. For example, during transcription, what happens to the protein surrounding the DNA? Does it break away while the DNA is serving as template for messenger molecules and

DNA OF MICROORGANISMS VS. CHROMOSOMES

FIGURE 7.10
Experiment demonstrating the universality of the genetic code. See text for explanation.

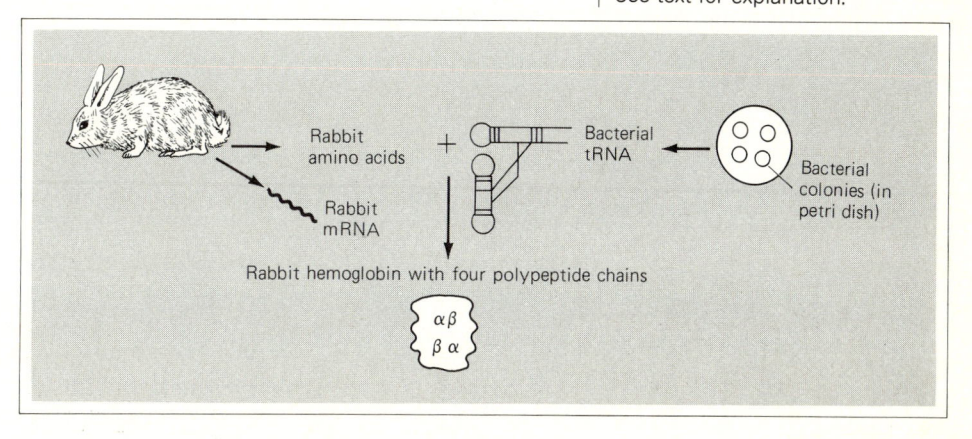

FIGURE 7.11
"Elephants are always drawn smaller than life, but an *Escherichia coli* always larger." [After Francois Jacob, Biology Today, CRM, Del Mar, Calif., p. 316.]

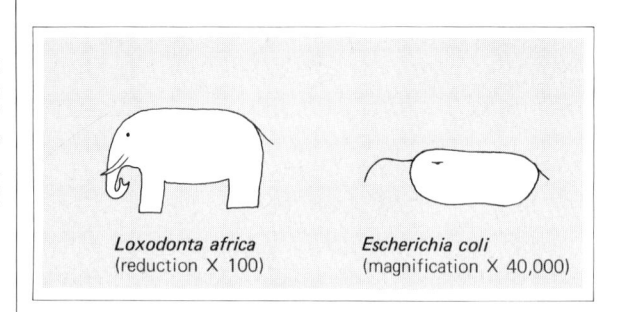

Loxodonta africa (reduction × 100)

Escherichia coli (magnification × 40,000)

then recombine after the message has been made? If so, what controls this?

IDENTIFICATION OF GENE LOCI ON CHROMOSOMES

Another area of difficulty is in the identification of the site of specific genes in higher organisms. Such identification is relatively easy in bacteria when a simple form of sexual reproduction, called *conjugation*, occurs between two bacterial cells. The cells lie side by side during conjugation, and with the formation of special structures called pili, the newly replicating DNA of one cell enters the second cell (Figure 7.12).

Jacob and Wollman, two French scientists, found that if they interrupted the mating procedure at different times after its initiation—using that sophisticated instrument of science, the electric blender—and observed the progressive genetic

FIGURE 7.12
Conjugation between two bacterial cells. The haploid state is restored at the next cell division of the recipient.

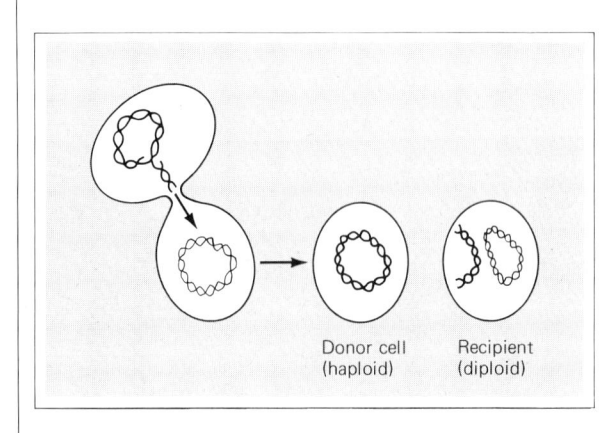

Donor cell (haploid)   Recipient (diploid)

changes in the offspring of recipients resulting from increasing amounts of DNA of the donors entering, they could determine the entire sequence of genes on the DNA of the bacterium.

As we saw earlier, there are no comparable techniques to allow the identification of specific genes on human chromosomes, since human cells do not undergo conjugation in tissue culture. Techniques such as the fusion of diploid cells in tissue culture (or somatic cell hybridization, as it is sometimes called) are beginning to allow the assignment of genes to specific chromosomes. However, their exact locations on the chromosomes are still unknown. And even though close to 80 genes have been linked to the X chromosome in humans, we have not yet been able to discern their actual position on the sex chromosome.

Difficulties such as these lead many geneticists to feel that total extrapolation from experimentation and data on microbial systems to complex systems (such as man's) is premature.

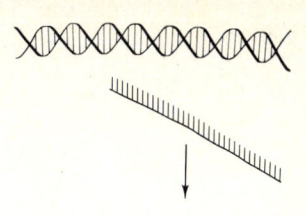

REFERENCES

Work cited:
Beadle, G. W., and E. L. Tatum, 1941: "Genetic Control of Biochemical Reactions in *Neurospora*," Proceedings of the National Academy of Science, **27**, 499–506. Also reprinted in J. Peters (ed.), *Classic Papers in Genetics*, Prentice-Hall, Inc., Englewood Cliffs, N.J., 1959.

The following two paperback volumes contain papers by almost every important contributor to the study of gene action (including those we have mentioned) written for the nonscientist; but some of the papers still require some background in chemistry or biology. Nonetheless, for the interested reader, they are highly recommended.

The Molecular Basis of Life: An Introduction to Molecular Biology, with introductions by Robert H. Haynes and Philip C. Hanawalt, *Readings from Scientific American*, W. H. Freeman and Company, San Francisco, 1968.

Facets of Genetics, with introductions by Adrian M. Srb, Ray D. Owen, and Robert S. Edgar, *Readings from Scientific American*, W. H. Freeman and Company, San Francisco, 1970.

BR 1

BR 2

BR 3

A

B

Chapter 8

The Regulation of Gene Action

The regulation of gene action is one area of genetics that is still not well understood. What specifically causes genes to become active or to be inactive? Even in one-celled organisms, not all the genes are active all the time. As we mentioned in the last chapter, this is also true in higher organisms. The very existence of different cell types reflects the fact that different genes are active in these cell types.

The first well-documented system of gene regulation in microorganisms was the genetic determination of three enzymes that cause the breakdown of lactose in *E. coli,* investigated by Jacob and Monod. The genes determining the structure and synthesis of the three enzymes were found to lie next to one another. Jacob and Monod called the genes that determine enzyme structure *structural genes.*

In sequence with these genes there appears to be another segment of DNA which does not code for protein, but which

REGULATION IN *E. COLI:* **THE LAC OPERON**

Chromosome puffs (labeled BR_1, BR_2, and BR_3) from polytene chromosomes of insect larvae. (A) Three large puffs on chromosome IV of salivary gland cells. (B) Autoradiography of the same puffs. [*Courtesy of Dr. W. Beermann, Max Planck Institute für Biologie.*]

exercises a control over the three by allowing transcription of mRNA for the three enzymes to take place. The DNA which controls their transcription is called the *operator* gene. A small region next to the operator called the *promoter* seems to be important for reacting with enzymes that cause transcription to occur.

But what turns the operator on and off? The operator is under the control of yet another segment of DNA called a *regulator* gene, which is situated apart from the operator and its structural genes. (Whatever happened to our simple beads on a string?) The regulator seems to specify a protein called a *repressor,* which then binds with the operator gene and renders it inactive. This prevents the enzymes bound to the promoter from progressing to the structural genes, so transcription cannot occur. The scheme is diagrammed in Figure 8.1. The operator, promoter, and structural genes together are called an *operon*. Since the system which we have described here involves the metabolism of lactose, this operon is commonly referred to as the "lac operon."

Occasionally, substances in the cytoplasm called *inducer* substances may bind with the repressor molecules. When this occurs, the operator gene is active and transcription will continue until the repressor is available. In Jacob and Monod's system, it was found that lactose interacts in this manner with the repressor. The structural genes then produce mRNA that determines the synthesis of enzymes which catalyze the breakdown of lactose. As the lactose is metabolized, repressor molecules are freed to bind to the operator, and transcription of the structural genes ceases.

Although the lac operon has been studied in great detail, it is not known if this system of gene control is applicable even to other genes determining the structure of enzymes of other metabolic reactions in *E. coli,* much less to other forms of life. For example, there is clear evidence that genes determining enzymes for other metabolic pathways are situated in separate parts of the DNA of bacteria. Their physical separation renders the idea of an operator gene controlling their transcription in a coordinated manner difficult to conceive. One might theorize the existence of a "common" repressor and operator that would act simultaneously on all the involved genes, but so far no evidence has been found to support this idea.

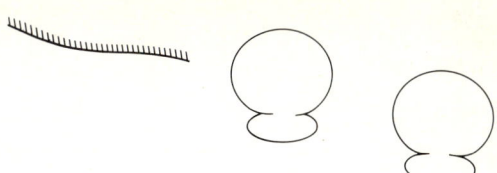

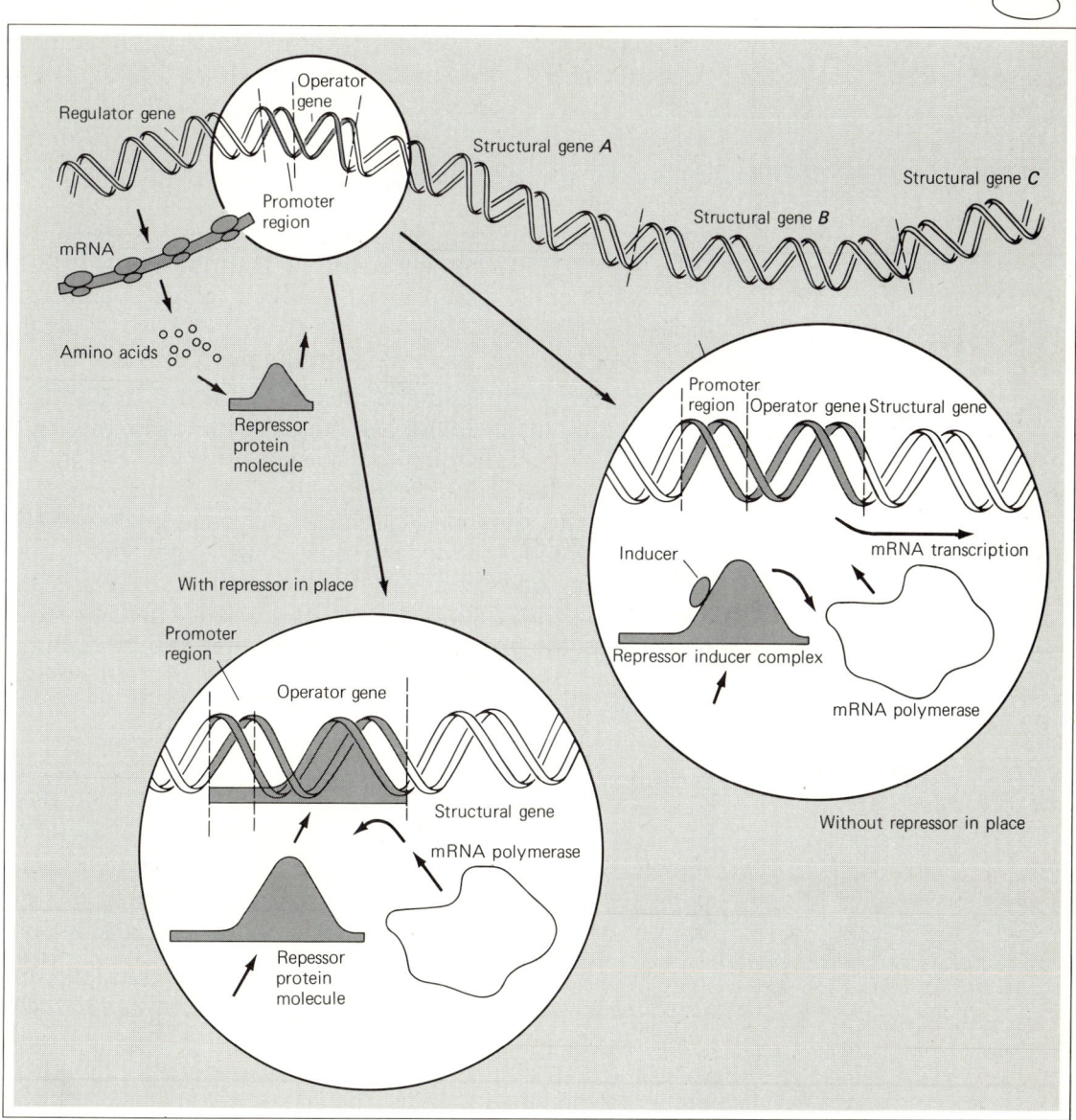

FIGURE 8.1
The basic scheme of gene regulation involving enzymes for the breakdown of lactose in the bacterium *Escherichia coli*. [*After Biology Today, CRM, Del Mar, Calif., 1972, p. 323.*]

A REDEFINITION OF "GENE"

The concept of the operon has contributed information which, with our knowledge of the double helix, now necessitates a review of our concept of the gene.

It now becomes very difficult to define the term *gene*. The old idea of beads on a string no longer holds, for we know that what determines a genetic trait is actually a segment of DNA which, being composed of subunits held together by chemical bonds, may break wherever these bonds are broken. Thus, the "gene" is, in fact, divisible, something which we shall discuss in detail in Chapter 12.

Furthermore, with Jacob and Monod's work, it is evident that not all segments of DNA have the same function, that is, to determine protein structure. For this reason, geneticists have coined new terms: *cistron* is the term used for the DNA segment that codes for a particular polypeptide chain,* and we have already defined an operon as a genetic system including operator and structural genes. A *muton* is the smallest unit that can undergo permanent change, and a *recon* is the smallest unit in the gene that can take part in crossing-over and recombination. Both recons and mutons are single nucleotides.

These terms are used primarily when molecular details of the system are known. For complex organisms such as man, in which understanding of such detail is almost nonexistent, the term *gene* is still used, although the concept of a gene's indivisibility has been shattered. In this book we shall continue to use the word *gene* to mean a unit of heredity, bearing in mind the new complexities which have been discovered.

The "one-gene–one-enzyme" theory that arose following Beadle and Tatum's classical studies on *Neurospora* has also had to be revised. It appears that the structural gene simply determines the structure of the polypeptide; the refinement of the polypeptide into a functional enzyme molecule occurs after the translation of the genetic message. Therefore, it might be better to think of "one gene, one polypeptide," but even this would not be entirely true in systems where operators and regulators are known to also play a role in the determination of a polypeptide chain, since more than one "gene" is involved.

---

* Note that *cistron* and *structural gene* may be used interchangeably.

An accurate definition of the term *gene,* then, would require an essay rather than a simple sentence or two! It is still a unit of heredity, but remember that there are now complex structural and functional aspects which could result in different meanings depending on the usage of the word. Definition aside, we have come a long way from Mendel's unit toward understanding the very basis of all life processes in just this century!

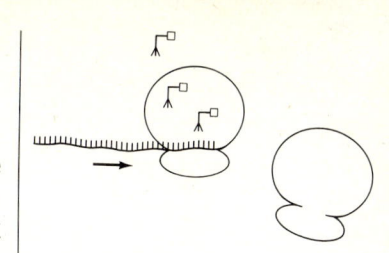

DEVELOPMENTAL GENETICS

Let us turn now to the question of whether the concept of the operon is applicable to higher organisms, and if not, what possible systems of gene regulation might exist. There is perhaps no area of study more relevant to discussions of gene regulation in higher organisms than that of developmental genetics.

Developmental genetics is the area of biology that deals with the control by genes of the differentiation and development of embryos of multicellular organisms. The basic question in this field is how the multitude of cells that arise from mitotic divisions of the fertilized egg cell and which are therefore genetically alike nonetheless become distinct cell types, differing in form and function. Since such differentiation is the result of some genes being active in some cells and not in others, the question of how differentiation and development occur is basically a question of how the activity of certain genes is regulated.

GENETIC POTENTIAL OF EMBRYONIC CELLS: CLONING

We know from the existence of identical twins and from Boveri's work on sea urchin embryos (see again page 33) that each embryonic cell possesses the entire chromosome complement, regardless of what the cell is destined to become in the adult. This implies that all cells possess greater potential than they ever attain.

In the 1950s, two embryologists, R. Briggs and T. King, developed a technique called *nuclear transplantation* which clearly demonstrated the genetic potential of embryonic cells. The nuclei from a number of frog egg cells are taken out with a micropipette and replaced with nuclei taken from the cells of an embryo (Figure 8.2). Although the cells of the embryo—at the point in development when they are enu-

150
The Regulation
of Gene Action

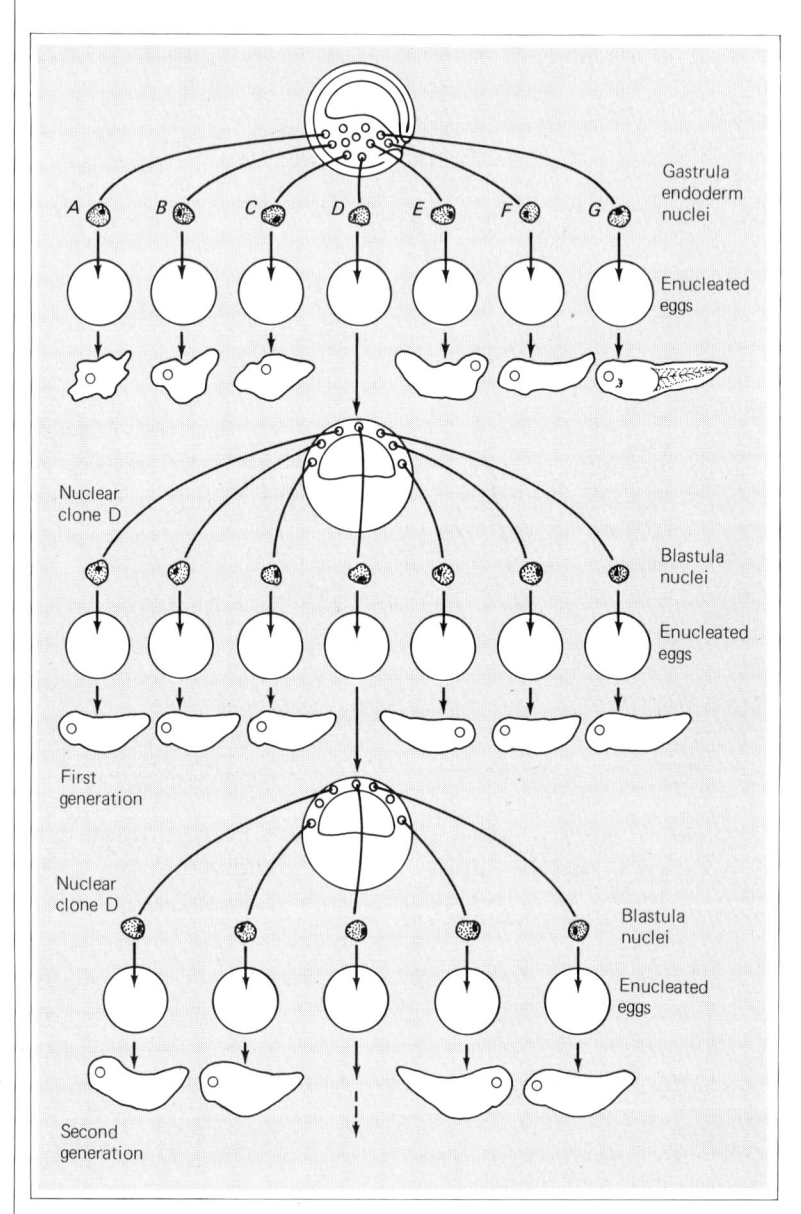

FIGURE 8.2
Nuclear transplantation in amphibians resulting in clones of genetically identical organisms, due to the use of nuclei from cells of the same individual. [*After N. T. Spratt,* Developmental Biology, *Wadsworth Publishing Company, Inc., Belmont, Calif., 1971, p. 435.*]

cleated—are already determined to be a particular cell type, Briggs and King found that the genes in the nuclei of many of these cells were still capable of total genetic activity. This was reflected in the fact that the recipient egg cells were able to develop into normal tadpoles and frogs, with all the different cell types normally found in frogs.

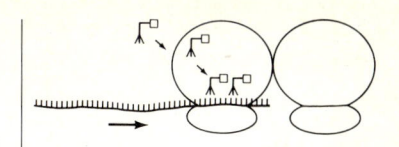

What made the nuclear transplantation technique of special interest was that it allowed the investigator to produce a number of *genetically identical organisms,* since the nuclei were taken from cells of the same animal. A group of such identical organisms is called a *clone,* and the term *cloning* is used to refer to the production of a clone. We shall discuss this further later in the book, since some publicity has been given to the possibility of cloning of humans.

Although cells of a tadpole evidently are capable of supporting complete development, experiments using nuclei from older embryos and tadpoles indicated that some changes that affect this potential take place during development. The older the individual from which the nuclei were taken, the less the recipient egg cell was able to develop normally through the entire process of metamorphosis. Also cells from different parts of the embryo differed in the degree of successful development attained after nuclear transplantation.

Experiments investigating gene products during development have confirmed that there is progressive change in gene action, as had been suggested by nuclear transplant experiments. One of the best-known examples of gene-product investigation comes from the laboratory of C. Markert and his co-workers, who have studied one enzyme in great detail. This enzyme, called LDH (lactate dehydrogenase), is found in all vertebrates and plays an important role in metabolism.

LDH is one of a number of enzymes known to exist in slightly different molecular forms. Because they are all still involved in the same chemical reaction, they are all referred to by the same name, LDH. Such enzymes are known as *isozymes.* In the case of LDH isozymes, four subunits (polypeptides) have been found, composed of two different kinds, designated *A* and *B,* which are determined by two different genes. There are, then, five possible isozymes of LDH: *BBBB, ABBB, AABB, AAAB,* and *AAAA,* and all five have been isolated from tissues of various experimental animals. For convenience of reference, the five types of LDH in the order mentioned have been designated LDH-1 through LDH-5.

PROGRESSIVE CHANGES IN PROTEINS DURING DEVELOPMENT: LDH AND HEMOGLOBIN

It was found that in mice the *BBBB* form of LDH, or LDH-1, is the only form present during oogenesis, indicating that at that stage there is activity of the *B* locus and relative inactivity of the *A* locus. Following fertilization, during the first half of the mouse's gestation period of around 20 days, there is a progressive suppression of *B*-gene activity and an increase in *A*-gene activity, reflected by a rise in the quantity of LDH forms containing the *A* subunit. Figure 8.3 shows the LDH isozymes found in heart tissue of mice at different stages of development. It is apparent that as the cells mature and differentiate, the *A* and *B* loci are not equally active. In the latter half of development, different tissues show divergence in the type of LDH present, and at the completion of development, the different tissues of the adult show great divergence in the relative amounts of the LDH isozymes present (Figure 8.4).

Hemoglobin in man provides another good example of gene regulation. Hemoglobin is a pigment in red blood cells which causes the cells to appear red; it is also the substance that allows red blood cells to effect the vital exchange of oxygen and carbon dioxide in the lungs and tissues. Like LDH, it is a complex protein composed of four polypeptide chains of which there are two kinds. Unlike LDH, however, two subunits, alpha (α) and beta (β), normally are present in equal amounts.

FIGURE 8.3
Changing patterns of LDH isozymes found in heart tissue of mice at different stages of development. The black spots represent presence of the various isozymes at different stages; the sizes of the spots represent relative concentrations. [*After C. L. Markert and H. Ursprung, Developmental Genetics, Prentice-Hall, Inc., Englewood Cliffs, N.J., 1971, p. 44.*]

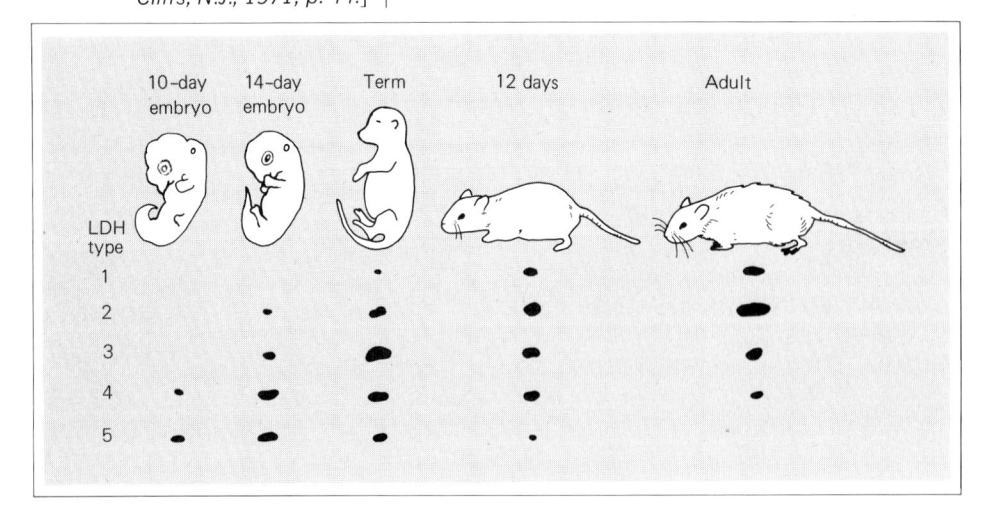

FIGURE 8.4
LDH types found in various tissues in the adult rat. [*From C. L. Markert and H. Ursprung, Developmental Genetics, (c) 1971. By permission of Prentice-Hall, Inc., Englewood Cliffs, N.J., p. 43.*]

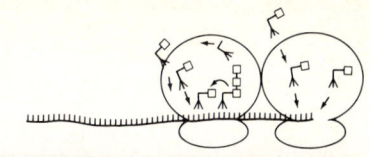

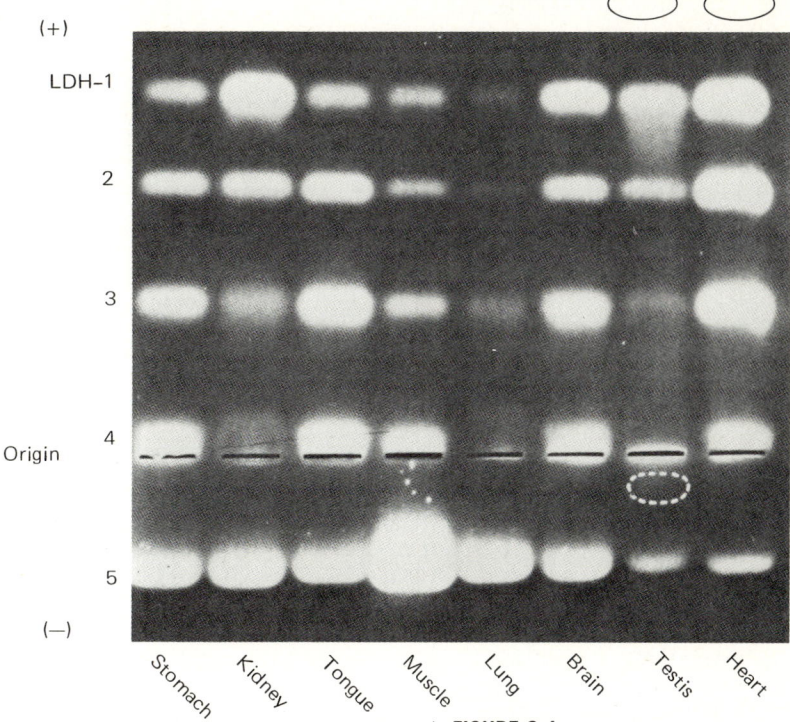

During fetal life the hemoglobin molecule of humans has, instead of beta, a third kind of polypeptide chain found only in the fetus, called gamma (γ). [There is, in fact, evidence that yet another type of hemoglobin exists in very early embryonic stages, containing alpha chains and what are called sigma (σ) chains. Not much is known about this hemoglobin because of the lack of the substance for study, although new and sensitive techniques are being developed to isolate it.] The three types of polypeptides, alpha, beta, and gamma, are determined by different genes. Here again is evidence that a particular gene, gamma, active during development, is in some way "switched off" as beta is "switching on," while the alpha gene is apparently active all along (Figure 8.5).

Presumably the change in protein types as exemplified by both the LDH and hemoglobin proteins is of some functional advantage to the organism. That is, as the individual

FIGURE 8.5
Schematic illustration of progressive change in hemoglobin subunits in human development, presumably due to the switching on and off of genes. $\alpha_2\gamma_2 = 2$ alpha chains + 2 gamma chains, $\alpha_2\beta_2 = 2$ alpha chains + 2 beta chains.

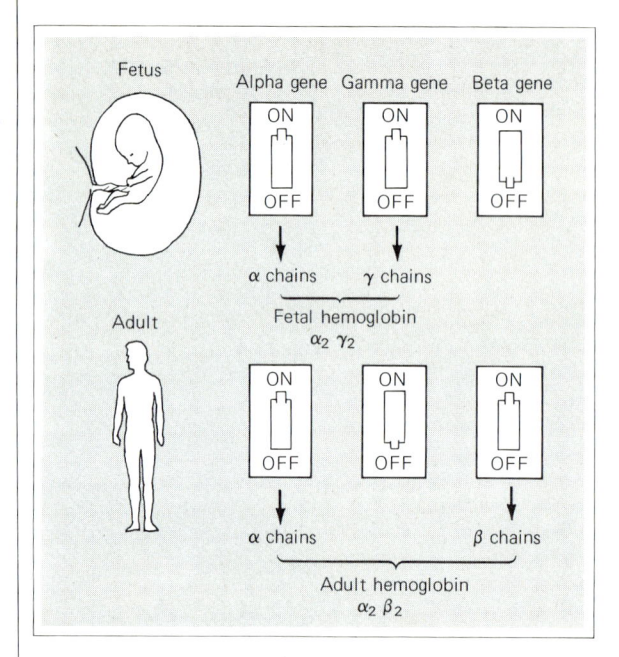

develops, its internal environment changes constantly, creating new metabolic demands that are better met by slight molecular changes in these vital proteins.

PROGRESSIVE CHANGES IN CHROMOSOME STRUCTURE DURING DEVELOPMENT: PUFFING

On the basis of our present knowledge about the steps in protein synthesis, one would expect that there should be detectable evidence for the switching on of genes in development, evidenced by studies on LDH and hemoglobin such as the synthesis of RNA, at particular active regions of chromosomes representing the transcription of messenger molecules. And in fact, such evidence exists, not in human chromosomes, for the reasons we have already discussed in previous chapters, but in chromosomes found in the salivary glands of larvae of certain insects like the gnat. (One of the most fascinating aspects of science is knowing of the obscure material which has been used in experimentation to provide important discoveries, and speculating how the investigator came to choose that particular system!)

In the salivary glands of these larvae there are found certain giant chromosomes which attain their enormous sizes through *polytenization,* a process in which there is replication of DNA without any separation of the chromatids. Consequently some polytene chromosomes have hundreds of chromatid fibers in them, making the chromosomes very easy to see under the light microscope. Figure 8.6 shows a micrograph of some polytene chromosomes.

Notice that there are distinct banding patterns on different chromosomes. Scientists found that the pattern of banding in different tissues was very similar, and provided evidence which led to the hypothesis that the banding correlated with the positions of various gene loci.

At different stages of development a phenomenon called *puffing* was seen to occur at different bands. The regions of puffing came to be known as *Balbiani rings* (see the introductory illustration to this chapter). On the basis of subsequent studies, the regions of puffing are now believed to be areas of

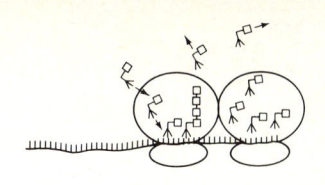

FIGURE 8.6
Polytene chromosomes found in insect larvae. [*Courtesy of Professor W. Beermann, Max Planck Institut für Biologie.*]

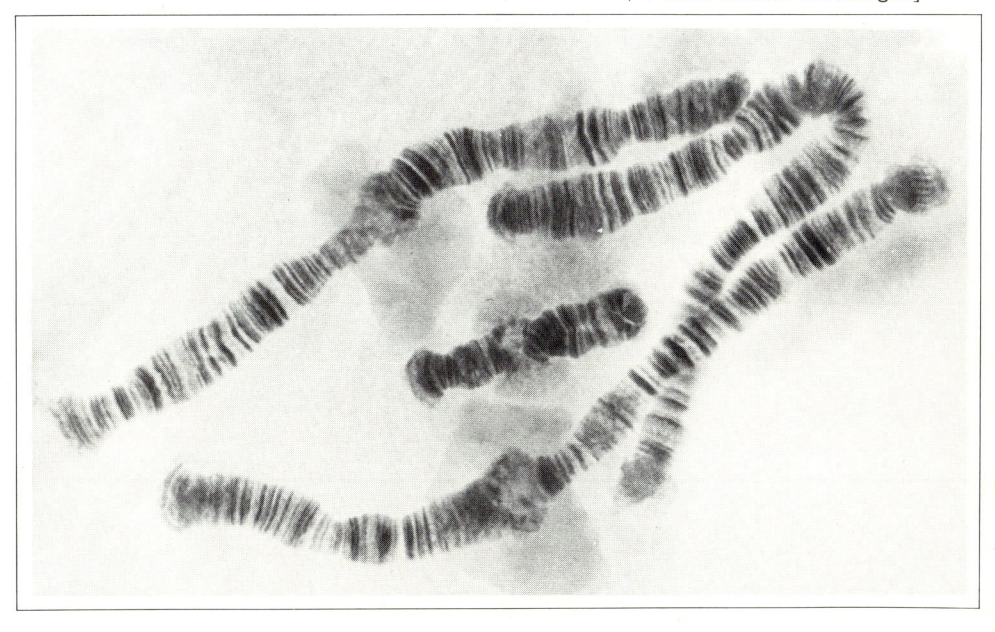

the chromosome at which the many strands of DNA are unwound in order for transcription of RNA to occur (Figure 8.7).

This idea was confirmed by experiments in which precursor molecules of RNA, that is, molecules which are the "building blocks" of RNA, injected into insect larvae were found concentrated in the regions of the puffs. The precursor molecules can be traced by making them radioactive and then using special techniques to trace the radioactive material. The concentration of newly synthesized RNA at the Balbiani rings is strong evidence that these are transcriptionally active loci.

It was also found that there was consistency in the appearance of puffs as the insect larvae continued to develop. In the metamorphosis of higher insects, the larva molts into the pupa and the pupa undergoes another molting to become the adult. Some puffs were seen to occur only in conjunc-

FIGURE 8.7
Illustration of the detailed nature of puffing. [*After W. Beermann and U. Clever, "Chromosome Puffs,"* copyright © *April 1964 by Scientific American, Inc. All rights reserved.*]

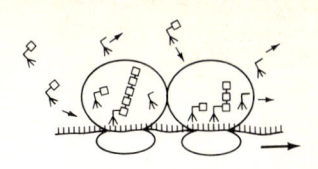

tion with the molting process, in some cases appearing regularly at the start of the molting process, in others after the process had begun. The interpretation of these particular puffs occurring at specific times is that they represent loci which determine the synthesis of some substance important to the development of the insect at that stage.

HORMONES AND GENE ACTION

In addition, it had been found previously that molting in insects is caused by a hormone known as ecdysone. If ecdysone is injected into insect larvae prior to the time that it is normally present, two specific puffs arise which are visible only at molting time. Apparently the presence of the hormone is the signal which causes these two genes to become active.

How this is accomplished remains a matter for conjecture. Perhaps the hormone directly activates the two genes; perhaps it serves to "de-repress" the two genes. Evidence of hormone-activated protein synthesis in mammals comes from studies that have shown that cortisone causes an increase in the synthesis of liver enzymes and that sex hormones cause an increase in protein synthesis in the sex organs. But again, how the increase is caused by the hormone is unknown.

GENE REGULATION IN COMPLEX ORGANISMS

Ever since the DNA-protein pathway and the genetic code were proved to be universal, scientists have searched for evidence indicating that the mechanism of gene regulation in higher organisms is similar to the one studied by Jacob and Monod for the lac operon in *E. coli*. The literature of recent years dealing with such studies on gene regulation in higher organisms is quite extensive; however, no formulation of an overall concept of the mechanism by which gene activity is regulated has been possible. Simply put, the existence of operons as defined in *E. coli* has not been found in higher organisms.

REGULATOR GENES

There are indications that some genes do govern the activity of other genes; for example, the "controlling elements" found in maize,* a system in which the presence of certain genetic elements can cause some loci to be active. The nature of this control has not been revealed.

> * See references to B. McClintock at the end of the chapter.

One system in mice which may prove to be of importance in the study of gene regulation in mammals involves a different aspect of hemoglobin synthesis. Earlier we discussed the globin, or protein portion, of hemoglobin; let us now consider some of the metabolic reactions that are known to occur during synthesis of the heme part of hemoglobin. Mice treated with a certain drug, DDC, show an increase in the synthesis of porphyrin, an intermediate substance in heme synthesis. The enzyme that appears to control the metabolic step limiting the *rate* of prophyrin synthesis is one called *ALA synthetase*. In other words, the drug is able to induce an increase in the synthesis of ALA synthetase, which results in an increase of porphyrin production in the liver.

Not all mice responded to DDC to the same degree, however. Different strains of inbred mice were found to be highly inducible, while other strains showed negligible response to the inducer drug. There must be genes in mice capable of regulating the activity of the gene determining ALA synthetase and its response to an inducer; however, the genetic system involved has not yet been discovered. Interestingly, the Harderian gland, a gland near the eye that is extremely high in porphyrin content, seems to have a different system of regulation, since it does not respond to DDC at all.* In other words, synthesis of the same chemical in two different tissues must be different in some way at the molecular level.

> * See references to J. Hutton and F. Margolis.

REGULATOR SUBSTANCES

Both the drug DDC and the hormone ecdysone can be considered examples of regulator substances. But to simply label them "regulator substances" without understanding *how* they cause activation of genes is by no means the solution to any problem. Other studies have indicated that proteins found in chromosomes of higher organisms may possibly regulate gene action in a manner analogous to repressor substances.

One group of proteins found in chromosomes is called *histones*. Given the proper mixture of DNA, substrates, and enzymes, the transcription of RNA can be induced in a test tube. If histones are added, the synthesis of RNA is substantially decreased. In peas, furthermore, the areas of DNA bound to histone in the chromosomes appear to be less active as templates for RNA synthesis than the areas of DNA bound to nonhistone proteins.

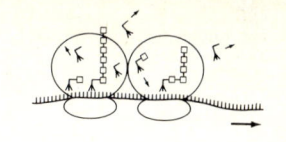

One of the difficulties in regarding histones as the regulators of gene action, however, lies in the fact that histones are generally homogeneous when extracted and studied chemically. How proteins which seem to be the same in different tissues, and even in different species, could have any specificity of gene control is a major conceptual problem to be overcome.

THE LYON HYPOTHESIS AND THE EFFECTS OF HETEROCHROMATIN ON GENE ACTION

A unique situation of gene inactivation is now widely held to exist in sex chromosomes of higher organisms such as humans and other mammals. Formulated by M. F. Lyon, a British geneticist, the hypothesis states that in the female, one of the X chromosomes is inactivated during early development, and only one of the two X chromosomes is actually genetically active. Which is inactivated is determined purely by chance.

There are two phenomena which led geneticists to investigate sex chromosomes and to formulate the so-called Lyon hypothesis. One is that males have only one X chromosome, yet obviously females have a double dosage of the proteins determined by genes on the X chromosome. As you will see in the chapter on mutation, and as we have implied in discussing Boveri's work, extra amounts or too little of genes and their products usually results in gross abnormalities. It was a curious situation, then, that the difference in quantity of *x*-linked genes was neither too little in males nor too much in females.

Furthermore, in studying cells under the microscope, scientists discovered a darkly staining piece of chromatin in the nuclei present only in females, never in normal males. This structure is called the *Barr body* (Figure 8.8) and has since been identified as one of the sex chromosomes. In fact, the presence or absence of the Barr body is a simple method used frequently to detect possible abnormalities in sex chromosome number. Again, this will be discussed in greater detail in the chapter on chromosomal mutations, but for now,

FIGURE 8.8
The Barr body in normal female cells (*A*) is absent in normal male cells (*B*). It is believed to represent an inactive X chromosome.

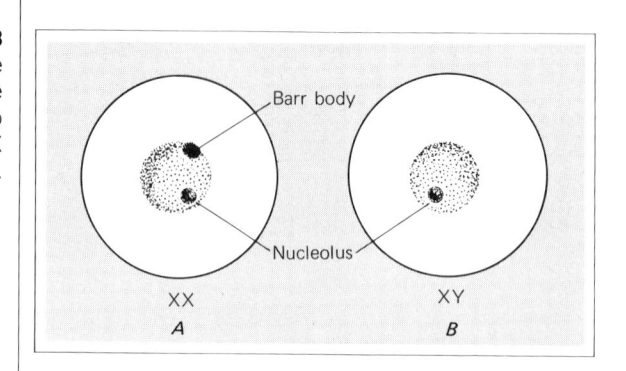

just remember that there is one fewer Barr body than the number of X chromosomes in a cell.

With these two phenomena in mind, the dosage differences and the existence of Barr bodies in female cells, the Lyon hypothesis was formulated. The Barr body is believed to represent an inactive X chromosome. Experimental evidence for the hypothesis exists, in that cells from females heterozygous for two codominant alleles determining two different isozymes of the enzyme G-6-PD, when grown in tissue culture, show the presence of only one of the two possible forms of the protein. This indicates that one of the two alleles is inactive in the cells, having been "turned off" early in development.

Other evidence indicates that the X is not totally inactive, but only partially so. There are still a number of questions to be worked out, but it is a fairly well-established phenomenon. One question is *how* is the chromosome, or parts of it, inactivated? It is believed that the physical state of the Barr body gives us a clue to the answer.

When chromatin is darkly staining, as is the Barr body, it is said to be *heteropyknotic,* and is called *heterochromatin.* This state is believed to be the result of a coiling of the chromosome which interferes with gene function. By "labeling" chromosomes with radioactive material, it has been found that areas of heterochromatin lag behind nonheteropyknotic areas (called *euchromatin*) during cell division, as does the entire X chromosome which forms the Barr body. Thus, the physical state of the chromosome may be involved with the capacity of the DNA to participate in both transcription and replication.

Furthermore, there is evidence that heterochromatic areas of chromosomes can actively affect neighboring genes. For example, if *Drosophila* females are heterozygous for a sex-linked recessive gene for white eye color they are phenotypically red-eyed, which is the wild-type dominant color. If, however, the two alleles are caused to be moved, or "translocated," next to a region of heterochromatin, the eyes of the heterozygotes become mottled, with patches of white and red. A number of such effects have been found, and they are called *variegated position effects.*

The variegation or mottling of eye color is believed to be due to the heterochromatin inactivating some of the genes normally determining red eye color. (There is no effect on the white allele, since white is caused by a "deletion" or loss of the DNA for eye color. What is not there cannot be inactivated. We shall speak more of deletions in Chapter 11.)

LEVELS OF ORGANIZATION IN MULTICELLULAR ORGANISMS

The above examples of gene regulation in higher organisms should be enough to convince you, as they have convinced most geneticists, that the regulation of gene action is one area in which microbial systems cannot be equated with systems in organisms as complex as that of man. Even species we tend to consider much less complex than man, such as the fruit fly, are multicellular systems with a number of different dimensions of organization not found in microorganisms, which necessitates changes from the operon system found in *E. coli,* if not different systems altogether. This does not imply, however, that the Jacob-Monod operon model is of no use in studies on higher organisms. Its very existence is a stepping-stone which guides geneticists in their endeavor to understand more complex systems.

CELLULAR COMPLEXITY

In microorganisms which have no true chromosomes but simply naked strands of nucleic acid, the genetic material is not bound to proteins, but is floating around in the cytoplasm. In our cells, the DNA is not only bound up in proteins of varying kinds, but is separated from the bulk of the cell by the nuclear membrane. The products of synthesis in a bacterial cell are right there, around the DNA. On the other hand, products of synthesis in human cells must cross through the nuclear membrane to the nucleoplasm if they are to serve as metabolic signals for gene action. In other words, there is more need for *intra*cellular communication in cells of higher organisms.

INTERCELLULAR INTERACTION IN DEVELOPMENT: EMBRYONIC INDUCTION

Our multicellularity adds another dimension of complexity to the picture. We know that in the development of embryos in complex organisms such as vertebrates, there is definitely an *inter*cellular communication which is absolutely essential for normal development. In other words, besides the cell's own genetic machinery there is control over the cell from external factors produced by other cells.

As an example of intercellular communication, let us follow the development of the vertebrate eye. In the first weeks of gestation, the area of the human embryo which eventually develops into the nervous system is tubular in shape. Anteriorly, where the brain develops, the tube is expanded. About the middle of the third week of gestation, two depressions appear laterally in the expanded areas and will eventually deepen into the forerunner of the eyes, the optic cups. About 10 days later, the ectoderm (the outer layer of cells covering the embryo) that is directly over the optic cups begins to thicken. This thickening eventually pinches off from the rest of the ectoderm and differentiates to become the lens of the eye. Figure 8.9 illustrates the major events in eye development.

In experiments on chick and amphibian embryos, embryologists found that if the optic cup is removed before the lens has formed, no lens ever appears where it normally would. The genes of the ectoderm cells that normally become active and direct the formation of proteins that lead to the differentiation of these cells into lens cells remain inactive. This ability of one group of embryonic cells to evoke the development of another group of embryonic cells is termed *induction*.*

Furthermore, it was found that if the optic cup is transplanted at the proper time to the belly region of the embryo,

---

* Incidentally, the use of the term *induction* by embryologists should not be confused with its use by molecular geneticists, who refer to induction as the removal of repression of a gene to allow the synthesis of a particular enzyme. An unfortunate, and, for the student, confusing problem in semantics.

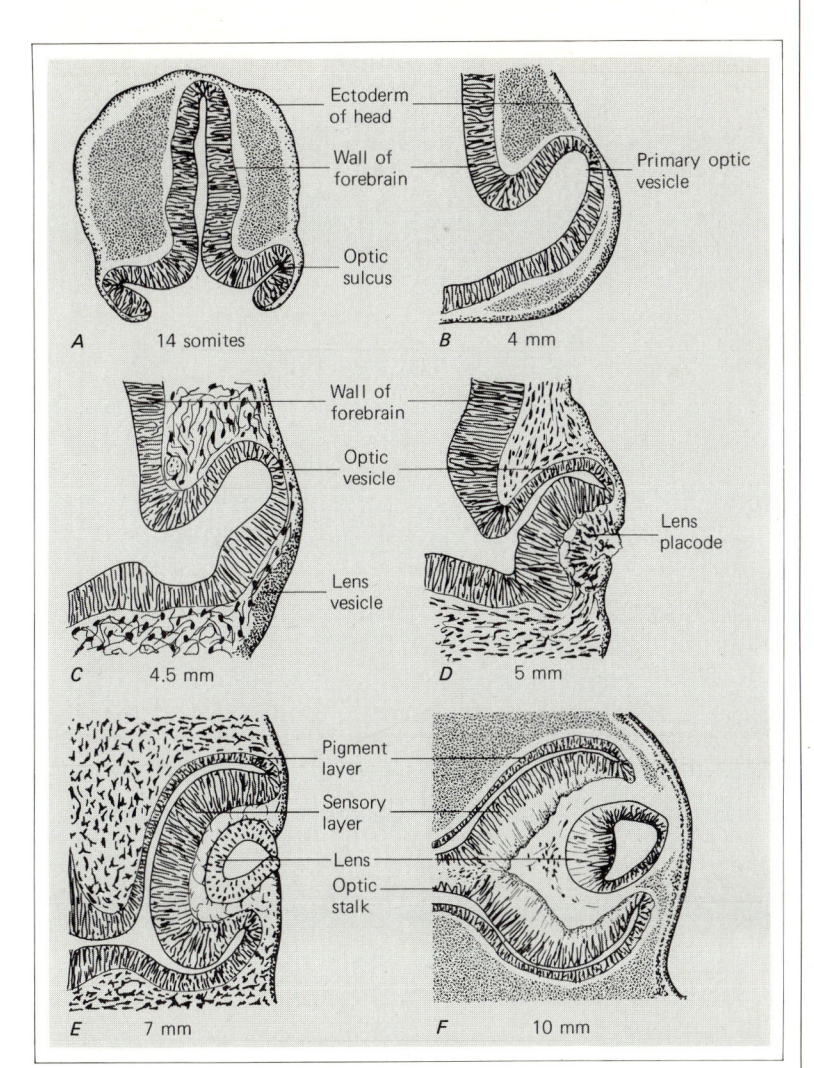

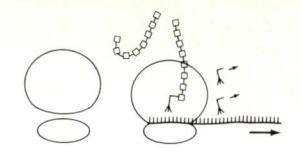

FIGURE 8.9
Early development of the human optic cup and lens. These are drawings of sections seen through the microscope. (A) Frontal view of developing head of an embryo. (B) to (D) Detail of one of the optic vesicles as it undergoes changes to become cup-shaped. (D) to (F) Development and differentiation of a portion of the tissue overlying the optic vesicle (ectoderm) into the lens of the eye. [From B. M. Patten, Human Embryology. Copyright © 1953 by McGraw-Hill, Inc. Used with permission of McGraw-Hill Book Company, New York, p. 398.]

the overlying ectoderm responds and forms a lens! This proves again the genetic potential of embryonic cells, since obviously no lens ever forms normally anywhere but at the level of the eye. (In case you should wonder, the eye thus induced is nonfunctional, since its connection with the brain has to be severed during transplantation!)

Experimentation has shown that this kind of intercellular communication is the rule rather than an exception during

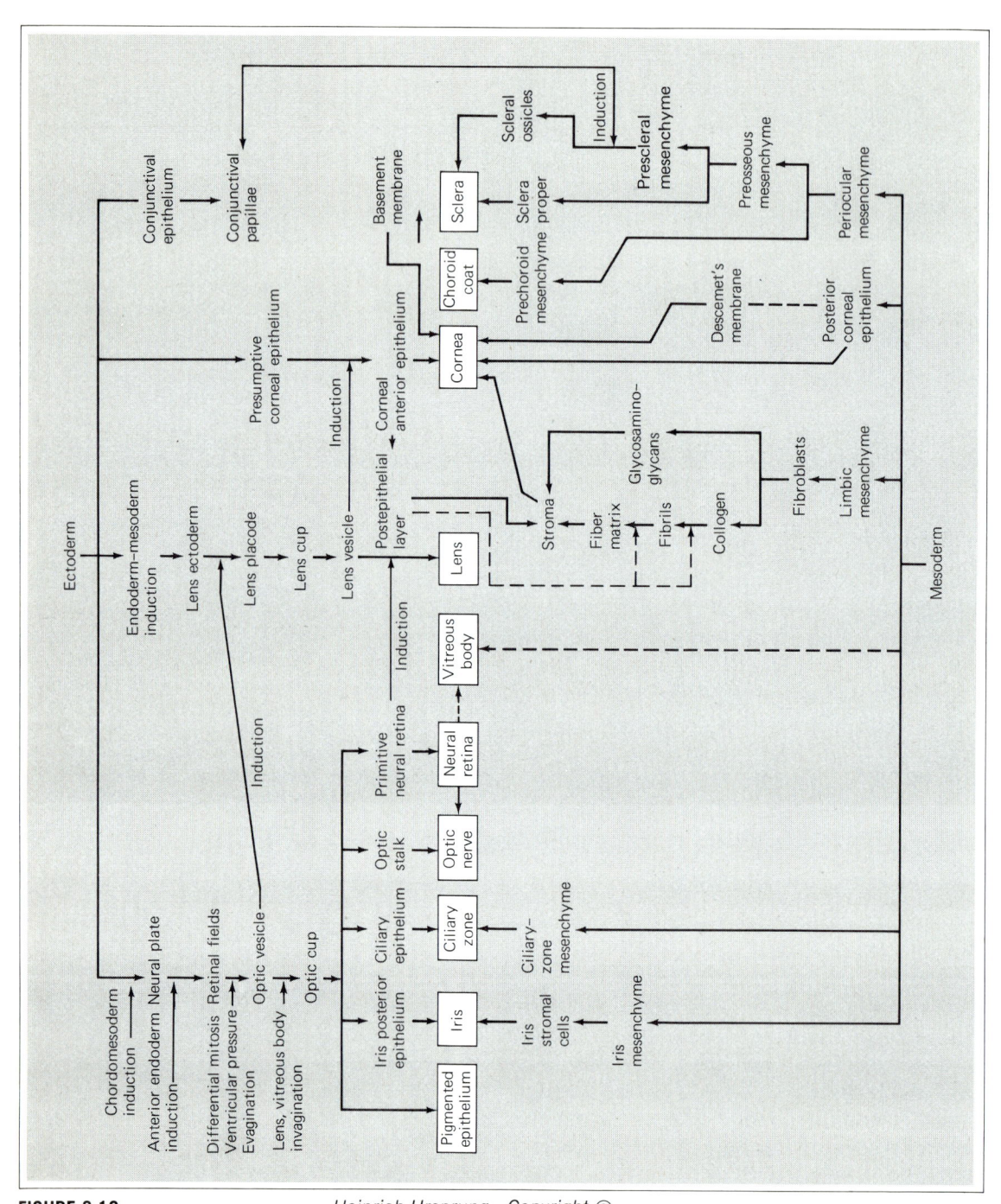

FIGURE 8.10
Interrelated events in the development of the vertebrate eye. [From "The Eye" by Alfred J. Coulombre, in Organogenesis, edited by Robert L. DeHaan and Heinrich Ursprung. Copyright © 1965 by Holt, Rinehart and Winston, Inc. Reprinted by permission of Holt, Rinehart and Winston, Inc.]

embryogenesis in higher organisms. In fact, the induction of the lens is only one of many events leading to normal development of the eye. Figure 8.10 is a summary of eye development compiled by A. Coulombre, and as you can see, the complexity of known events in the formation of one single structure in the human embryo is staggering, and we have discovered only a few.

What is the nature of the intercellular communication that must exist to cause induction to occur? This has been studied in another system of induction in the embryo, the induction of kidney formation by the ureter, which develops first. The components of this system were grown in tissue culture, separated by special filters that have openings too small for cells to pass through. The fact that induction nonetheless occurs indicates that some chemical is transmitted from one tissue to the other and must act on the genes of the cells being induced to develop in a particular manner. What the substance is has not been determined, but it appears to be a relatively large molecule.

INTERACTION BETWEEN DEVELOPING ORGAN SYSTEMS

In addition to the communication between cells in induction, the normal development of higher organisms involves a relationship between different organ systems. An example of this is the known dependence of bone development on normal muscle development. In order for bone to develop its proper shape, the stresses supplied by muscles and tendons on the embryonic bone are essential. Without the proper stresses, the bones will still develop but the shape will be abnormal.

There is a condition in mice which illustrates this point very well. The condition is caused by homozygosity for a recessive gene called muscular dysgenesis (*mdg*) which interferes with normal muscle differentiation. When the homozygous recessive individuals are born, no normal striated muscle can be found anywhere. There are a number of skeletal abnormalities such as short jaw, abnormal sternum, and abnormal vertebral column, which are suspected to result from the absence of normal muscle during development (Figure 8.11). Very similar skeletal deformities in newborn humans have been reported by Banker et al., also apparently resulting from widespread muscle deficiencies.

CAUSAL ANALYSIS OF GENE ACTION IN DEVELOPMENT

It should be apparent from this brief introduction to the fascinating problems of developmental biology that events in the progressive formation of the embryo from the one-celled zygote in higher organisms are very much interrelated. One event causes another to begin, and so on. This causal relationship has led to use of the term *causal analysis* to describe the manner in which students of development in higher organisms must trace the steps leading to the "progressive unfolding" of the mature form. In microorganisms, such tracing is relatively simple to do; in man and other higher organisms it is, as you can imagine, quite difficult because of the different levels of organization that we have just briefly discussed.*

*See reference to Gluecksohn-Waelsch, 1951.

Because of the interrelationship between developmental processes, causal analysis, simple as it may sound, frequently presents the investigator with baffling associations to make. For example, there are two loci in the mouse, the *W* locus and *steel (Sl)*, which are known to be inherited in simple Mendelian patterns as single genes, yet each of these causes abnormalities in pigmentation, germ cell formation, and the normal red blood cell picture.

The developmental relationship among the three diverse characteristics determined by the two genes seems to derive in part from the effects of the *W* gene that is indigenous to blood-forming tissue, and in part from effects of the *Sl* gene on neighboring cells, such as stromal cells in the spleen, an organ that serves as a source of blood cells. There are many examples of syndromes of diverse abnormalities due to the inheritance of a single mutant gene, and the term *pleiotropy* is used for this phenomenon.

On the other hand, to further complicate the analysis of gene action in development, recall that a number of different genes are frequently involved in the determination of *one* particular phenotypic trait. The more one learns about genes and development, the more difficult the situation becomes!

OTHER PROBLEMS

In addition there are problems of a technical nature in working with complex systems. For example, there is great difficulty in manipulating mammalian embryos which develop within the mother. Developmental geneticists therefore

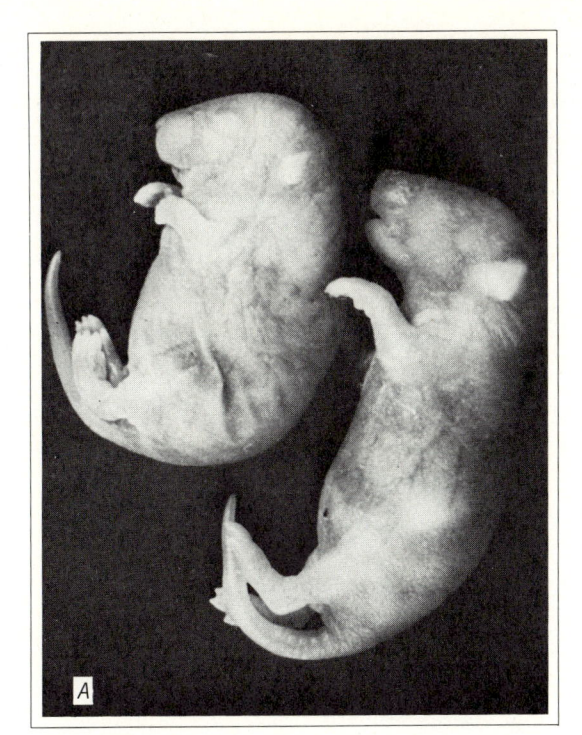

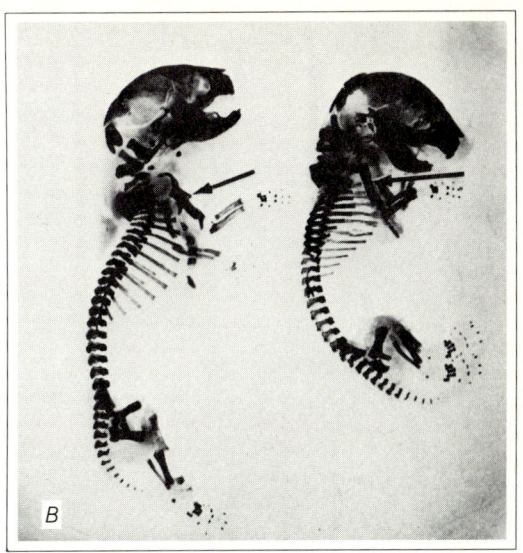

FIGURE 8.11
(A) Normal newborn mouse on the right; on the left, a mutant newborn homozygous recessive for a lethal mutation, *mdg*, which prevents the development of skeletal muscles. The lower jaw is foreshortened and the mutant shows extensive edema. (B) Stained skeleton of normal newborn (left). Homozygous *mdg* newborn skeleton (right) shows abnormal jaw, ribs, cervical vertebrae and absence of a bone projection on the humerus (arrow). Many of the skeletal abnormalities are attributed to the absence of muscle tissue during development. [*From A. C. Pai, "Developmental Genetics of a Lethal Mutation, Muscular Dysgenesis (mdg) in the Mouse,"* Developmental Biology, **11**:82–92, 1965.]

"treasure" mutations such as the *mdg* gene, which in effect perform experiments interfering with normal development in mammals that could not be done experimentally with present techniques without killing the embryos.

A very good example of a scientifically useful mutation is Danforth's short tail (*Sd*), an autosomal dominant mutation which interferes with the development of normal ureter formation. The relationship mentioned before between the ureter and kidney formation, discovered in lower vertebrates such as amphibians and birds, could not be confirmed experimentally in mammals. But this mutant gene which, among other effects, interferes with normal development of ureters, indeed showed that without normal growth of the ureter, no kidney is induced in mammals (Figure 8.12).

A number of helpful mutations have been found in various laboratories. Even so, technical difficulties we encounter in analyzing the earlier stages of abnormal development

FIGURE 8.12
Abnormal ureter and kidney formation in mutant mice illustrating the inductive relationship between the two structures in development. [*After S. Gluecksohn-Waelsch, 1945. "The Embryonic Development of Mutants in the Sd Strain in Mice,"* Genetics, **30**:29–38.]

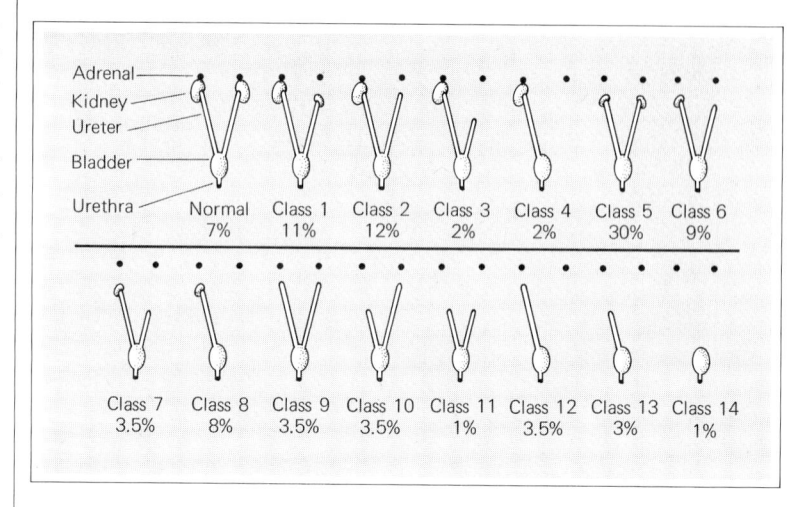

frequently result in causal analysis reaching an impasse long before we approach primary gene action, that is, the event directly determined by the particular gene or mutation.

The field of embryology has long been burdened with the frustration of only being able to describe causal events in development without truly understanding the basis of how they occur; consequently, embryology stands to gain a great deal from progress in molecular genetics. This is, in fact, one of the most active areas of biological research today.

PRESENT STATUS OF GENE REGULATION

Gene regulation is still largely characterized by great gaps in our knowledge, and, as is obvious from our discussion above, the more complex the system the greater the gaps. How soon scientists can break through these barriers remains to be seen, but the past accomplishments of molecular geneticists lends to optimism that the future holds great promise for our understanding of basic life processes.

Because of the universality of the double helix and the genetic code, many still feel that the scheme for gene regulation in complex organisms will be found to be basically the same as that in microorganisms with, of course, modifications to account for the greater complexity.

With this in mind, many scientists who have made significant contribution to our understanding of gene action in microorganisms are now turning their attention to cells of higher organisms, realizing the challenge posed by the great complexity of organization of multicellular beings, and that what is true in *E. coli* is not necessarily always true in the elephant!

REFERENCES

The different mutations in mice mentioned in Chapter 8, *W, Sd, Sl,* mdg, all are discussed in:

E. L. Green (ed.), 1966: *The Biology of the Laboratory Mouse,* 2d ed., McGraw-Hill Book Company, New York.

In addition, the following review article and titles cover most of the important work in developmental genetics:

Gluecksohn-Waelsch, S., 1951: "Physiological Genetics of the Mouse," *Advances in Genetics,* **4:**1–51.

Markert, C. L., and H. Ursprung, 1971: *Developmental Genetics,* Prentice-Hall, Inc., Englewood Cliffs, N.J.

Spratt, N. T., 1971: *Developmental Biology,* Wadsworth Publishing Company, Inc., Belmont, Calif.

In addition, specific works cited are listed as follows, although the layman will probably find them too advanced:

Banker, B. Q., M. Victor, and R. D. Adams, 1957: "Arthrogryposis Multiplex Due to Congenital Muscular Dystrophy," *Brain,* **80:**319–334.

Beermann, W., and U. Clever, 1964: "Chromosome Puffs," in *Facets of Genetics,* W. H. Freeman and Company, San Francisco.

Briggs, R., and T. King, 1952: "Transplantation of Living Nuclei from Blastula Cells into Enucleated Frogs Eggs," *Proceedings of the National Academy of Science,* **38:**455–463.

Gross, S. R., and J. J. Hutton, 1971: "Induction of Hepatic Aminolevulinic Acid Synthetase Activity in Strains of Inbred Mice," *Journal of Biological Chemistry,* **246:**606–614.

Jacob, F., and J. Monod, 1961: "Genetic Regulatory Mechanisms and the Synthesis of Proteins," *Journal of Molecular Biology,* **3:**318–356.

Lyon, M. F., 1962: "Sex Chromatin and Gene Action in the Mammalian X-chromosome," *American Journal of Human Genetics,* **14:**135–148.

Margolis, F., 1971: "The Regulation of Porphyrin Biosynthesis in the Harderian Gland of Inbred Mouse Strains," *Archives of Biochemistry and Biophysics,* **145:**373–381.

McClintock, B., 1950: "The Origin and Behavior of Mutable Loci in Maize." Reprinted in J. Peters (ed.), *Classic Papers in Genetics,* Prentice-Hall, Inc., Englewood Cliffs, N.J., 1959.

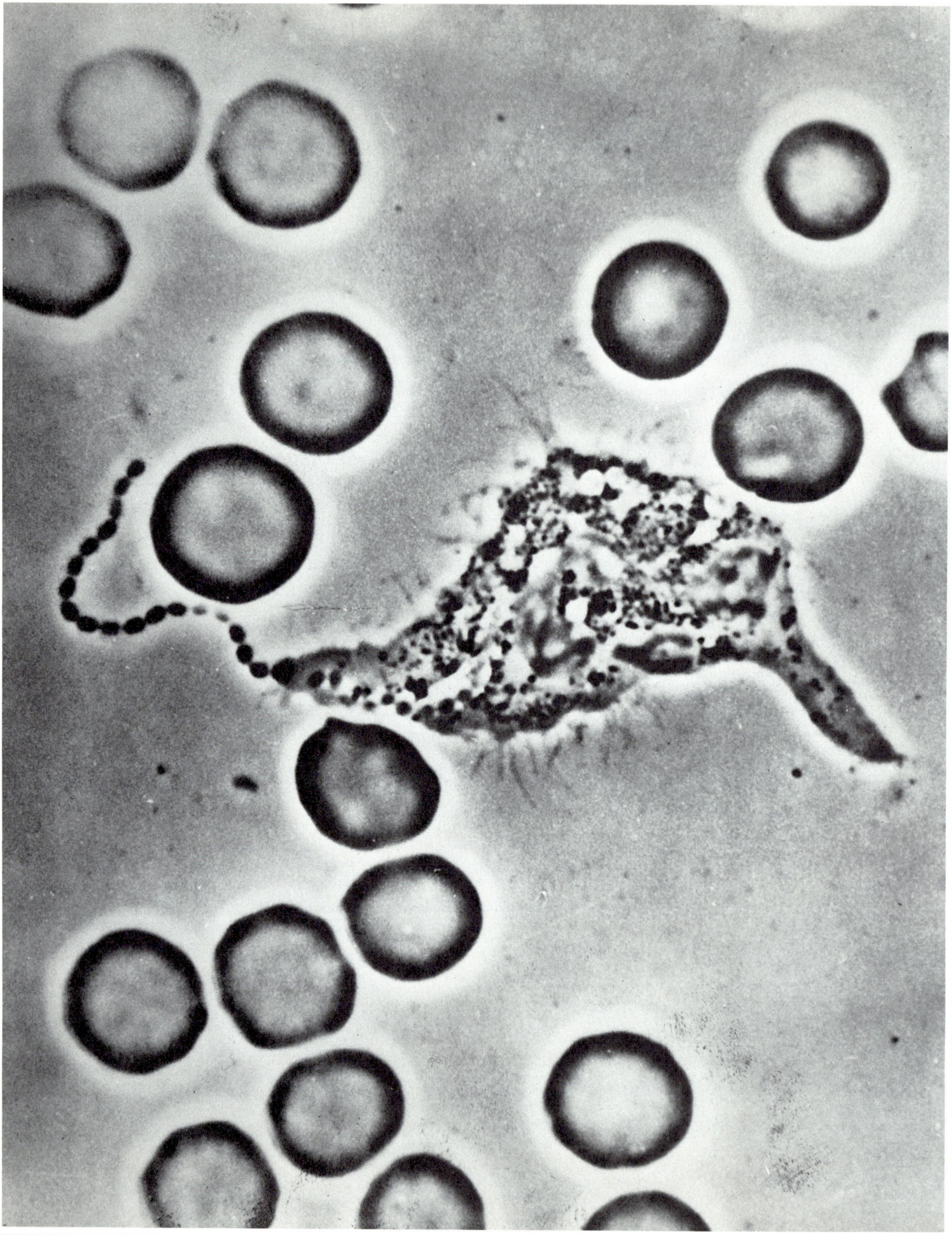

Chapter 9

The Genetics of Immune Reactions

There is perhaps no more dramatic example of the uniqueness of every individual than the tragic failure of attempts at organ transplantation. The terms *compatibility* (meaning acceptance of transplants) and *rejection* (meaning failure of transplants) are now familiar to all who have followed surgeons' desperate attempts to replace failing vital organs.

The ability of our bodies and cells to resist invasion by foreign objects, be they cellular, viral, or chemical, is the subject of interest for the increasingly important field of *immunology*. Like all life processes, the basis for immune reactions such as the failures of transplants can be traced to the genetic constitution of those involved. Because of the profound influence of molecular genetics on all cellular studies, there is now an area of genetics known as immunogenetics that combines the two disciplines.

WHAT IS AN IMMUNE REACTION?

To the layman, the term *immune reaction* generally brings to mind common situations such as allergies and the use of vaccinations to render a person immune to some disease.

A human white blood cell ingesting a chain of streptococci (bacteria). [*Courtesy of Public Relations, Pfizer Laboratories.*]

These are immune reactions, of course, but so are blood-type incompatibility and transplantation rejection. Also, recent evidence indicates our susceptibility to cancer is closely related to our immune mechanism. In fact, so many conditions can be attributed to immune reactions that the elucidation of the mechanism for immune reactions has one of the highest priorities in research in medicine and genetics today.

ANTIGENS AND ANTIBODIES

Basically an immune reaction involves the interaction of two factors, an outside agent that is unacceptable to the special cells of the body involved in immune reactions, and the body factor that responds to the invader. The "foreign" factors are called *antigens,* and are usually organic substances or cells of some sort. When they enter the body, they elicit a reaction from the specialized cells of the host which results in either the production of special proteins called *antibodies* or the proliferation of cells called *phagocytes* that literally attack the antigens.

When antibodies are involved, a number of different kinds of reactions can occur: if the antigen is in solution, reaction with antibodies can cause the antigen to precipitate out of solution; or if cells are the antigenic agent, antibodies can cause them to break open, or lyse; finally, antibodies can cause the antigens to agglutinate, or stick together in clumps.

ACQUIRED AND PASSIVE IMMUNITY

It is well known that there are many diseases to which humans become immune after having been exposed to the disease organism. This is because the presence of that particular organism causes the cells to produce specific antibodies against it, and to "remember" the antigen so that if it infects the individual again, it will be destroyed before causing symptoms of the disease again. This is called *acquired immunity.* Immunologists have used this phenomenon to protect humans against diseases before they encounter the disease organism. By using killed microorganisms which can no longer cause disease but which still retain their antigenic qualities, our bodies can be induced to produce antibodies against the organism as if we had naturally been exposed to it. Immunity developed in this manner is referred to as *artificial immunity;* it is a form of acquired immunity.

On the other hand, sometimes immunity can be transferred by transferring antibodies that have been made in one individual into a second individual. This *passive immunity* does not last very long in contrast to acquired immunity, which generally lasts many years, and in some cases a lifetime. Passive immunity occurs naturally, such as in transfer from the first milk (colostrum) of nursing mothers to the newborn. It can also be introduced artificially by injection of antibodies to confer immunity for some diseases, although this technique is being used less and less frequently.

The definitions of *antigen* and *antibody,* then, are circular: an antigen is a factor that elicits the production of antibodies; and an antibody is the protein, found in blood plasma, which is produced by cells of the body in reaction to the presence of an antigen, and which specifically reacts to that antigen. Because they circulate in the plasma of blood, antibodies which participate in immune reactions are called *circulating antibodies.*

CIRCULATING ANTIBODIES: BLOOD GROUPS AND RESISTANCE TO DISEASE

A familiar example of the kind of immune reaction involving circulating antibodies is the incompatibility of certain blood types for transfusion. The technique of transfusion was not used to any extent until the nineteenth century, and even then its use was limited because of frequently serious effects. Patients receiving blood might go into shock and, if the reaction was severe, die as a result of the transfusion.

THE ABO BLOOD GROUP

In 1900, an Austrian physician, K. Landsteiner, discovered that there were in fact different "types" of blood in different people. He found that if the wrong types were mixed together, a clumping of red blood cells would ensue, and this was what caused the deaths in earlier attempts at transfusion. This reaction was correctly interpreted to be an antigen-antibody reaction.

THE IMMUNE BASIS TO ABO BLOOD-TYPE REACTIONS It has since been established that blood-type incompatibility is indeed an immune reaction. In discussing codominance in Chapter 5, we stated that what makes a person type A is a particular substance on the surface of his red blood cells. We can now

TABLE 9.1
A summary of the genetic basis of the ABO blood types

BLOOD TYPE	GENOTYPE	ANTIGEN	ANTIBODY
A	$I^A I^A$, $I^A i$	A	Anti-B
B	$I^B I^B$, $I^B i$	B	Anti-A
AB	$I^A I^B$	A, B	None
O	ii	None	Anti-A, Anti-B

restate this by saying that a person with blood type A has the A antigen on the surface of his red blood cells, as determined by the codominant gene I^A. Similarly, type-B individuals have the type-B antigen, determined by the gene I^B; type AB people have both antigens, and type-O people have neither A nor B antigens.*

* Actually, there are two kinds of type-A antigen, A and A¹. For clarity, however, we shall discuss the two as one type, type A.

THE GENETIC BASIS OF ABO BLOOD TYPES The clumping reaction, or agglutination, occurs when different blood types are mixed because type-A individuals have anti-B antibody molecules present in their blood and type-B people have anti-A antibodies. Type-O individuals are found to have both anti-A and anti-B antibody molecules. On the other hand, type-AB individuals have no antibodies to the ABO group of antigens at all. Table 9.1 summarizes the genetic basis of the ABO blood types.

Notice that if type-A blood were introduced into type-B recipients, the A antigen on the surface of the red blood cells would react with the anti-A antibody molecules in the blood of the recipient, and the result would be agglutination. The same principle holds true for all the possible combinations of blood types, as seen in Table 9.2.

TABLE 9.2
Effects of mixing different types of blood

RECIPIENT TYPE	DONOR TYPE				
	A	B	AB	O
A	−	+	+	−
B	+	−	+	−
AB	−	−	−	−
O	+	+	+	−

\+ indicates agglutination; − indicates no reaction.

Because AB individuals do not have any antibodies, they can receive, with certain precautionary measures, all types of blood without unfavorable reaction, and so sometimes are called "universal recipients." Notice, however, that they cannot donate their blood to any type except AB. On the other hand, type-O blood may be given to most individuals without serious reaction because of the absence of any antigens to elicit antibody reaction. Type-O people are called "universal donors," and can receive only type-O blood.

THE ORIGIN OF THE CIRCULATING ANTIBODIES What is the source of the ABO antibodies? This is one of the many questions for which no clear answer is available. Most antibodies that appear in reaction to antigens are produced by special white blood cells from bone marrow, called *plasma cells*. What is perplexing about the ABO antibodies is that although plasma cells are known not to be able to produce antibodies until some time after birth, the ABO antibodies are already present at birth. This is termed *natural immunity*.

A clue to the origin of natural antibodies may come from studies of the immune system of birds. In chickens there is a structure, a derivative of the intestines, called the bursa of Fabricius that is the source of circulating antibodies in that species. Immunologists hypothesize that a similar, as yet undiscovered, structure also exists in man.

THE Rh FACTOR

Another well-known antigen carried by the red blood cells is the blood group known as the Rh, or rhesus, factor, so named because it was first found in the red blood cells of rhesus monkeys. People who are Rh-positive (Rh+) have a particular Rh antigen on the surface of their red blood cells which is different from the ABO antigens and is determined by different genes. Rh-negative (Rh−) individuals do not have the Rh antigen.

One basic difference between the ABO and Rh systems is that Rh antibodies are not natural antibodies; that is, they are not present at birth, but are synthesized in response to the presence of the Rh antigen. Obviously, people who are Rh-positive do not synthesize anti-Rh antibodies, since this would cause agglutination of their own cells, but Rh-negative people react strongly to Rh antigens with the production of antibodies.

TABLE 9.3
Genetic basis of the Rh factor

BLOOD TYPE	GENOTYPE	ANTIGEN	ANTIBODY
RH+	*DD* or *Dd*	+	None
RH−	*dd*	None	+

THE GENETIC BASIS OF THE Rh BLOOD TYPES The genetic determination of the Rh factor is not entirely clear. There is evidence that more than one pair of genes are involved, but for simplicity, we shall depict the genetic system as one pair of alleles. It has been found that the Rh+ condition is dominant to Rh−, and thus we can use the letter *D* to symbolize the dominant gene, and *d* for the recessive gene (see Table 9.3).

THE IMMUNE BASIS FOR Rh INCOMPATIBILITY Interest in the Rh factor stems from an incompatibility between the blood of Rh− women and that of their Rh+ fetuses. Although fetal and maternal blood do not come into direct contact in humans because of the placental barrier, some fetal red blood cells do manage to enter the bloodstream of the mother. The presence of Rh antigens on the surface of these cells causes maternal white blood cells, the plasma cells, to produce anti-Rh antibodies that *can* cross the placental barrier. When these antibodies enter the fetal bloodstream, they cause agglutination of fetal red blood cells. The fetus then suffers from a blood disorder known as erythroblastosis fetalis, an anemia which stems from an insufficient number of red blood cells to effect the exchange of gases (Figure 9.1). Depending on the severity of reaction, the disease may be fatal to the baby.

Notice that the disorder arises only if the mother is negative and the baby positive; it does not occur nor is there harm done to the mother if the mother is positive and the baby negative. The reason is that even if the negative baby produces antibodies against his mother's cells and these molecules enter the mother's bloodstream, there is proportionally much more of the mother's blood than of the baby's antibodies. Therefore, the effect is negligible. For this same reason, a negative mother can produce enough antibodies to overwhelm the production of red blood cells in the fetus.

TECHNIQUES TO MINIMIZE Rh INCOMPATIBILITY The condition was a real medical problem a few years ago, but several tech-

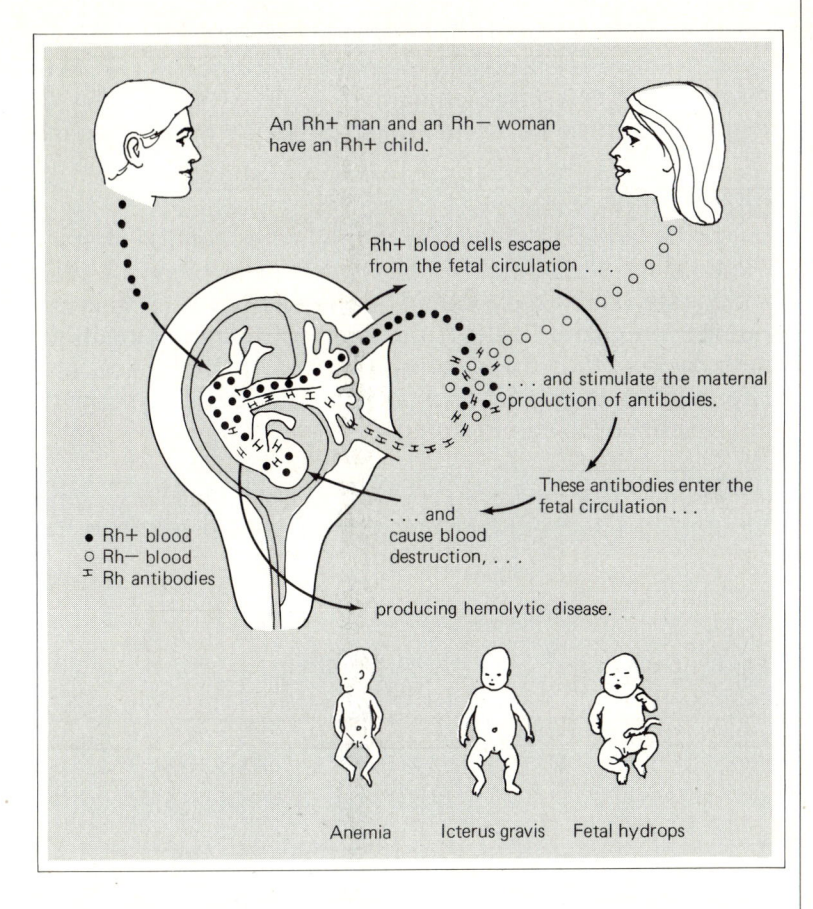

FIGURE 9.1
Rh incompatibility between an Rh-negative mother and an Rh-positive fetus results in a blood disorder, erythroblastosis fetalis. [*After Rh by E. L. Potter. Copyright © 1947 by Year Book Medical Publishers. Adapted by permission.*]

niques now available can minimize the danger to the Rh+ child of an Rh− mother. One method is to perform whole-body transfusion, and today this can be done even before the child is born, if necessary. If the level of antibodies rises dangerously in the mother's blood before term, the fetus can be transfused with Rh− blood by the insertion of a needle through the womb into the fetal abdomen. The Rh− cells will then be absorbed into the bloodstream of the fetus with no apparent ill effects on either mother or child.*

Another method is the use of the anti-D antibody. One of the problems with Rh incompatibility is that once the production of anti-Rh antibodies has begun, the reaction of an Rh− woman against her Rh+ offspring will be progressively more severe with each subsequent pregnancy. One reason is that

* Eventually the child will revert to Rh+ blood.

with the physical trauma of birth, a large number of fetal red blood cells are released into the mother's bloodstream. This normally elicits a large amount of antibody production in response, and the next fetus would be faced with the handicap of existing antibody molecules already in the maternal bloodstream.

When the anti-D antibody is administered to the mother immediately after birth, it causes fetal red blood cells to agglutinate before the mother's own cells begin to react, so the next pregnancy will not be faced with large amounts of antibody from the time of conception (Figure 9.2). Of course, care must be taken that no more than the needed amount of anti-D antibody is administered.

ERYTHROBLASTOSIS FETALIS FROM ABO INCOMPATIBILITY One would expect that ABO incompatibility should result in the

FIGURE 9.2
Schematic illustration of the effect of injecting an artificial anti-Rh antibody into an Rh− mother who has just given birth to an Rh+ child. The antibody will react with fetal red blood cells before maternal cells become immunologically active.

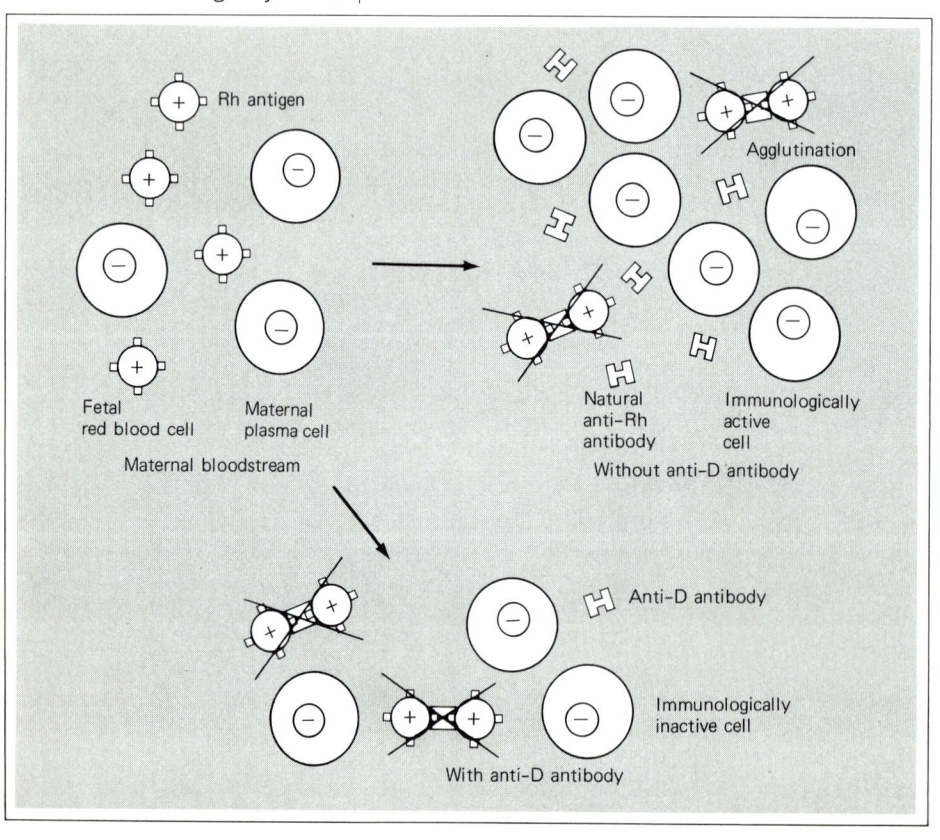

same kind of immune reaction between mother and child. For example, a type-O female should cause problems with the development of any offspring other than type O. In fact, the most common cause of erythroblastosis fetalis *is* ABO incompatibility; however, the Rh incompatibility is of more medical concern because the Rh antibody passes through the placenta more easily than the A and B antibodies, and thus causes the strongest reaction.

By "strong" we mean the severity of the reaction between antibody and antigen. There actually are many different blood groups determined by different genes, but most of them result in such weak immune reactions that there is no need to ensure matching types in transfusion, or to worry about incompatibility between mother and child in pregnancy. In these "weak" reactions, the antigen does not cause a strong antibody reaction, and blood cells are consequently not agglutinated. The anemia resulting from Rh incompatibility frequently is so severe as to cause death from heart failure, whereas the reaction from ABO incompatibility results in a much less severe anemia that rarely leads to death.

In fact there is some evidence that ABO incompatibility may protect the fetus against the development of anemia due to Rh incompatibility. It is thought that ABO antibodies in the mother's serum destroy fetal red blood cells as soon as they reach the mother's bloodstream, thus removing them before the mother's immune system can begin to react against the Rh antigens. However, it is interesting that most of the cases of anemia resulting from ABO incompatibility involve mothers who are type O and children who are type A.*

* See references to Lawler and Lawler and to Race and Sanger at the end of the chapter.

LEGAL IMPLICATIONS OF BLOOD TYPING

Because of the number of blood groups other than ABO and Rh, it would be virtually impossible, if all groups were considered, to find two people (aside from identical twins) with identical blood types, because for each blood group a number of alleles exist. This genetic heterogeneity in a population, that is, the existence of different alleles in a population, is sometimes called *genetic polymorphism*.

Because the combination of blood types in every individual is unique, blood typing has been used for purposes of identification, and as evidence in many legal disputes such as paternity cases. For example, if a woman accused a man of being the father of her child, and the blood types involved were the following, what would your decision as a geneticist be if you were called in consultation? Mother = type A; man = type AB; child = type O. In this case, the man is clearly not guilty, since the child would have inherited one or the other of his two codominant alleles, and therefore could not be homozygous recessive. The mother's genotype is therefore $I^A i$, and the father, whoever he is, also must be at least heterozygous for the recessive *i* allele.

Actually, use of the blood type can only prove innocence, not guilt. Had the man's blood type been type O, the only conclusion that should be drawn is that he *could* have been the father. Some of the other blood groups of the individuals concerned should be analyzed to provide more information.

IMMUNITY TO DISEASE ORGANISMS

Circulating antibodies are also of great importance in the resistance of higher organisms to infection by microorganisms. The presence of infectious agents which act as antigens causes the production of antibodies. The effect of antibodies on the microbes can vary, in some cases causing agglutination much in the same way as blood incompatibility, in other cases causing *lysis,* or disintegration, of the foreign cells.*

* The science of immunology actually began with the study of the body's reaction to disease organisms. A discussion of the history and application of principles of immunology in medicine is beyond the scope of this book, but it is an absolutely fascinating chapter in the development of scientific thought, and the interested reader is encouraged to turn to the relevant references at the end of the chapter.

Another poorly understood aspect of our immune system against disease organisms is an *innate immunity* which different species have toward microorganisms known to cause disease in other species. For example, we are unaffected by organisms which cause distemper in dogs; similarly, dogs are completely unaffected by measles virus. The differences in susceptibility no doubt reflect genetic differences between species, but in what way is unknown.

IMMUNE REACTIONS: PROBLEMS OF CELL-MEDIATED TRANSPLANTATION

We began our discussion of immune reactions by stating that there are two major kinds, one involving circulating antibodies. The second kind is a cell-mediated reaction, which is responsible for problems encountered in the transplantation of tissues and organs from one organism to another.

The presence of transplanted tissues and organs elicits not the production of antibodies by plasma cells, but rather the

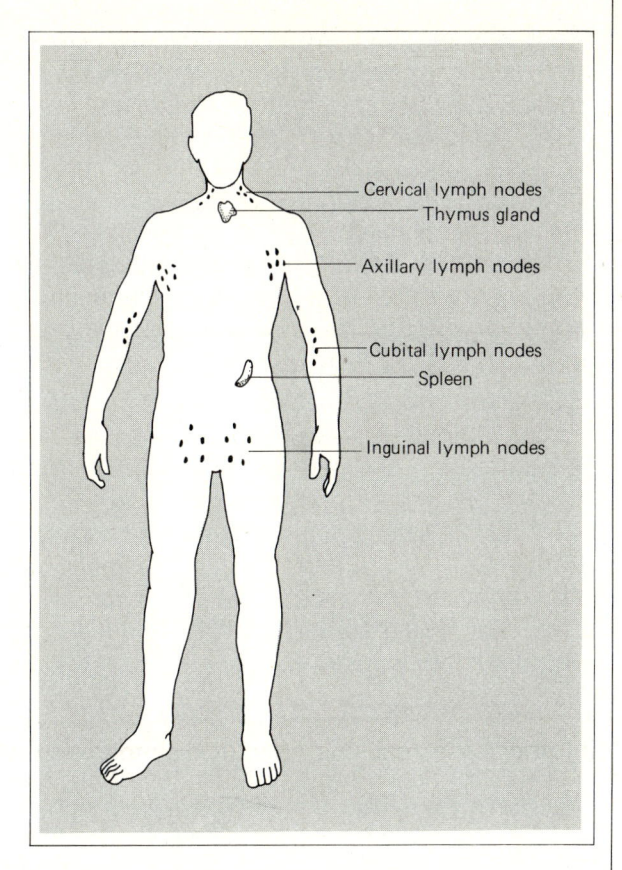

FIGURE 9.3
The location of the spleen and some important lymph nodes, sources of lymphocytes, in the human body.

- Cervical lymph nodes
- Thymus gland
- Axillary lymph nodes
- Cubital lymph nodes
- Spleen
- Inguinal lymph nodes

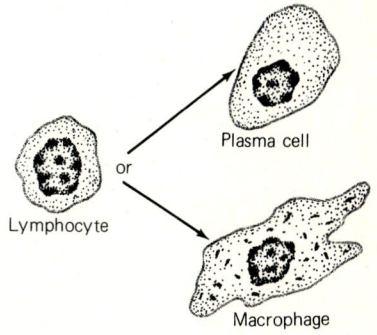

FIGURE 9.4
Lymphocytes are capable of differentiating either into plasma cells which produce antibodies or into macrophages which are phagocytes and can consume bacteria. [*After A. A. Maximow and W. Bloom,* A Textbook of Histology, *7th ed. Copyright © 1957 by W. B. Saunders Co., Philadelphia.*]

proliferation of phagocytes, cells that literally "eat up" the foreign cells. These cells develop from white blood cells called lymphocytes, so named because they occur in great numbers in our lymph nodes. (See the opening illustration to this chapter for an actual photograph of a phagocyte consuming bacteria.) The lymphocytes originate in the thymus gland, and in early life migrate from the thymus to various organs of the body which later serve as the source of these vital cells, such as the lymph nodes and spleen (Figure 9.3).

The versatility of these cells is very interesting in terms of gene action, because given the proper stimulus, the lymphocyte can develop into either a phagocyte or a plasma cell that is capable of antibody production (Figure 9.4). Which of these forms the lymphocytes take is a reflection of differential gene action.

HISTOCOMPATIBILITY LOCI

The ability of lymphocytes to recognize what is foreign and what is self appears to be determined by the presence of "individuality markers" (antigens) present on the cell surface of all nucleated cells. (The distinction is made between nucleated and nonnucleated cells because mature human red blood cells, which do not have nuclei, do not elicit lymphocyte reaction, although they do possess antigens on their surfaces which cause circulating antibodies to react.) These markers are different in all individuals because they are determined by a number of different genes that are known as *histocompatibility loci*.

C. C. Little and G. Snell of the Jackson Laboratory in Bar Harbor, Maine, pioneered studies on the genetic basis of transplantation compatibility using inbred strains of mice. To date, at least 15 different histocompatibility loci have been found, producing markers or antigens of differing strengths. The antigen causing the most severe immune reaction is determined by the *H-2* locus, a multiple allelic system for which some 20 different alleles, all presumably codominant, have been found.

The exact number of histocompatibility loci in humans has not been determined, but there are several, including one strong one that is believed to be the counterpart of the *H-2* locus in mice. This complex locus in man is called *HL-A*, and is believed to have two subregions, each of which has at least two dozen alleles. The large number of possible forms of these genes, which is a good example of genetic polymorphism, causes problems when transplantations of tissues and organs are attempted between individuals who are not identical siblings.

PROBLEMS OF HUMAN ORGAN TRANSPLANTS

The great variability which exists in the human population renders it highly unlikely that the kinds of individuality markers present on the cell surfaces of any two people are genetically alike. Since the recent attempts at heart transplantation are between a recipient and donor who are not, of course, chosen solely on the basis of histocompatibility type but rather by fate since the donors are usually victims of sudden violence, it is not surprising that few of the transplants have lasted more than a few years. Most have been rejected within weeks.

To overcome this problem, the patients who are to receive organ transplants have been treated with chemicals to destroy the immunologically active lymphocytes. However, this method is unsatisfactory on two counts. First, some of the drugs used produce very unpleasant side effects such as hallucinations. Second, the absence of the body's main line of defense against infection leaves the patient susceptible to severe and possibly fatal attacks from organisms that would normally cause minor illness, such as a cold virus. In addition, the repression induced is only temporary.

A recently developed technique that may help solve this problem is one which allows scientists to actually remove the markers from the surface of the cells and to introduce them in a solubilized form to a recipient. Somehow, and the reason is not known, the fact that they are not attached to cell surfaces seems to confuse lymphocytes that would normally proliferate and attack the markers. In a solubilized form, the markers seem to be accepted, and subsequent tissue transplants that carry the same antigens are not as quickly rejected. This method is still in experimental stages, but holds promise for the future.

Another area of active research in circumventing the problems of tissue incompatibility is the construction of artificial organs composed of nonantigenic material. There have already been successful replacements of structures such as blood vessels with plastic tubing. Of course, artificial organs also have the advantage of being available in quantity, whereas the availability of real tissues and organs is limited.

There are a few areas of the human body which are called "immunologically privileged" sites because, for poorly understood reasons, the rejection of foreign tissue is decreased in severity, or absent altogether. One example of such a site is the cornea of the eye. The majority of cases of corneal transplants are successful. Another such privileged region is the anterior chamber of the eye. One theory for these exceptions is that there is less blood or lymph circulating through

THE PHYSICAL NATURE OF THE ANTIBODY MOLECULE

What are antibodies, and how are they determined? These basic questions are being actively investigated now by both medical and molecular genetics researchers. It has been established that antibodies are a group of proteins in the plasma of blood that are also called *gammaglobulins*. There appear to be a number of different kinds of immunoglobulins, which have been designated by different letters of the alphabet. They all have a general structure, and we shall describe one of the most common types, immunoglobulin G, to illustrate the general nature of an antibody molecule.

The antibody molecule appears to be a beautifully symmetrical one composed of four polypeptide chains linked by disulfide bonds, so called because two atoms of sulfur are involved. Two of the chains are large and identical to each other, and are called the "heavy" chains; the other two are

FIGURE 9.5
The generalized structure of immunoglobulin G. CHO represents a side chain of carbohydrate molecules. [*After Biology Today, CRM, 1972, Del Mar, Calif., p. 450.*]

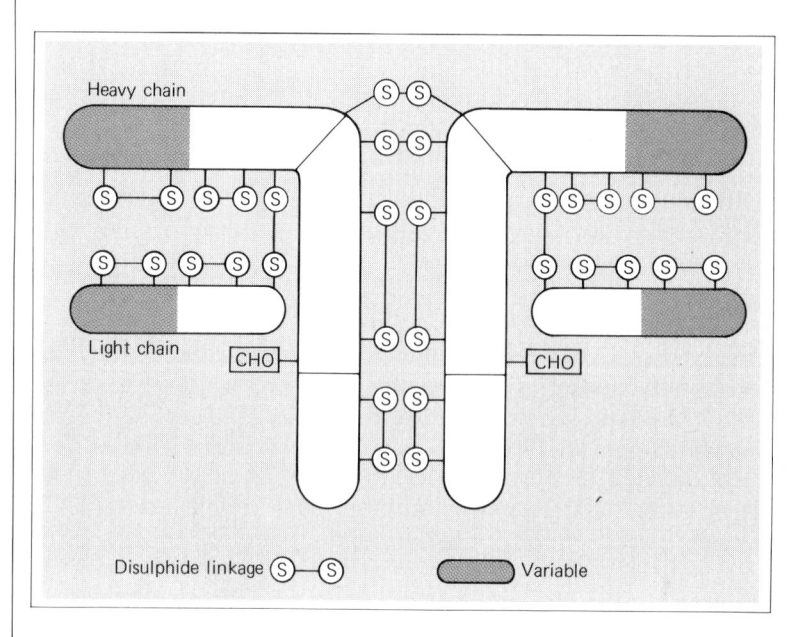

also identical to each other and are called the "light" chains. Figure 9.5 gives a schematic illustration of this molecule, which is one of the most important our cells are capable of producing.

From what we have discussed so far, it should be obvious that the ability of cells to synthesize proteins is absolutely essential for the life of the cell and the well-being of the entire organism. In this regard, did you ever stop to think why malnutrition is almost always accompanied by disease? Now that you know a little about our immune mechanism, the relationship should be apparent: malnutrition itself is not a disease-causing condition, but without a proper diet which includes proteins that are broken down to amino acids upon digestion, there would be no amino acids for cells to utilize in synthesizing their own proteins such as antibodies. The result is a weakened response to disease organisms.

THE SPECIFICITY OF ANTIGEN-ANTIBODY REACTIONS

If the antibody molecule is basically as described above, what then accounts for the specificity of antigen-antibody reaction, for as we mentioned before, it has long been known that for each antigen, a specific antibody molecule exists? The techniques developed by molecular biologists in analyzing the amino acid sequence of proteins have been of great help in answering this question. In the analysis of the amino acid sequence of the antibody molecule, it appears that a portion of each light and heavy chain never varies in amino acid sequence; however, the remaining portion of each chain does indeed show variation from antibody to antibody. (See again Figure 9.5.)

Presumably the variability of the sequence of amino acids in some way causes a specific configuration of the antibody molecule, so that the molecule then reacts with an antigen possessing a complementary configuration. The present theory is that in this way a "lock-and-key" type of reaction occurs, with each antibody molecule having two sites of reaction with antigen molecules (Figure 9.6).

GENETIC DETERMINATION OF ANTIBODY MOLECULES

The genetic determination of antibody molecules has been more difficult to discover than has their biochemistry. By

FIGURE 9.6
Schematic illustration of the "lock-and-key" theory of antigen-antibody reaction.

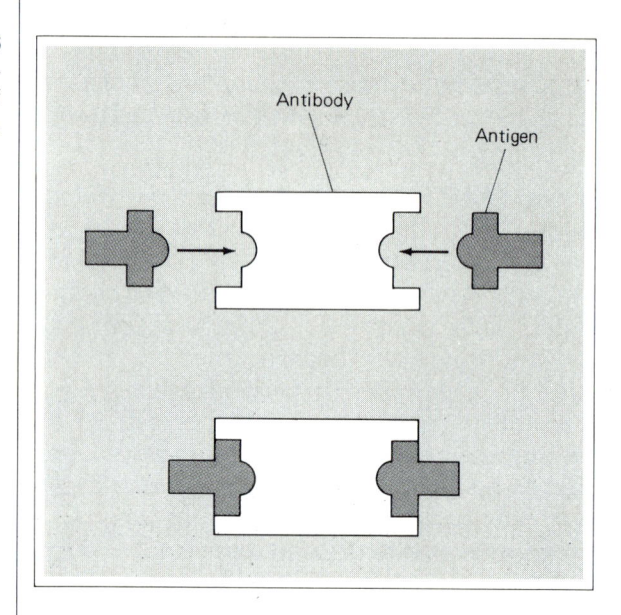

being able to respond specifically to each antigen that enters the body, every individual has the potential to produce a staggering number of different antibody molecules, considering not only all the different microorganisms that can infect us but also all the various organic substances and plant materials to which many people react. Therefore, it is difficult to conceive that there is a gene for every different immunoglobulin molecule produced.

Without any real experimental evidence that would allow geneticists to decide which situation actually exists, a number of different theories on antibody determination have arisen. One, known as the "multiple-germ-line theory," states that the constant parts of the antibody molecules are determined by a few alleles, and that many different genes code for the variable portions of both the light and heavy chains. Presence of a particular antigen would activate certain genes in any plasma cell.

Another theory states that the variability results from a few uniquely unstable genes which "hypermutate" in the plasma cells themselves, creating many different kinds of plasma cells, certain ones of which are then stimulated to divide actively in the presence of the antigen for which they can produce antibodies. This is known as the "clonal" or "cell selection" theory. Which, if either, of these two theories is actually the case has not been determined. Figure 9.7 illustrates the clonal theory.

Another immune phenomenon that intrigues scientists is how plasma cells recognize what is foreign and what is "self." In other words, if our cells are antigenic to others, why are they not antigenic to ourselves? Sir Macfarlane Burnet, who has been one of the foremost proponents of the clonal selection theory, has hypothesized that as the embryo develops, and as plasma cells appear which could produce antibodies against the antigens of the embryonic cells, there is an immediate response so severe that those particular plasma cells are destroyed. Consequently, only those cells capable of producing antibodies against antigens *other* than those encountered within the individual are allowed to survive. These then become our main weapons in reacting to the presence of foreign agents.

THE RECOGNITION OF SELF

FIGURE 9.7
Schematic illustration of the clonal selection theory. [*After* Biology Today, *CRM, 1972, Del Mar, Calif., p. 444.*]

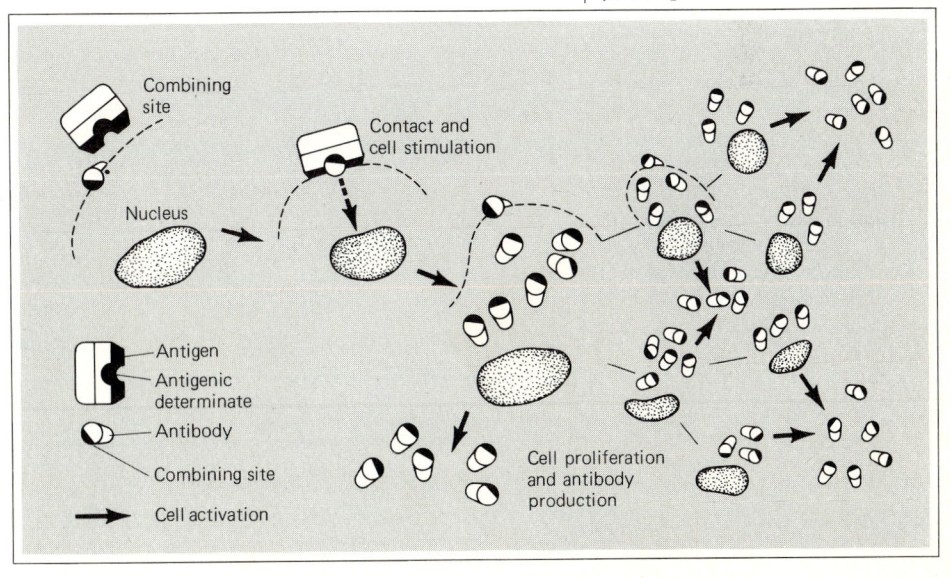

INDUCTION OF TOLERANCE

Some evidence to support Burnet's theory is the loss of an organism's ability to react against an antigen once exposed to it at a very early stage of life. This situation is known as *induced tolerance*. Normally, because of antigenic differences, transplantation of skin from mice of one inbred strain to another is not successful. However, if cells from mice of strain *A* are injected into embryonic or newborn mice of strain *B*, then these strain-*B* mice will accept skin grafts from strain-*A* mice as adults. According to Burnet's theory, the introduction of strain-*A* cells must presumably destroy strain-*B* plasma cells that normally would have reacted to foreign *A* cells later in life (Figure 9.8).

One might conjecture on the possible application of this phenomenon to humans to overcome our problems of organ transplant rejection. If tolerance can be induced in humans, one can envision the advantages offered by the induction of tolerance between, say, siblings, so that in the future, organs such as kidneys (of which we have two, and need only one) may be transplanted, if necessary, from one sib to another with total success.

FIGURE 9.8
The induction of tolerance in genetically different strains of mice (*CBA* and *A*) by introducing cells of one strain into another at early stages in life. The recipients then tolerate transplants from the donor strain as adults. [*After Rupert Billingham and Willys Silvers*, The Immunobiology of Transplantation, © *1971. Adapted by permission of Prentice-Hall, Inc., Englewood Cliffs, N.J., p. 134.*]

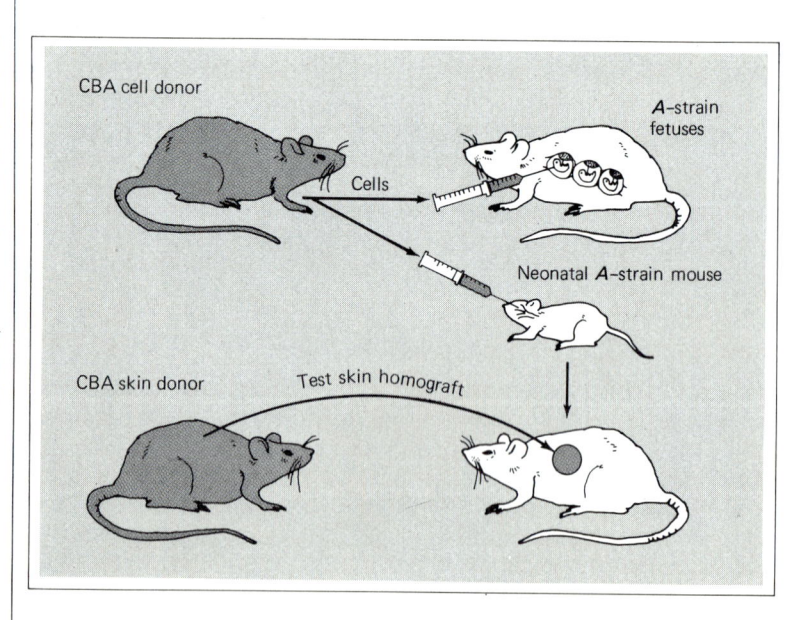

AUTOIMMUNE DISEASES

Actually, there do exist disease conditions in humans and experimental animals which seem to result from the failure of the immune cells to recognize certain "self" substances so that they turn on their own cells as if those cells were foreign. Such conditions are called *autoimmune diseases,* and include conditions such as hemolytic anemias, Lupus erythematosus, and rheumatoid arthritis. Intensive research is continuing on the causes for these presumed aberrations of the immune process.

Recently, evidence has been accumulating which indicates that some viral infections can cause a person's immune mechanism to destroy his own cells, possibly by forcing the infected cells to produce new and "foreign" substances which attract antibodies. The relationship between this phenomenon and autoimmune diseases is at present under investigation.*

* See reference to Notkins and Koprowski at the end of the chapter.

OTHER PROBLEMS IN IMMUNOLOGY

The wide scope of biological processes which fall under the heading of immune reactions points up the great importance of this vital area, to which genetics has already contributed a great deal of information. But as you have probably noticed, we have far to go before we understand some of the complex situations which exist, partly because we understand so little of some of the basic concepts of immune reactions.

For example, the confusion that results in the immune system when "individuality markers" are introduced in solubilized form points out one of the basic questions that has not yet been answered: what causes something to be antigenic? Factors known to affect the antigenicity of molecules include size, rigidity of structure, amount of exposure, and as in the case of the markers, association with the cell membrane when exposed to cells of the immune system.

Another poorly understood phenomenon is "immunological memory," a common characteristic of immune reactions. This term refers to the fact that once a cell has been exposed to an antigen, a second exposure results in a faster, stronger reaction. For example, if a skin graft is made between two incompatible strains of mice, it will be rejected, say, in a week's time. If the same graft is made again shortly after the first, the rejection may occur in half the time, presumably because of the presence of immunologically active cells elicited by the first graft. However, a number of different factors are known to affect immunologic memory, such as the age of the individual and the quantity of antigens present.

These are just a few of the unsolved questions of immune reactions confronting scientists. There are many other aspects we have not touched upon because they are beyond the scope of this book. In the next chapter we shall center our attention on yet another area to which researchers are devoting intense effort: the relationship between viruses and cancer, and the possible role played by our immune system in resisting both viral infections and cancer.

REFERENCES

There are not many sources on immunogenetics written for the nonscientist. Most of the principles discussed here can be found in the more extensive general biology texts such as:
Biology Today, 1972, CRM Books, Del Mar, Calif.

A good general article:
Burnet, Sir Macfarlane, 1961: "The Mechanism of Immunity," *Scientific American,* **204**:58–67.

An article on the production of disease by immune reactions:
Notkins, A. L., and H. Koprowski, 1973: "How the Immune Response to a Virus Can Cause Disease," *Scientific American,* **288**:22–31.

One paperback on transplantation phenomena with a clear and simply written approach:
Billingham, R., and W. Silvers, 1971: *The Immunobiology of Transplantation,* Prentice-Hall, Inc., Englewood Cliffs, N.J.

A small book, written on a basic level:
Lawler, S. D., and L. J. Lawler, 1971: *Human Blood Groups and Inheritance,* 3d ed., St. Martin's Press, New York.

A classic, more advanced book on blood groups is:
Race, R. R., and R. Sanger, 1962: *Blood Groups in Man,* 4th ed., Oxford University Press, Blackwell, England.

A more comprehensive volume, but also rather technical:
Abramoff, P., and M. LaVia, 1970: *Biology of the Immune Response,* McGraw-Hill Book Company, New York.

For the history of immunology:

Dubos, R., 1962: *The Unseen World,* Rockefeller University Press, New York.

———, 1960: *Pasteur and Modern Science,* Anchor Books, New York.

DeKruif, P., 1953: *Microbe Hunters,* Harcourt Brace Jovanovich, New York.

See also references to texts in history of science at the end of Chapter 1.

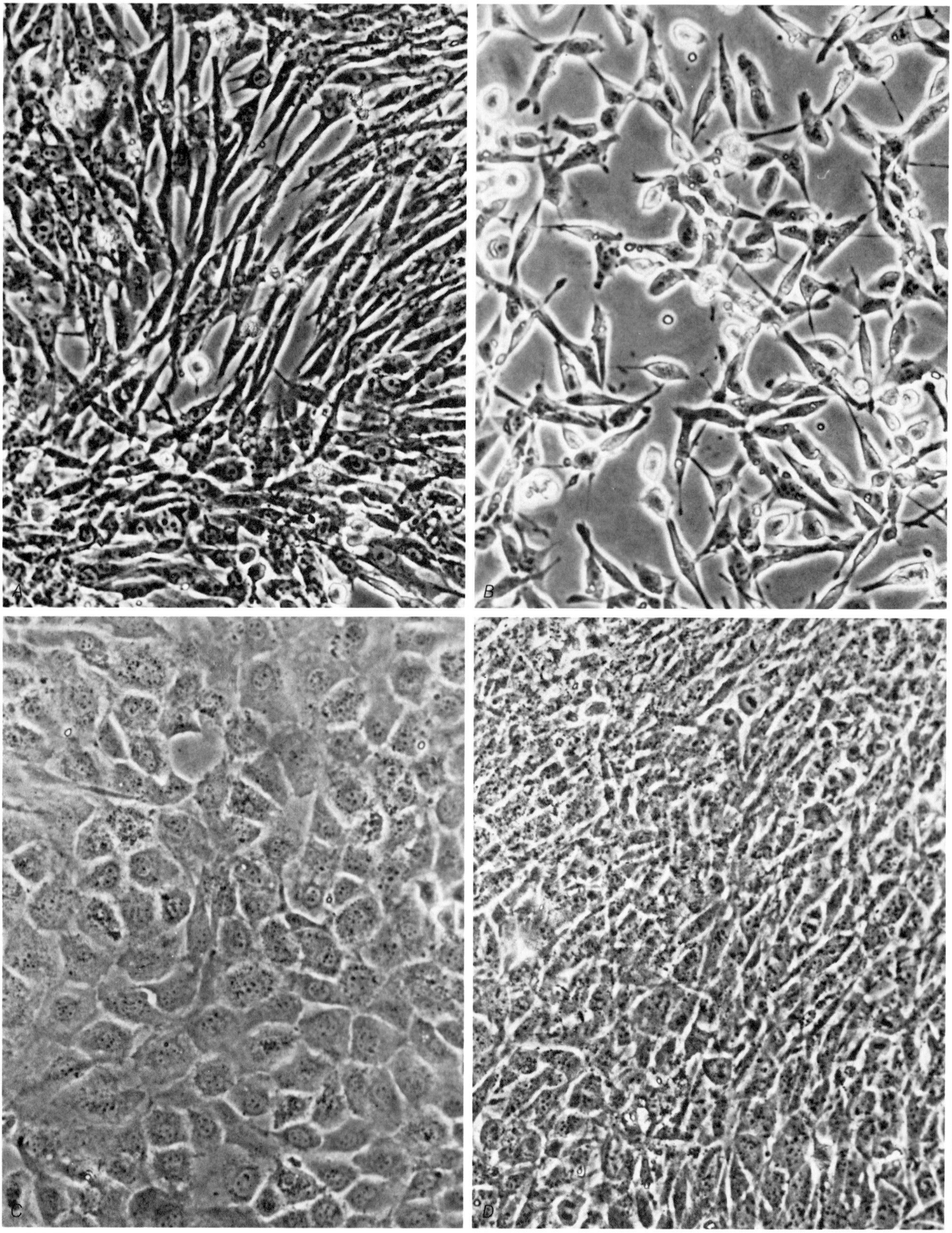
A
B
C
D

Chapter 10

The Genetics of Viruses and Cancer

The field of immunology is fascinating and, because of its far-reaching implications, essential to our understanding of many life processes. Techniques developed by molecular biologists, together with concepts of immunology, promise to be major factors in helping to conquer many medical problems, including one of our most dread diseases, cancer.

The contributions by Little and Snell to our understanding of the genetic basis of transplant rejection began as studies of cancer. They attempted to transplant tumors from one mouse to another and, in doing so, discovered that rejection was due not to the tumor-causing agent but to differences in what was eventually identified as the *H-2* locus. More recently, studies on the relationship between cancer and immune phenomena have focused on the ability of the cell to prevent infection by microorganisms—specifically, by tumor viruses.

For example, experiments have shown that many cancers in experimental animals are caused by viruses. Since viruses are antigenic, the field of immunology may contribute greatly to our understanding of the processes involved in the change

(*A*) Normal hamster cells in tissue culture. (*B*) Hamster cells transformed by polyoma cancer viruses. (*C*) Normal mouse cells. (*D*) Mouse cells transformed by SV-40 cancer viruses. [*Courtesy of Dr. Renata Dulbecco, Imperial Cancer Research Fund Laboratories, London.*]

from normalcy to malignancy. This knowledge, in turn, may help us to prevent or control cancer and other virus-caused diseases.

GENERAL CHARACTERISTICS OF VIRUSES

The existence of viruses was first postulated in the nineteenth century, when scientists discovered infectious agents that could not be trapped by filters with pores the size of cells. With modern microscopes we can now see these tiny particles. Ironically, although they often create havoc for the health of host cells and organisms, viruses have proved to be extremely useful for genetic studies.

All viruses are obligate parasites. That is, they cannot survive independently but must parasitize host cells in order to reproduce. Because the ability to reproduce is one of our criteria for an organism to be classified as living, viruses are considered to be subliving particles.

Generally speaking, viruses are composed of an inner core of nucleic acid; some contain DNA and others RNA. The nucleic acid is surrounded by a protein coat. See again Figure 6.6. Some of the common viruses shown there are known to parasitize animal cells, including those of humans, and others to parasitize bacteria.

Much is now known about the life cycles of viruses, primarily of those which prey on bacteria, commonly referred to as *bacteriophages* or simply *phages*. Scientists are now extending this knowledge to the study of cancer, as evidence has accumulated which points to viruses in animal cells being the basic cause for changes of cell characteristics to malignancy. Because there are some differences between bacteriophages and animal viruses, their effects on host cells will be considered separately. And because the details are more clearly understood in bacteriophages, we shall discuss these first.

BACTERIOPHAGES

* R. S. Edgar and R. H. Epstein, "The Genetics of a Bacterial Virus," and R. W. Horne, "The Structure of Viruses," in *The Molecular Basis of Life*, W. H. Freeman and Company, San Francisco, 1968, pp. 145–152 and 153–163.

A number of bacterial viruses have been studied and their development and genetics clearly analyzed.* Scientists have discovered that three possible routes may be taken by the virus following infection of bacteria.

One of the routes is the production of new viral particles following infection of the bacterium. These new viral particles will be released upon the *lysing,* or breaking open, of the host cell (Figure 10.1). The stage of the phage life cycle during which it is actively reproducing in the host cell is known as the *vegetative stage.*

Upon lysis of the host cell, viruses occasionally carry pieces of cellular DNA from the first host to another host cell. If the second host cell is not lysed, the foreign cellular DNA may be incorporated into its genome, causing a genetic change.

LYTIC PHASE

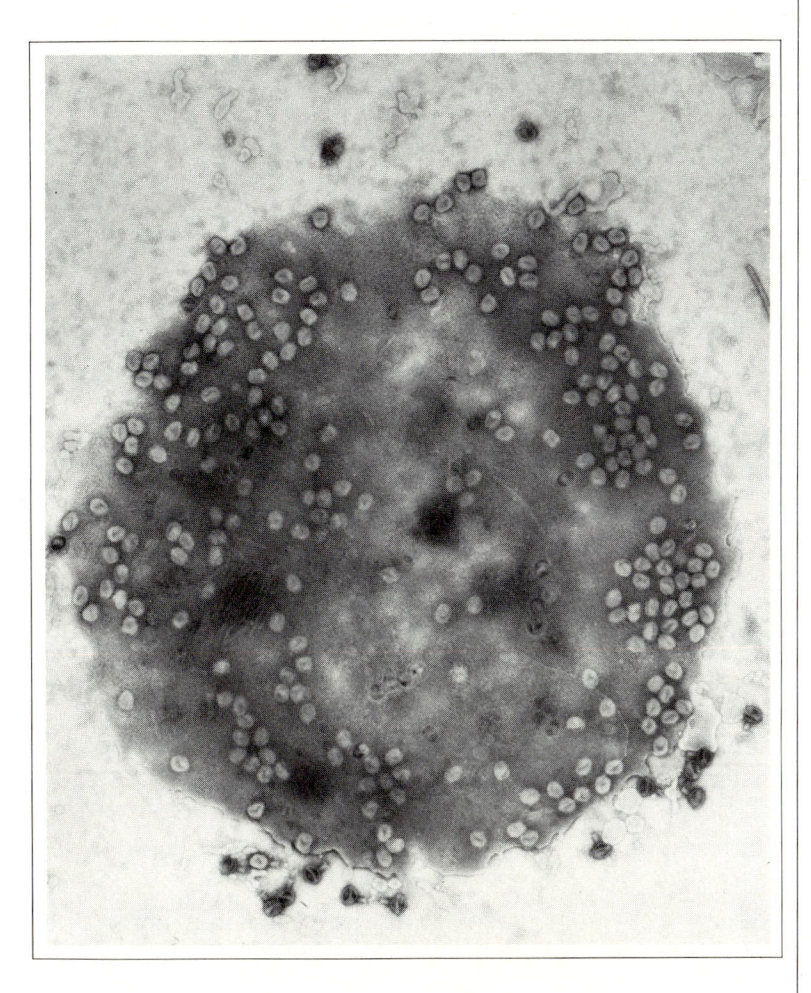

FIGURE 10.1
Lysis of a bacterial cell and release of newly formed viruses. [*From Eugene Rosenberg,* Cell and Molecular Biology. *Copyright © 1971 by Holt, Rinehart and Winston, Inc., New York. Reprinted by permission of Holt, Rinehart and Winston, Inc., p. 63. Photo by Louis Cañedo.*]

196
The Genetics
of Viruses and Cancer

FIGURE 10.2
Schematic illustration of the process of transduction, in which viruses carry genetic material from one host cell to another.

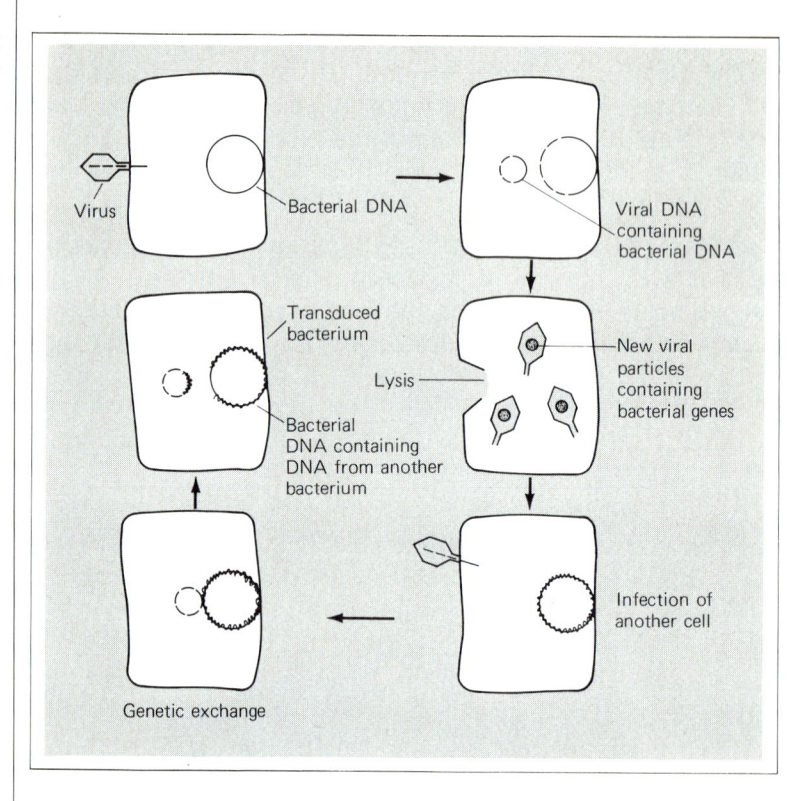

This virus-mediated form of gene transfer between cells is known as *transduction* (Figure 10.2).

TEMPERATE PHASE

A second possible consequence of bacterial virus infection is, as noted above, not lysis at once but a period during which the viral nucleic acid may remain dormant in the cell. If it is DNA, the viral nucleic acid may become incorporated into the cell's DNA so that as the cell divides, each daughter cell contains a piece of viral DNA. The virus is said to be in its *temperate phase* and is called a *prophage*. The bacterial host is referred to in this situation as being *lysogenic*. Figure 10.3 illustrates the temperate phase of the bacteriophage life cycle. Note that bacterial and viral DNA exist naturally as closed circles.

In some cases the presence of viral DNA does nothing to change the characteristics of the host cell. In fact, as shown in Figure 10.3, during cell division the viral particle can sometimes be lost from a daughter cell, and the line of cells derived from the "cured" cell will be free of viral DNA. On the other hand, the dormant nucleic acid may become active, and the cell will produce new viral particles as described earlier, resulting in lysis.

Geneticists have taken advantage of the relatively simple genetic constitutions of viruses, their short generation span, various life cycles, and the numbers of progeny available for

CONTRIBUTIONS
OF BACTERIOPHAGES
TO GENETICS

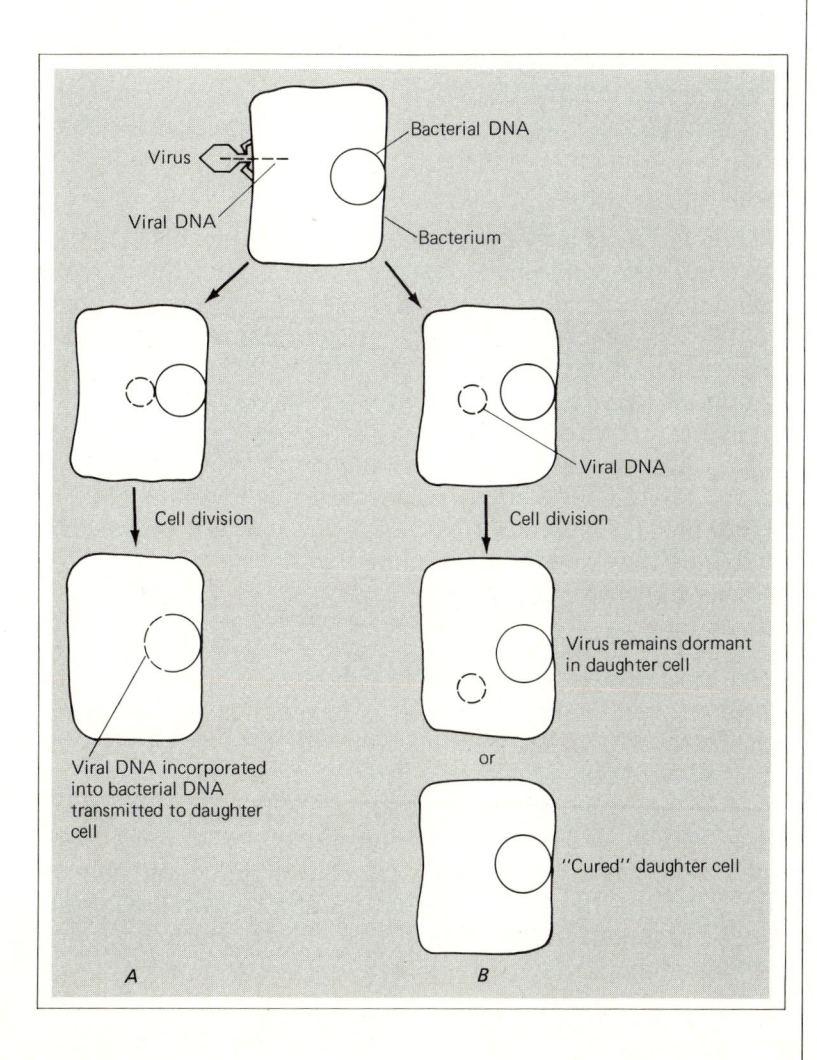

FIGURE 10.3
Possible fates of temperate phage. In one case (*A*) the viral DNA is incorporated into the host DNA. When this happens, the virus is transmitted to daughter cells following division. Another possibility (*B*) is that the virus is not incorporated into the bacterial genome and may be lost during cell division.

analysis to study many basic aspects of genetics. Mutation rate, the fine structure (molecular detail) of genes in viruses, and the mapping of genes on the viral nucleic acid all have been studied in great detail (see Chapter 12 and Figure 12.8). Furthermore, interest in bacterial viruses has led to the discovery of viruslike genetic particles which are becoming increasingly important in genetic research.

VIRUSLIKE GENETIC PARTICLES
EPISOMES

These elements are identical to bacterial viruses, with one exception: they do not have protein coats. F. Jacob and E. L. Wollman gave the name *episome* to these DNA particles, which can exist independently in the cytoplasm or become incorporated into the bacterial nucleic acid. In fact, as Stent has pointed out, "According to Wollman and Jacob's definition, . . . temperate phages, . . . thus belong to the class of bacterial episomes."*

* G. S. Stent, *Molecular Genetics*, W. H. Freeman and Company, San Francisco, 1971, p. 415.

Among the most interesting and important types of episomes are the resistance transfer factors, or RTF. These have been found to cause host bacterial cells to be resistant to a number of antibiotics, such as streptomycin and sulfonamides. In fact, it is believed that the RTFs are the most important factors causing some bacteria to be resistant to antibiotic treatment. Apparently many genes are involved, since resistance is to several antibiotics at once. The mechanism for the resistance remains unknown. Although the RTFs are not found in all bacteria, the obvious health problems they pose are attracting the attention of scientists all over the world.

PSEUDOVIRIONS

With their understanding of the nature of viruses and extra chromosomal elements such as episomes, molecular geneticists are now experimenting with introducing genes into cells of higher organisms in a variation of bacterial transduction. In fact, scientists have produced a viruslike particle composed of mouse embryo genes surrounded by a virus protein coat, which has shown itself capable of infecting and introducing the mouse genes into human cells. These particles are called *pseudovirions*. We shall discuss the implications of this experiment in Chapter 16.

ANIMAL VIRUSES
ABORTIVE INFECTION

Like phage, animal cell viruses are DNA- or RNA-containing organisms which are obligate parasites. There is evidence that some animal viruses, like temperate phage, can enter a host cell without affecting it in any way and can then be lost from it. This is referred to as an *abortive infection*.

PRODUCTIVE INFECTION

On the other hand, *productive infection* does result in the virus being reproduced. However, animal cell viruses, especially RNA viruses, do not always cause lysis of the host cell when they reproduce. They manage to leave the cell without breaking it open. For example, a study of influenza viruses has shown that when the viral particles are reproduced, the inner component of the virus, called the *nucleocapsid,* is assembled inside the host cell. It then becomes associated with a part of the host cell membrane which contains viral proteins presumably made by the cell and deposited in the membrane following infection. The cell membrane has been seen to form a protein envelope around the nucleocapsid by a form of budding (Figure 10.4).

LYSIS

Not all viruses follow the same pattern as influenza viruses. Some, especially DNA viruses, do cause lysis. The polyoma tumor virus (see Figure 6.6), for example, when introduced into cultures of embryonic mouse cells almost always causes lysis of the host cells.*

* J. Ponten, *Spontaneous and Virus Induced Transformation in Cell Culture,* Springer-Verlag of New York, New York, 1971.

TRANSFORMATION

Besides abortive and productive infections, a third possible consequence of animal virus infection, and one of great concern to medicine, is *transformation* to neoplasia (tumor formation), either malignant or benign.

TRANSFORMATION IN ANIMAL CELLS: CANCER

It must be pointed out here that transformation by animal viruses is distinct from genetic transformation. You may recall that genetic transformation is the process by which a cell is genetically changed after receiving nucleic acids from another cell (see page 115).

Transformation in animal cells as a result of viral infection refers to a number of changes in cell function which result in

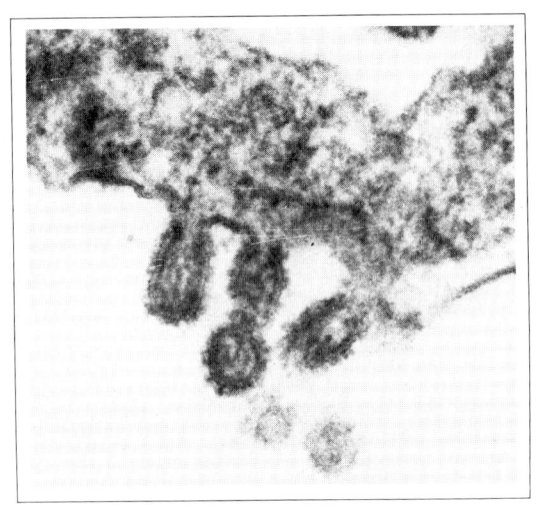

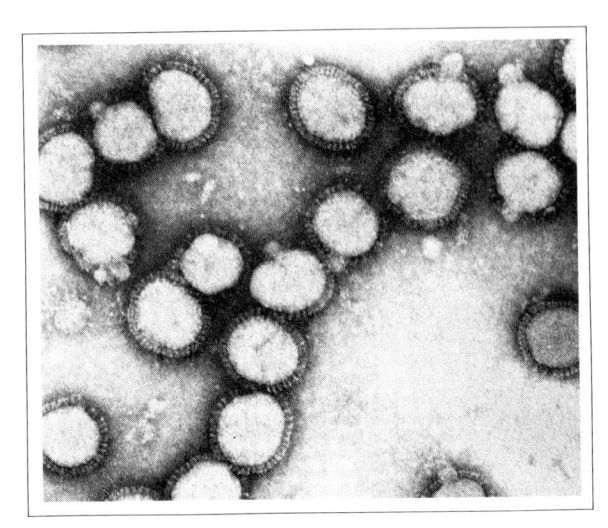

FIGURE 10.4
Formation of an influenza virion by a process involving budding of a portion of the cell membrane. (A) Influenza virions in the process of budding at the cell surface. (B) Influenza virions fully formed. [*From R. W. Compans and P. W. Choppin, in K. Maramorosch and E. Kurstak (eds.), Comparative Virology, Academic Press, Inc., New York. By permission.*]

our calling the cell cancerous. For the remainder of the chapter, our use of the term *transformation* shall refer to cancer and not to genetic transformation. There are three major criteria by which we recognize a transformed cell: (1) uncontrolled multiplication of the cell, (2) infinite multiplication, and (3) changes involving the cell membrane which lead to, among other phenomena, a loss of "contact inhibition" in tissue culture.

UNRESTRAINED AND INFINITE GROWTH

It is common knowledge that cancer cells multiply unceasingly. The reasons for this growth are unknown, though it is thought to reflect the action of some part of the viral genome. Along this line, scientists have found that in cells infected with a DNA tumor virus known as SV 40, complete SV 40 genome equivalents appear to be present in the dividing cells. No infectious viruses are produced, however, and the viral genes are transmitted to the daughter cells. In addition, studies on the polyoma virus have shown that cells may lose the viral DNA by excretion or by the breaking down of the viral nucleic acid. When this happens, cells revert from the transformed to the normal state.

Infinite growth is also characteristic of transformed cells. It has been found that cells grown in tissue culture do have a finite life span. After a series of cell generations, which vary in number from cell line to cell line, the culture will die. However, transformed cells are capable of dividing well beyond their normal limit; some scientists feel that they can survive indefinitely.

These two characteristics of transformed cells—unrestrained and infinite growth—are important to biologists not only from the point of view of curing a dread disease. They also involve a very important basic trait of all living cells: the regulation of cell division, or growth. If we can determine the mechanism for unlimited growth by studying cancer cells, we can then infer the nature of the normal mechanism. The infinite capacity for growth of transformed cells may also some day shed light on aspects of aging.

CHANGES IN THE CELL MEMBRANE OF TRANSFORMED CELLS

Another difference between normal and transformed cells can be seen in their recognition of contact with other cells in tissue culture. When normal cells come into contact, there is a cessation of movement; the cells do not glide over one another. This is called *contact inhibition.* In this manner, normal cells usually form one-cell-layer aggregations which are fairly well organized. Cancer cells, on the other hand, seem to lose the ability to recognize contact with other cells. Instead they glide over one another, and as a consequence, overlap and form irregular clumps. (See the introductory illustration to this chapter.) This loss of contact inhibition may reflect the ability of cancer cells to spread throughout an organism.

If you think about it, normal cells in complex organisms such as humans generally remain in place: liver cells stay in the liver, lung cells in the lungs. This may seem like a facetious statement because it is so obvious, but we really are not cognizant of the forces that keep cells in their normal position. Whatever they are must be altered in the change to malignancy, because cancer cells do migrate through the system. The term used by doctors to describe this movement is *metastasis.*

The change in movement is believed to reflect some as yet undiscovered change in the cell membrane. This heretofore largely unappreciated cell organelle has begun to attract much attention, as a result of studies on immune reactions and cancer. Figure 10.5 shows a schematic illustration of some proposed theories on the organization of the cell membrane. Different protein molecules, including transplantation markers, exist in various states of association with the lipoprotein that forms the matrix of the cell membrane. ("Lipo" refers to molecules of lipids or fats, lipoproteins to complex proteins containing fats.)

It is becoming apparent that the cell membrane plays more than a passive role as a semipermeable substance holding the cytoplasm and other cell structures together. It also actively participates in a number of biological processes, including immune responses (as a result of the individuality markers mentioned in Chapter 9) and the relationship between neighboring cells. Both of these functions are directly related to studies on malignancy.

VIRUSES IN HUMAN CANCERS

One of the most puzzling aspects of cancer in humans is that although cancer viruses from chickens were first isolated

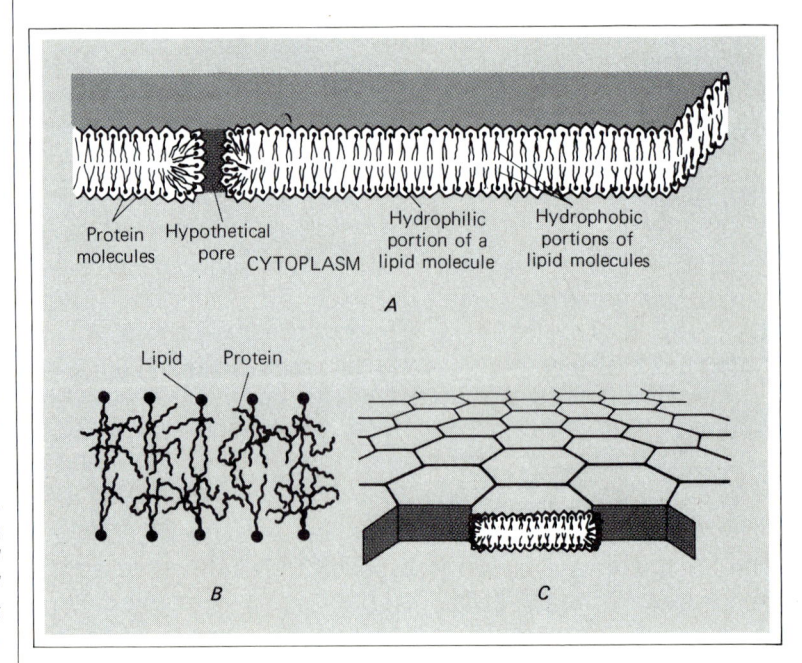

FIGURE 10.5
Schematic diagrams of the theories on the structure of the cell membrane. [After Alex B. Novikoff and Eric Holtzman, Cells and Organelles. Copyright © 1970 by Holt, Rinehart and Winston, Inc., New York. Adapted by permission of Holt, Rinehart and Winston, Inc.]

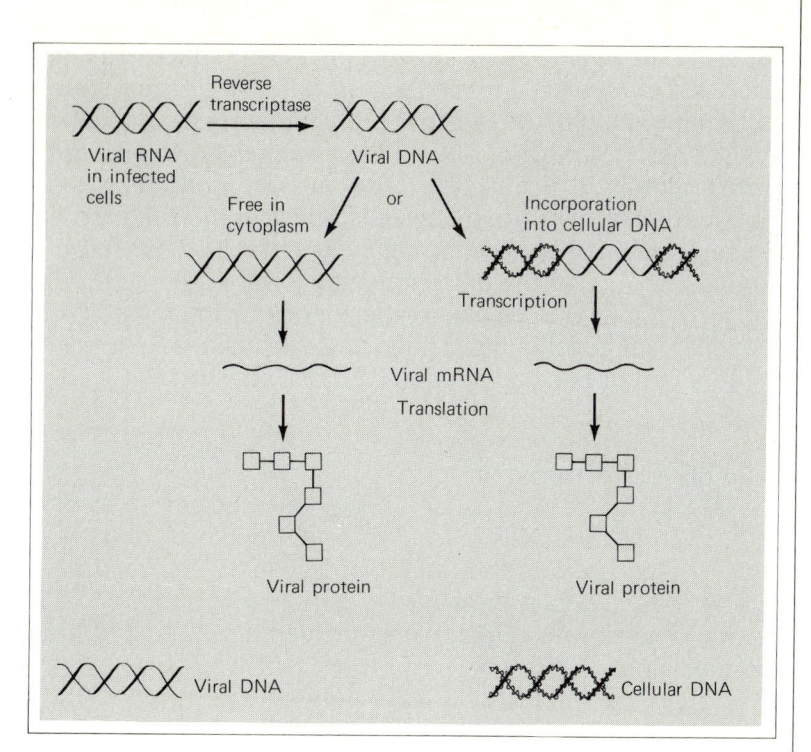

FIGURE 10.6
Schematic representation of how RNA viruses are thought to cause viral protein formation in cells, which then are transformed into cancer cells.

more than sixty years ago by Peyton Rous, and since then cancer-inducing viruses have been isolated from a number of experimental animals including primates, no whole virus has been found in human tumors. Some viruslike tumor particles have been found, but none that could be positively identified as viruses with protein coats, which are known to cause cancer when introduced into other human cells. Most of these oncogenic particles are RNA rather than DNA. The term *oncogenic* refers to a capacity to cause transformation.

The discovery of an enzyme, reverse transcriptase, in 1970 was of great interest, for it allowed scientists to deduce how the genetic machinery of the cell could be taken over by viruses or viruslike particles containing RNA. This enzyme catalyzes the transcription of viral DNA from RNA, producing active viral genes which presumably cause transformation of the host cell. Figure 10.6 illustrates the process of DNA transcription from RNA.

Since all oncogenic RNA viruses were found to cause the synthesis of reverse transcriptase in host cells, it was originally thought that the presence of this enzyme indicated malignancy. However, since then, studies have been reported showing the presence of the enzyme in normal, nontransformed human lymphocytes.* Still, it is most likely that viruses or viral genetic information is related to cancers in humans. Although the details are still to be worked out, scientists feel that viruses cause malignancy by forcing cells to produce viral genetic information and substances that result in the aberrant behavior characteristic of cancer cells.

* P. E. Penner, L. H. Cohen, and L. A. Loch, "RNA-Dependent DNA Polymerase: Presence in Normal Human Cells," *Biochemical and Biophysical Research Communications,* **42:**1228, 1971.

DIFFICULTIES IN CURING CANCER

From our discussion of immunological principles in Chapter 9 it is easy to see why cancer, once it has begun, is so difficult to cure. The cells are not foreign and therefore do not elicit a response from the white blood cells. Also, any drugs that would kill these transformed cells would probably kill normal cells as well. Herein lies the dilemma faced by modern medicine.

CANCER-CAUSING AGENTS AND INNATE RESISTANCE

Recent evidence indicates that probably every individual has been in contact with cancer viruses and is probably harboring the viral genes in some cells. Why, then, does not everyone fall victim to various forms of cancer? The answer is that a number of factors seem to be involved in an individual's susceptibility to the disease.

CARCINOGENS

One factor is that in some cases, the viruses appear to need the stimulus of external factors to become activated. A number of such factors, called *carcinogens,* are known, including radiation and chemicals such as cigarette smoke, mustard gas, and asbestos. The increase of these factors in our environment is believed to be related causally to the increase in cancer within the last few decades.

INTERFERON

For another thing, it has been established that a form of immune response is elicited when cells are infected by viruses. That is, once cells have been infected without being lysed, they become resistant to further infection by the same type of virus. This natural defense seems to be due, at least in part, to the production of a protein called *interferon*. Interferons are not the same as antibodies in that no specific interferon is produced in response to a particular virus. Many different viruses cause the cell to produce the same interferon. However, interferons are species-specific. In other words, chick interferon does not protect rabbits from viral infection, although monkey interferon seems to give some protection to human cells.

The virus nucleic acid is believed to induce the production of interferon. How this occurs is not understood. It is known that viral RNA in its double-stranded form, which occurs during multiplication, immediately causes the synthesis of interferon; how DNA viruses induce interferon production is not known. The presence of double-stranded RNA may de-repress genes which code for the interferon protein. Once formed, the interferon prevents multiplication of viral nucleic acid. In fact, it has been found that interferon produced by one cell will enter another rendering it immune to viral infection by preventing either the duplication of viruses that may infect the cell or synthesis of viral products (Figure 10.7).

When interferons were first isolated, there was great excitement. There appeared to be a real chance for scientists to

FIGURE 10.7
The mechanism of natural resistance to viral infection by production of interferon begins with the entrance into a cell of a virus containing single-stranded RNA (1,2). The RNA replicates to form double strands (3–5) which cause the cell to produce interferon (5). Interferon produced by the infected cell can then enter another cell and prevent replication of RNA from viruses following subsequent infections (6–10). [After "The Induction of Interferon," by Maurice R. Hilleman and Alfred A. Tytell, copyright © July 1971 by Scientific American, Inc. All rights reserved.]

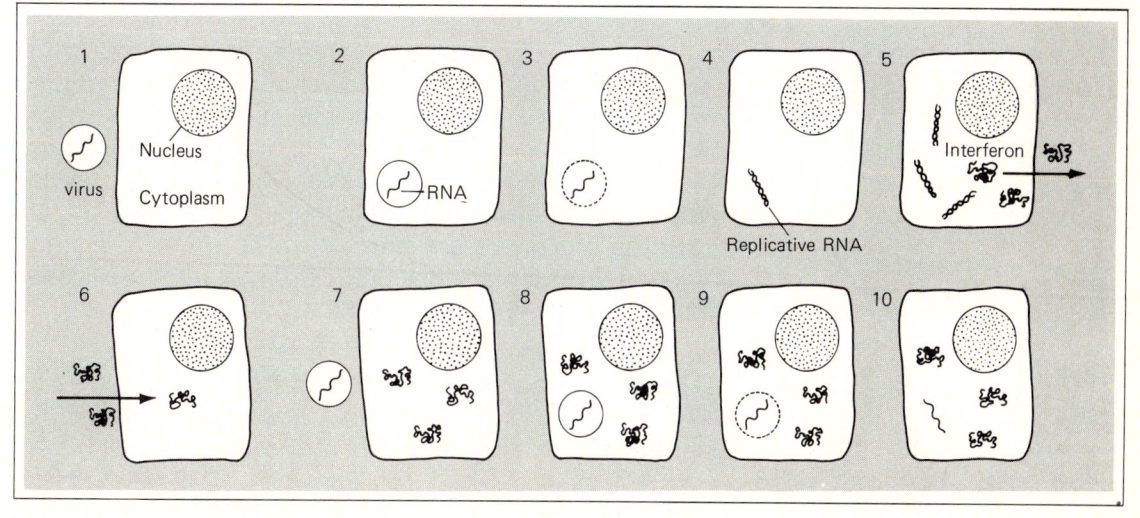

develop a preventive measure or cure for any viral disease, including cancer, by introducing interferon into the organism. However, two great obstacles have been encountered. (1) Interferons are produced in such minute quantities as to render it impossible to "harvest" usable quantities from human cells grown in tissue cultures. (2) Animal interferons are not effective on human cells.

Present efforts are concentrated on synthetic substances, which can be made in substantial quantities and may induce cells to produce interferon. A synthetic polynucleotide called *poly I:C* has been found to do just this in experimental animals. A study conducted on cancer patients* revealed that injection of poly I:C did increase interferon production in 14 of 20 patients; however, no significant improvement in their condition resulted from the greater interferon production. The scientists involved feel this may be due to the fact that the study did not go beyond one or two injections and that continuous induction of interferon might provide the protection needed.

Work is continuing along these lines, with many problems to be resolved. For example, will continuous induction of interferon cause deleterious side effects? What is the most effective dosage? It is hoped that our increased understanding of basic gene action and immune responses will one day break down the obstacles preventing us from exploiting natural defense mechanisms.

* M. R. Hilleman and A. A. Tytell, "The Induction of Interferon," *Scientific American*, **225**:14, 26–31, 1971.

GENETICALLY DETERMINED DIFFERENCES IN RESISTANCE TO CANCER

Experiments on inbred strains of mice have indicated that the susceptibility or resistance to various forms of cancer is genetically determined. For example, certain strains of mice show high incidence of spontaneous leukemia. Other strains are very resistant to spontaneous leukemia but will develop the disease if treated with carcinogens. This suggests that the disease organisms are present in the resistant animals but for some reason are inactive. What form the resistance takes is not clear, but it is inherited.

If our resistance to cancer is indeed genetically determined, it is not surprising that we differ in resistance just as we differ in hair color or height. This is not to say, however, that if one member of a family has contracted cancer, all his kin will inevitably be similarly affected. Resistance is probably polygenic, and the distinct role of carcinogens renders prediction impossible. One of the perplexing aspects of cancer studies is that we can never be sure whether the malignancy is due to the patient's genetic makeup or to exposure to a carcinogen. Often, by the time the condition is manifested, the initial contact with the carcinogen — if in fact a carcinogen is involved — is long past.

PRESENT THEORIES ON THE ORIGIN OF CANCER

A number of theories exist to explain the transformation of normal to cancer cells. We shall discuss three which are most prevalent today, although the validity of any remains to be proven.

THE PROVIRUS

One is the theory of the *provirus*, proposed by H. Temin, one of the discoverers of reverse transcriptase. This theory states that after a cell is infected by a tumor virus, the nucleic acid of the virus (the provirus) becomes incorporated into the DNA of the cell. The cell is now genetically able to produce viral RNA and proteins. Under the influence of various factors, the cell could either produce more viruses to spread to other cells or could itself become abnormal and continue to divide, giving rise to large numbers of cancer cells, each containing the viral DNA. (See Figure 10.3A.)

ONCOGENES

A second theory, proposed by R. Huebner and G. Todaro of the National Cancer Institute, postulates that cancer virus genes were incorporated into our cells millions of years ago during our early evolutionary history. Such genes, which they term *oncogenes,* are normally repressed and are activated only under the influence of carcinogens, such as radiation, chemicals, and viruses. This theory therefore assumes (1) the potential of all cells to become cancer cells and (2) a common mechanism for all cancers.

THE PROTOVIRUS

The third and latest hypothesis, also proposed by H. Temin, is the *protovirus* theory, which holds that pieces of genetic information (the protovirus), presumably the result of previous infections, exist in cells and have the potential to form cancer viruses if the *correct assortment* of genes is present.

Evidence to support each of these theories exists, but not enough to enable scientists to formulate one definite scheme of carcinogenesis.

CHROMOSOME ABNORMALITIES AND CANCER

One of the most common defects found in cancer cells is in the chromosome content. Both abnormal structure and number of chromosomes have been found in most neoplastic tissue. In fact, so many kinds of chromosome abnormalities have been found in cancer cells that it is generally impossible to relate any specific abnormality to a particular cancer.

One type of chromosome abnormality that *has* been found associated with a particular cancer, chronic myeloid leukemia, is found in chromosome 22 which is missing part of the long arm. This particular abnormality, called the *Philadelphia chromosome,* has not been found in any other disease condition (Figure 10.8). The cause of the chromosome loss is not known, but since it occurs only in the white blood cells of these leukemia patients, it is not an inherited condition.

The existence of abnormal chromosomes or abnormal numbers of chromosomes in most malignant cells may be a *result* of the wild cell divisions of cancer cells rather than the *cause* of malignancy. In support of this, it has been found that cancer viruses do cause chromosome abnormalities in the process of transforming normal cells in tissue culture to cancer cells.

With regard to chromosome abnormalities and neoplasia, J. Hamerton, a cytogeneticist, has summarized the situation:

It would thus seem unimportant whether the chromosome alteration came before, after, or at the same time as the neoplastic transformation. What is important is that the production of an altered genome in the tumor provides a population of cells from which lines capable of unlimited growth in an ever-changing tumor environment

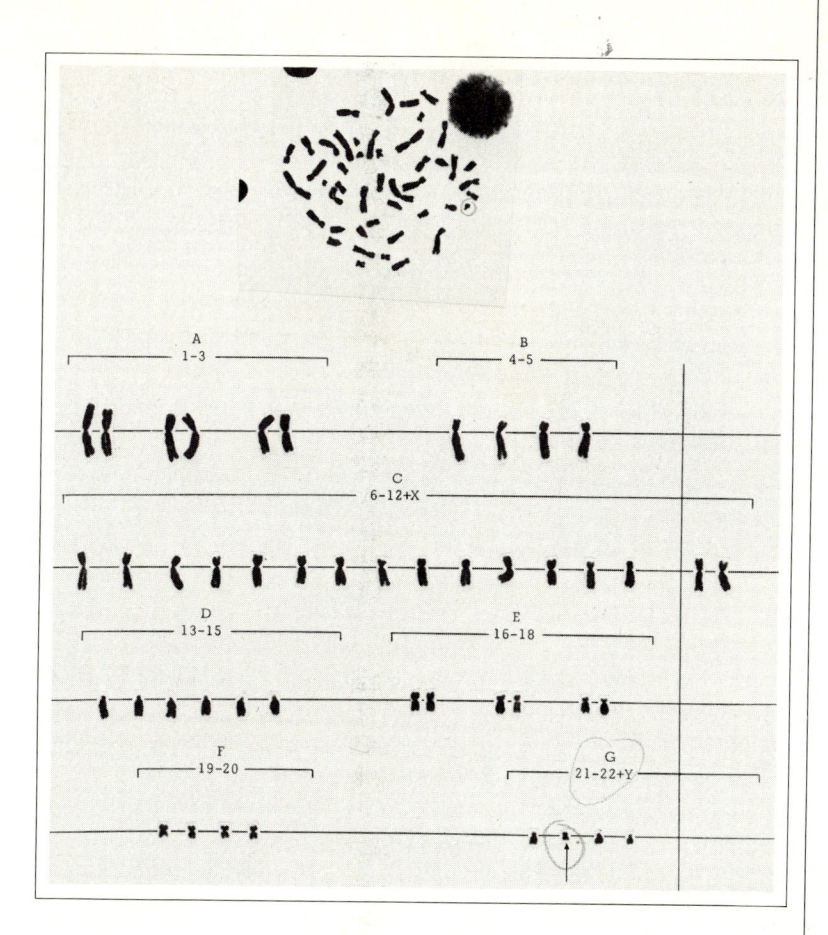

FIGURE 10.8
Photograph of the metaphase plate of a cell from a female patient with chronic myeloid leukemia. The karyotype reveals a deletion of one of the chromosomes in the G group (arrow). [*Courtesy of Emilie Mules and the Chromosome Laboratory, University of Virginia.*]

can be selected. All the evidence suggests that the chromosome and neoplastic change occurred at the same time, or at least very close to each other, and that the changed chromosome complements immediately became important in the progression of the tumor.*

* J. L. Hamerton, *Human Cytogenetics*, vol. 2, Academic Press, Inc., New York, 1971, p. 439.

It is to be hoped that the brave men and women who expose themselves to possible infection in their studies of viruses and cancers will produce a breakthrough in the near future. We must caution, however, against overoptimistic proclama-

A CURE FOR CANCER?

tions made recently in news media that science is on the verge of conquest. The problems, as we have seen, are many and great. We are certainly headed toward a solution of many of them, but it is unlikely that all forms of virus-caused diseases, including cancers, will be preventable or curable in the near future.

REFERENCES

A detailed and technical discussion of chromosomes and cancer can be found in:

Hamerton, J. L., 1971: *Human Cytogenetics,* vol. 2, Academic Press, Inc., New York.

All of the following are good review articles understandable to the layman:

Culliton, B. J., 1971: "Reverse Transcription: One Year Later," *Science,* **172:**926–928.

———, 1971: "Cancer Virus Theories: Focus of Research Debate," *Science,* **177:**44–47.

Harris, M., 1970: "Interferon: Clinical Application of Molecular Biology," *Science,* **170:**1068–1070.

Hilleman, M. R., and A. A. Tytell, 1971: "The Induction of Interferon," *Scientific American,* **225:**14, 26–31.

Schmeck, H. M., 1972: "Immunology: Doctors Use It in Fight Against Cancer," *The New York Times,* July 30, pp. 1 and 36.

Temin, H. M., 1972: "RNA-Directed DNA Synthesis,' *Scientific American,* **226:**10, 24–33.

A technical article on the presence of reverse transcriptase in normal cells:

Penner, P. E., L. H. Cohen, and L. A. Loch, 1971: "RNA-Dependent DNA Polymerase: Presence in Normal Human Cells," *Biochemical and Biophysical Research Communications,* **42:**1228.

A technical article on the resistance transfer factor:

Davies, J. E., and R. Rownd, 1972: "Transmissible Multiple Drug Resistance in Enterobacteriaceae," *Science,* **176:**758–768.

Technical discussions on animal cell viruses:

Compans, R. W., and P. W. Choppin, 1971: "The Structure and Assembly of Influenza and Parainfluenza Viruses," in K. Maramorosch and E. Kurstak (eds.), *Comparative Virology,* Academic Press, Inc., New York.

Howatson, A. F., 1971: "Oncogenic Viruses: A Survey of Their Properties," in K. Maramorosch and E. Kurstak (eds), *Comparative Virology,* Academic Press, Inc., New York.

Pontén, J., 1971: *Spontaneous and Virus Induced Transformation in Cell Culture,* Springer-Verlag, New York Inc., New York.

Good review articles on bacterial viruses can be found in *The Molecular Basis of Life,* W. H. Freeman and Company, San Francisco, 1968. This is a compilation of articles from *Scientific American,* among them:

Edgar, R. S., and R. H. Epstein: "The Genetics of a Bacterial Virus," pp. 145–152.

Horne, R. W.: "The Structure of Viruses," pp. 121–129.

Wood, W. B., and R. S. Edgar: "Building a Bacterial Virus," pp. 153–163.

CHROMOSOME STUDY

CHROMOSOME LABORATORY
University of Virginia Hospital
Charlottesville, Va. 22901

NAME:

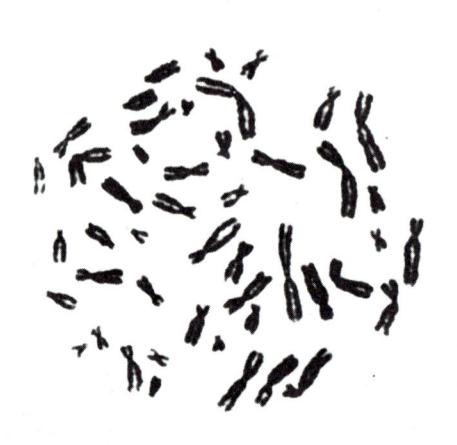

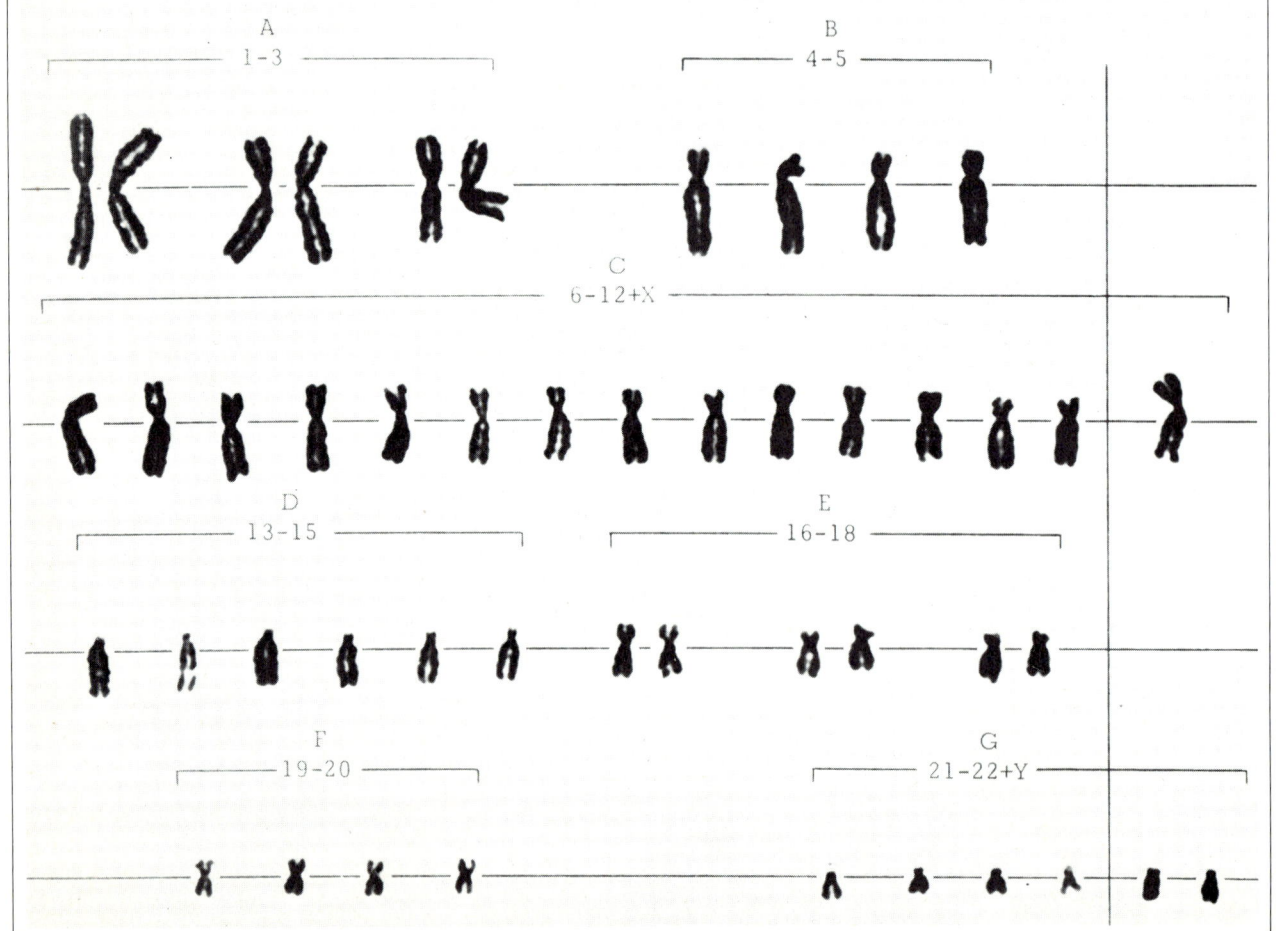

A
1-3

B
4-5

C
6-12+X

D
13-15

E
16-18

F
19-20

G
21-22+Y

Chapter 11

Chromosomal, or Gross, Mutations

With our numerous references to mutant conditions, you are now well aware that the genetic constitutions of organisms do not remain static, but undergo changes. If you recall, we mentioned in Chapter 1 that one of the criteria biologists use to distinguish living from nonliving matter is whether the subject is capable of mutation. Although the terms *mutant* and *mutation* usually imply a change for the worse, the very existence of a species depends entirely on its ability to mutate.

It is necessary, then, for students of biology and certainly of genetics to study the types of mutations that exist and their importance for living organisms. This chapter will deal with the first topic, namely, the different types of mutations which are known to exist in man and other organisms.

First we should note that geneticists tend to classify as mutations only those genetic changes that occur within the genome of an individual. Thus, genetic changes derived from the acquisition of foreign genetic material, such as occurs in bacterial genetic transformation and conjugation, are usually not called mutation. Cells which are changed in

Metaphase plate and karyotype of a cell from a man with XYY sex-chromosome complement. A controversy existed at one time, associating this genetic constitution with acts of violence and other antisocial behavior. [*Courtesy of the Chromosome Laboratory, University of Virginia.*]

this manner are usually referred to as *transformed cells,* although the changes are every bit as permanent and transmissible as mutations that occur within the genome. Nor is the transformation of animal cells to malignancy considered mutation. Presumably the changes characterizing neoplasia are due to genetic information other than that of the cells, namely, to genes of the oncogenic viruses.

GERMINAL AND SOMATIC MUTATION

Geneticists are primarily concerned with mutations which occur in the gametes, as these are then transmitted to future generations. Mutations found in eggs and sperm are referred to as *germinal mutations;* they would affect the offspring of the mutant individual but not the individual himself. On the other hand, since we define mutation simply as a change within the genes of a cell, mutations can occur in any cells of the body—not only in germ cells, but in somatic cells as well; these are known as *somatic mutations.* Somatic mutations affect only the individual and are not transmissible to future generations.

CHROMOSOMAL MUTATIONS

What, then, are some of the alterations of the genetic material that result in altered phenotypes? Mutations can involve either entire chromosomes or large portions of chromosomes and are known as *gross* or *chromosomal mutations.* Or they may change the sequence of nucleotides within individual genes and are then known as *point mutations.* In this chapter we shall discuss chromosomal mutations.

Generally, there are two categories of chromosomal mutations: those which result from aberrant numbers of chromosomes and those involving changes in the structure of individual chromosomes.

ABERRATIONS OF CHROMOSOME NUMBER

Numerical aberrations of chromosomes may involve the entire set, a condition called *polyploidy,* or individual chromosomes, called *aneuploidy.* The latter is indicated by the suffix *somy*. Thus, an individual whose karyotype shows an extra chromosome 21 is called a *trisomy* 21. An individual possessing only one rather than a pair of 21 chromosomes would be *monosomic*.

Although polyploidy is known to be a viable condition in other organisms, such as some plants, it seems to be fatal to humans and has been found only in aborted fetuses. There are, however, a number of conditions found in humans resulting from aneuploidy of both autosomes and sex chromosomes, some of which may be known to you.

DOWN'S SYNDROME (MONGOLOID IDIOCY)

One of the most familiar conditions in man resulting from aneuploidy is trisomy 21. Originally studied by Langdon Down in 1866, it had been termed mongoloid idiocy or mongolism because of certain facial characteristics that suggested resemblance to oriental features, although as L. S. Penrose, a well-known British geneticist has wryly observed, "Conversely, the Asiatics regard them as looking like Europeans." The condition is now referred to as *Down's syndrome,* which diplomatically removes the nuances of associating mental retardation with any particular group of people. (See Figure 11.1.)

The presence of an extra copy of one of the smallest chromosomes results in a wide variety of abnormalities in almost every area of the body, differing in combinations and degree of severity from patient to patient. One of the most constant characteristics which allows doctors to identify Down's syndrome in very young children, even in newborn babies, is the lines found on the palms of both hands, which are known as *dermatoglyphic patterns.* Without going into detail, the presence of a "simian crease" (Figure 11.1C) and a tendency for every finger to have a loop in the print region, rather than whorls or arches, distinguish the Down's child from a normal child. Because of various defects, such as in the cardiovascular system, the life expectancy of Down's syndrome patients is shorter than normal. One study sets the life expectancy at sixteen years.

As we mentioned in Chapter 2, the incidence of Down's syndrome increases dramatically with maternal age (see Fig-

FIGURE 11.1
(A) Down's-syndrome patient with typical facial characteristics. [*Courtesy of Dr. Irene Uchida, McMaster University Medical Center.*] (B) Typical "simian crease" on the palms of Down's-syndrome patients. [*Courtesy of Emilie Mules and the Chromosome Laboratory, University of Virginia.*] (C) Typical weakness and lack of muscle tone in Down's-syndrome patients.

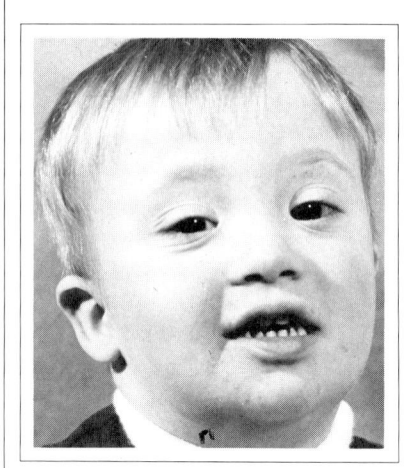

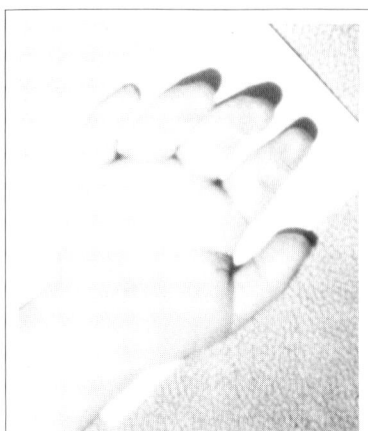

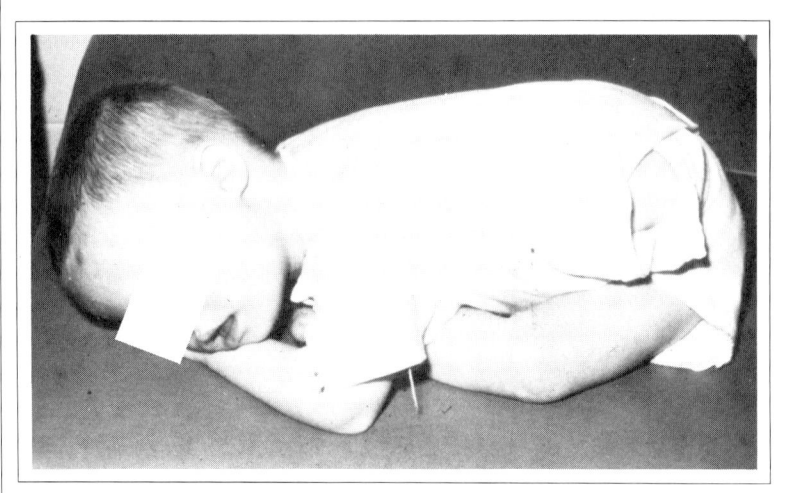

ure 11.3). This increase is thought to result from the age of the egg cell that is fertilized to produce the zygote. The human female is born with an estimated 400,000 gametes and will not produce any new ones in her lifetime. Therefore, if a woman is forty-five years old, her germ cells are also forty-five years old. Increasing age is believed to result in a tendency for a particular aberration of meiosis, called nondisjunction, to occur. There is no such correlation of the incidence of Down's syndrome with paternal age.

Nondisjunction can occur before either the first or the second meiotic division. The basic event in nondisjunction is just as the term implies: the chromosomes do not separate, do not "disjoin." This results in aneuploidy in the resulting daughter cells.

Gametes containing the extra chromosome will produce a Down's-syndrome patient. The nullisomic (missing a particular chromosome altogether) gametes are presumably nonfunctional or lethal, as no monosomy of autosomes has been found in living individuals. If nondisjunction occurs in germ cells that are normal, it is called *primary nondisjunction*. If,

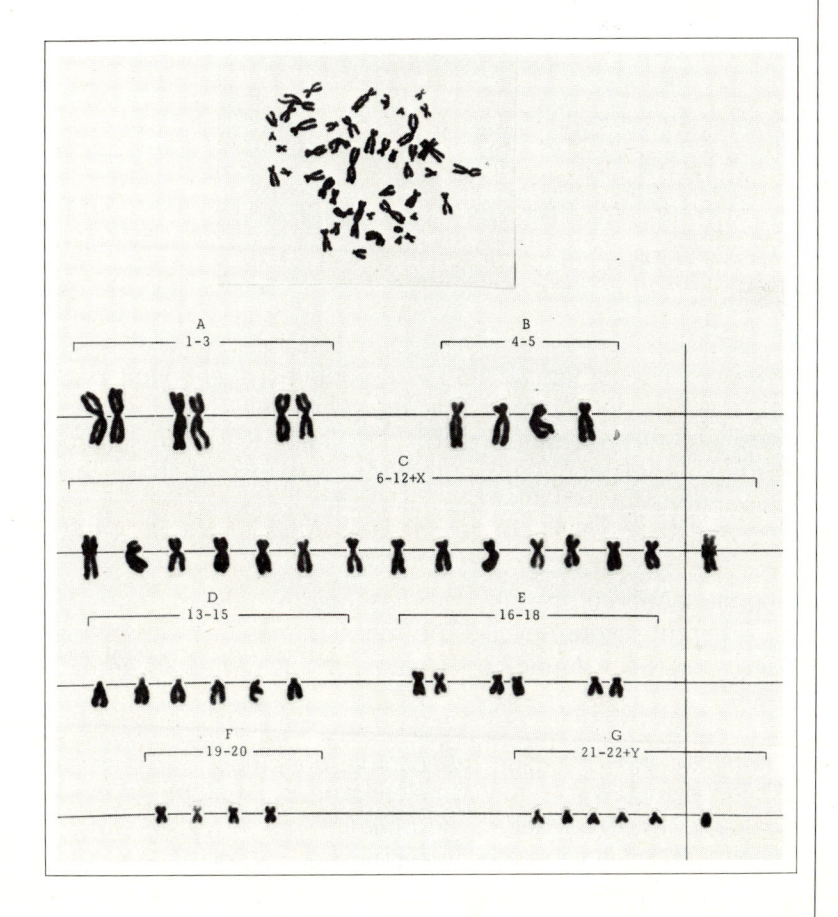

NONDISJUNCTION

FIGURE 11.2
Metaphase plate and karyotype of a trisomy 21 individual with Down's syndrome. [*Courtesy of Emilie Mules and the Chromosome Laboratory, University of Virginia.*]

FIGURE 11.3
The incidence of Down's syndrome at birth. [*After J. L. Hamerton, Human Cytogenetics, Academic Press, Inc., New York, 1971, p. 202.*]

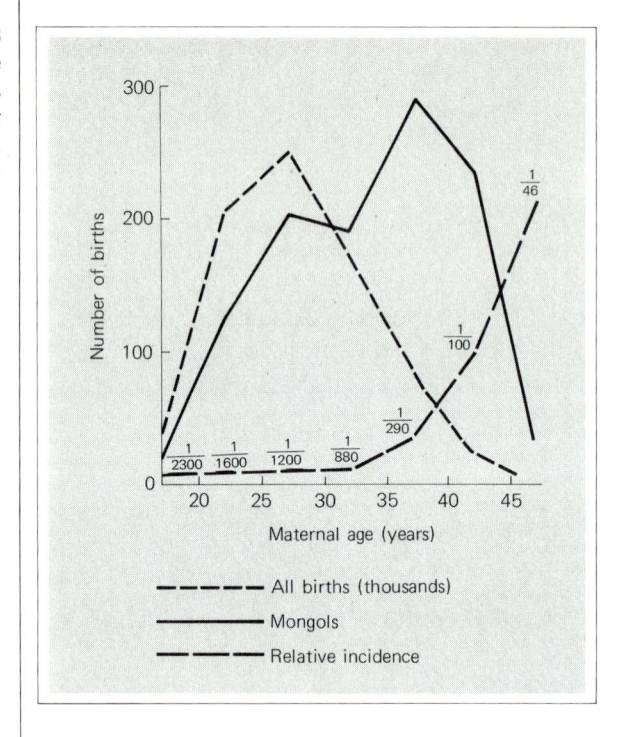

however, a trisomic individual reproduces, the nondisjunction resulting in aneuploidy in the germ cells is called *secondary nondisjunction*. Figure 11.4 illustrates the results of primary nondisjunction, Figure 11.5 those of secondary nondisjunction.

SEX-CHROMOSOME ABERRATIONS

Nondisjunction of the sex chromosomes results in a number of different syndromes. The only monosomic condition which has been found to be viable involves the X chromosome.

TURNER'S SYNDROME The monosomic condition, depicted as XO, is called *Turner's syndrome*. Figure 11.6 is a photograph of a patient with Turner's syndrome and her karyotype, which shows one fewer chromosome than normal in the C group.

FIGURE 11.4
Diagrams illustrate the results of (A) nondisjunction at the first meiotic division and (B) the second meiotic division, which give rise to cells containing an extra copy of a chromosome (DS) or missing a chromosome (L). Since monosomy of an autosome has not been found in humans, the gametes marked L are presumed to be nonviable.

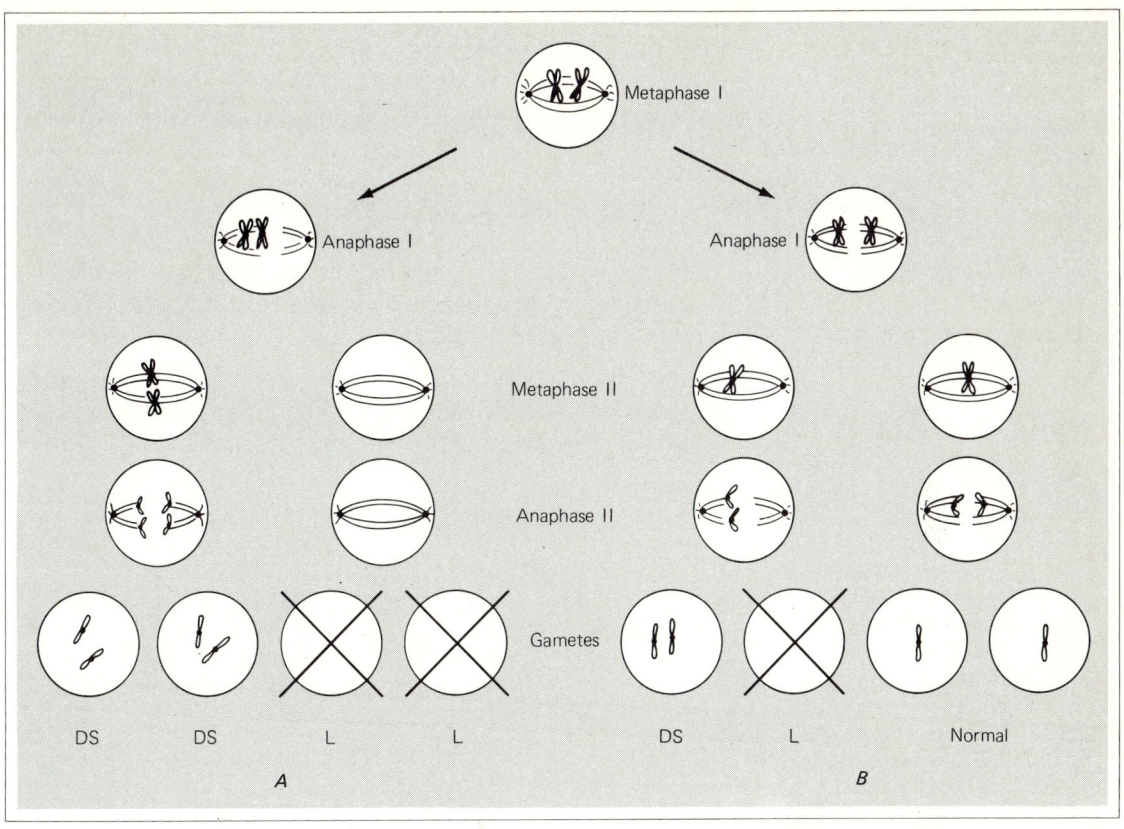

Such individuals are usually phenotypically female, although the gonads are abnormal and the patient is sterile. Thus, another term for Turner's syndrome is *gonadal dysgenesis*. The patients tend to be short, and secondary sex characteristics in the adults, such as breast development, are slight. Other abnormalities frequently associated with Turner's syndrome include webbing of the neck and cardiovascular abnormalities. Some studies have indicated slight mental retardation; others have not shown any association of the syndrome with significant mental impairment.

The following diagrams illustrate the manner in which "O" gametes (missing a sex chromosome) that would produce

Turner's syndrome individuals could result from nondisjunction of the sex chromosome:

Nondisjunction in the male

XY ↙ XY ↘ O gametes

Nondisjunction in the female

XX ↙ XX ↘ O gametes

In both cases, if the O, or nullisomic, gamete were involved in fertilization with an X-bearing gamete from the other parent, the resulting zygote would be XO.

Although monosomy of the X chromosome is the most common aberration associated with gonadal dysgenesis, a number of Turner's syndrome patients have been studied

FIGURE 11.5
Secondary nondisjunction is the term used for the process resulting in aneuploidy when germ cells of a trisomic individual, such as a Down's syndrome patient, undergo meiosis. Half the gametes would be expected to result in trisomic individuals following fertilization by normal gametes.

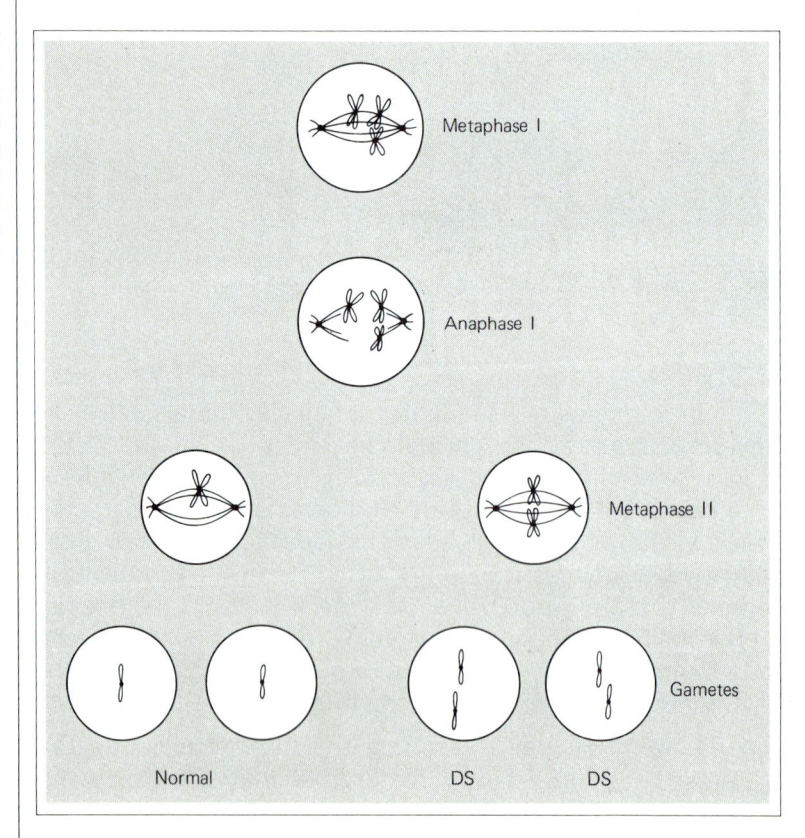

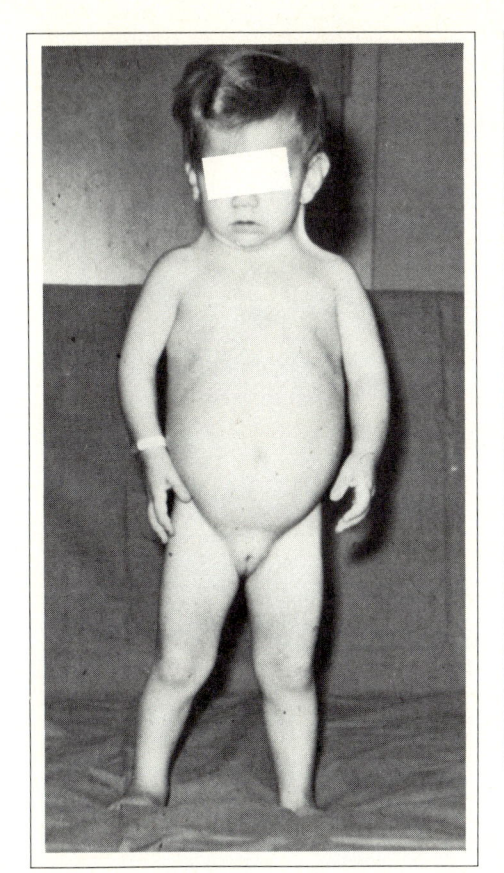

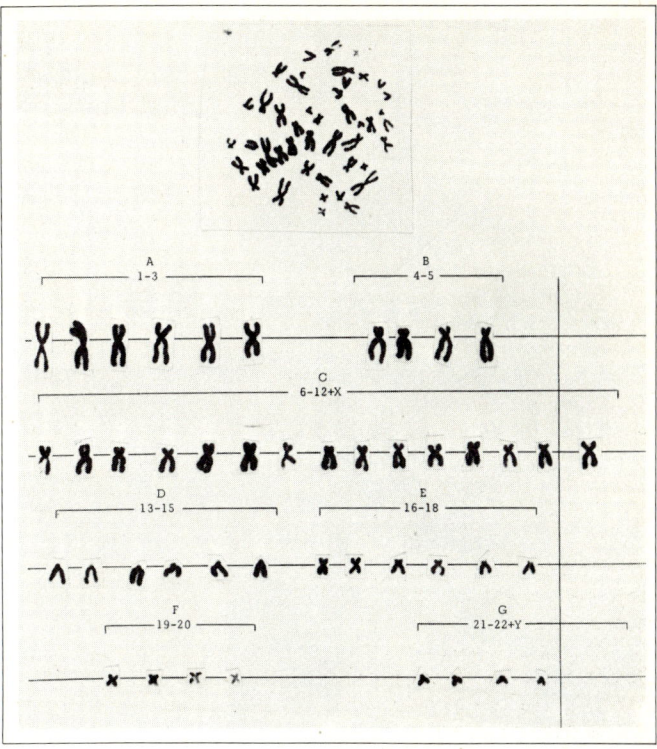

FIGURE 11.6
(A) Turner's syndrome: female patients are usually shorter than average, with webbed neck, abnormal ovarian development, and sterility. (B) Karyotype of the patient in (A). [Courtesy of Emilie Mules and the Chromosome Laboratory, University of Virginia.]

* M. A. Ferguson-Smith, "Karyotype-Phenotype Correlations in Gonadal Dysgenesis and Their Bearing on the Pathogenesis of Malformations," *Journal of Medical Genetics*, **2**:142–155, 1965.

whose karyotype reveals the correct number of chromosomes but who have abnormal sex chromosomes. One report gives evidence that a deletion of the short arm of the Y chromosome results in gonadal dysgenesis. This finding may indicate some homology of the short arm of the Y chromosome with parts of the X chromosome.*

Since nondisjunction produces gametes with extra sex chromosomes, one would also expect trisomy of the sex chromosomes if the gametes containing an extra sex chromosome took part in fertilization. Conditions resulting from trisomy of the sex chromosomes in fact occur even more frequently than Turner's syndrome.

FIGURE 11.7
(A) Klinefelter's syndrome: Male patients are usually taller than average, with a tendency toward breast development, abnormal testicular development, and sterility. (B) Karyotype of the patient in (A) [*Courtesy of Emilie Mules and the Chromosome Laboratory, University of Virginia.*]

KLINEFELTER'S SYNDROME If a sperm from nondisjunction were to fertilize a normal X-bearing ovum, and if an XX ovum rising from nondisjunction in oogenesis were to be fertilized by a normal Y-bearing sperm, the zygotes in each case would be as follows:

$$XX \times Y \rightarrow XXY \qquad XY \times X \rightarrow XXY$$

This imbalance in the number of sex chromosomes leads to a series of abnormalities known collectively as *Klinefelter's syndrome* (Figure 11.7).

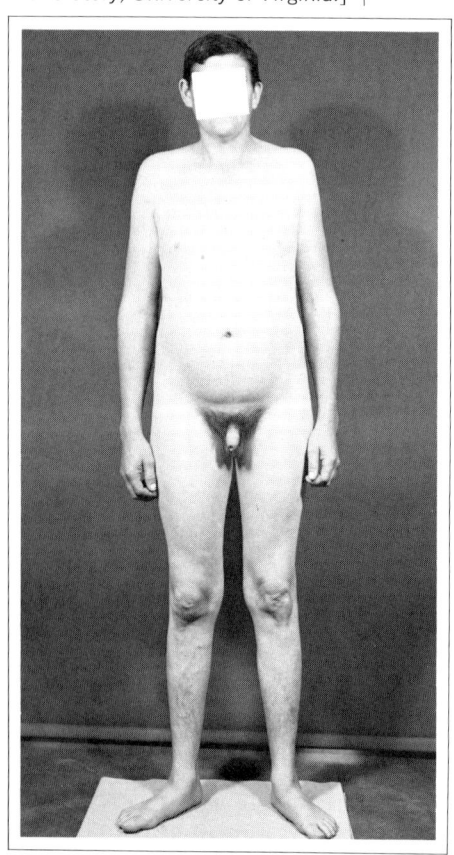

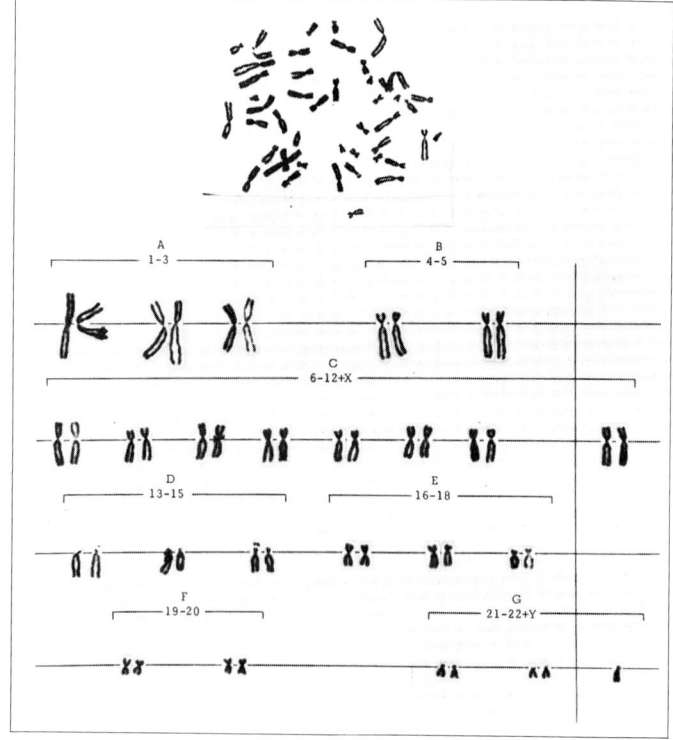

As you see from the karyotype, such individuals have forty-seven chromosomes, including a Y and two X chromosomes. They are generally tall, unusually long-limbed, sterile due to underdeveloped gonads, and tend toward breast development. Mental retardation is often associated with Klinefelter's syndrome, although the degree of impairment varies.

XXX "SUPERFEMALES" Fertilization of XX germ cells by X-bearing gametes results in XXX persons, or "superfemales." There is no characteristic phenotype associated with this aneuploid condition, as the level of intelligence and development of the reproductive system vary from normal to subnormal.

XYY SYNDROME Another type of abnormal sperm is created by the nondisjunction of Y chromatids prior to the second meiotic division. The result is a spermatid without sex chromosomes and a spermatid with two Y chromosomes (Figure 11.8). Fertilization involving either of the two abnormal sperms results in an imbalance of sex chromosomes in the zygote. The O sperm leads to Turner's syndrome when it fertilizes a normal X-bearing egg. The YY sperm results in an interesting condition, XYY, which has been referred to in the past as the *criminal syndrome,* since at one time it was thought that persons possessing the XYY complement tended to commit acts of violence.

The association of the XYY chromosome complement with aggressive behavior stems from some studies in the mid-1960s of patients in mental institutions and prisons. These studies indicated that a relatively large percentage of inmates (3 percent) were XYY. This correlation was strengthened when a number of notorious crimes were found to have been committed by XYY individuals. (It was also erroneously reported that Richard Speck, the man who murdered eight student nurses in Chicago, was an XYY. This has since been disproved.) A recent review by Kessler and Moos[*] of the literature on the XYY syndrome points out that actually only one characteristic was most consistently found in XYY males: they were usually taller than average. No other physical or mental trait could be found to characterize this aneuploidy. For example, XYY males were found to possess intelligence

[*] S. Kessler and R. H. Moos, "The XYY Karyotype and Criminality: A review," *Journal of Psychiatric Research,* **7**:153–170, 1971.

FIGURE 11.8
Nondisjunction of sex chromosomes results in the formation of sperm that would give rise to XYY individuals and to Turner's-syndrome individuals. The longer chromosome is the X chromosome; the shorter one is the Y chromosome.

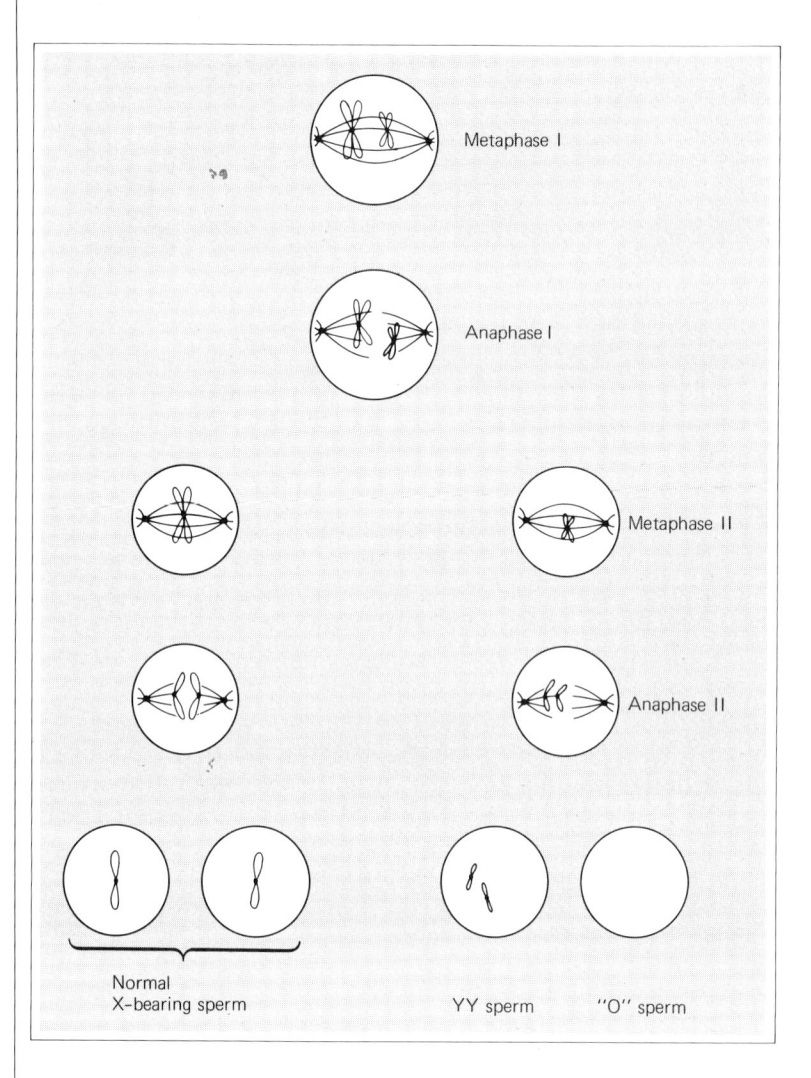

*Ibid.

levels ranging from "grossly retarded to substantially above average. An IQ score of 125 has been found in one XYY male."* Most cases seem to be free of abnormalities of the reproductive system which are found in the other sex chromosome aberrations described above.

The relationship between antisocial behavior and the XYY condition was used as a defense in the 1960s in some cases of criminal violence. Because of his genetic makeup, the defendant was said to be uncontrollably aggressive and therefore not accountable for his actions. Thus Daniel Hugon, a Frenchman convicted of murdering a prostitute, was given a lighter than usual sentence, and in Australia, a man was actually acquitted of murder on this ground. However, the fact that a man is genetically XYY is no longer acceptable as a legal reason for acquittal, for the following reasons.

First, although some studies have indicated a higher incidence of XYY among institutionalized criminals as compared to the noncriminal population,* many nonviolent XYYs are known to exist. Since XYY individuals do not usually have any physical impairments comparable to those associated with Klinefelter's or Turner's syndrome, many of them are probably unaware of their chromosomal aberrations and are leading normal lives.

Second, when one deals with behavior, it is difficult, if not impossible, to determine whether behavior patterns are due to genetic constitution or to environment. We shall speak more of this in a later chapter, but it is clear that some of the studies linking XYY to criminality do not consider the individual's background. For example, the fact that some are unusually tall may cause them to feel isolated from their contemporaries and thus lead to antisocial attitudes. To verify this hypothesis, one would then have to conduct studies of the effect of height on normal (XY) individuals. Such complexities make behavior analysis extremely difficult.

Finally, we must be aware of the implications involved in claiming an abnormal genetic constitution as the basis for aberrant behavior and violent acts. Such individuals would have to be treated as medical rather than social problems. They should then be exempt from possible capital punishment and be institutionalized for life, since there is no way (at present, at least) to change their chromosome complement.

* E. B. Hook, "Behavioral Implications of the Human XYY Genotype," *Science,* **179:**139–149, 1973.

OTHER SEX-CHROMOSOME ABNORMALITIES

Aneuploidy resulting from nondisjunction has been found in many different forms with regard to the sex chromosomes. (Incidentally, although there are indications that nondisjunction of sex chromosomes is related in some of these cases to

TABLE 11.1
Sex-chromosome aneuploidy in humans

	SEX PHENOTYPE	FERTILITY	NUMBER OF BARR BODIES	SEX-CHROMOSOME CONSTITUTIONS
Normal male	Male	+	0	XY
Normal female	Female	+	1	XX
Turner's syndrome	Female	−	0	XO
Klinefelter's syndrome	Male	−	1	XXY
XYY syndrome	Male	+	0	XYY
Triple X syndrome	Female	±	2	XXX
Triple X-Y syndrome	Male	−	2	XXXY
Tetra X syndrome	Female	?	3	XXXX
Tetra X-Y syndrome	Male	−	3	XXXXY
Penta X syndrome	Female	?	4	XXXXX

Source: V. McKusick, *Human Genetics,* 2d ed., Prentice-Hall, Inc., Englewood Cliffs, N.J., 1969, p. 19.

the mother's age, the correlation is hardly as clear-cut as with Down's syndrome.) Table 11.1 shows some of the gross aberrations and their effects on the phenotype of the individuals.

TRANSSEXUALS AND HERMAPHRODITES

Persons with sex chromosome abnormalities should not be confused with *transsexuals,* those who undergo surgery for a change in sexual phenotype. The latter are usually genetically and phenotypically normal but have deep psychological problems accepting their given sex.

Still other individuals are true intersexes in the sense that they possess both ovaries and testes. These are called *hermaphrodites.* The abnormality arises from the presence of primordia for both sets of reproductive organs in the first few weeks of embryonic life. Normally, one or the other set regresses, depending on the genotype of the individual's sex chromosomes. For reasons not understood, both sets partially develop in hermaphrodites, many of whom have genetically normal sex-chromosome complements.

Some hermaphrodites show the presence of cells containing both XX and XY chromosomes, a condition referred to as *genetic mosaicism.* This arrangement may be caused by a double mitotic nondisjunction in some cells of a male. Another possibility is dispermic fertilization of a secondary oocyte, in which an X-bearing sperm and a Y-bearing sperm would unite with the nucleus of the egg and a polar body before the latter is forced out. These two possibilities are illustrated in Figure 11.9.

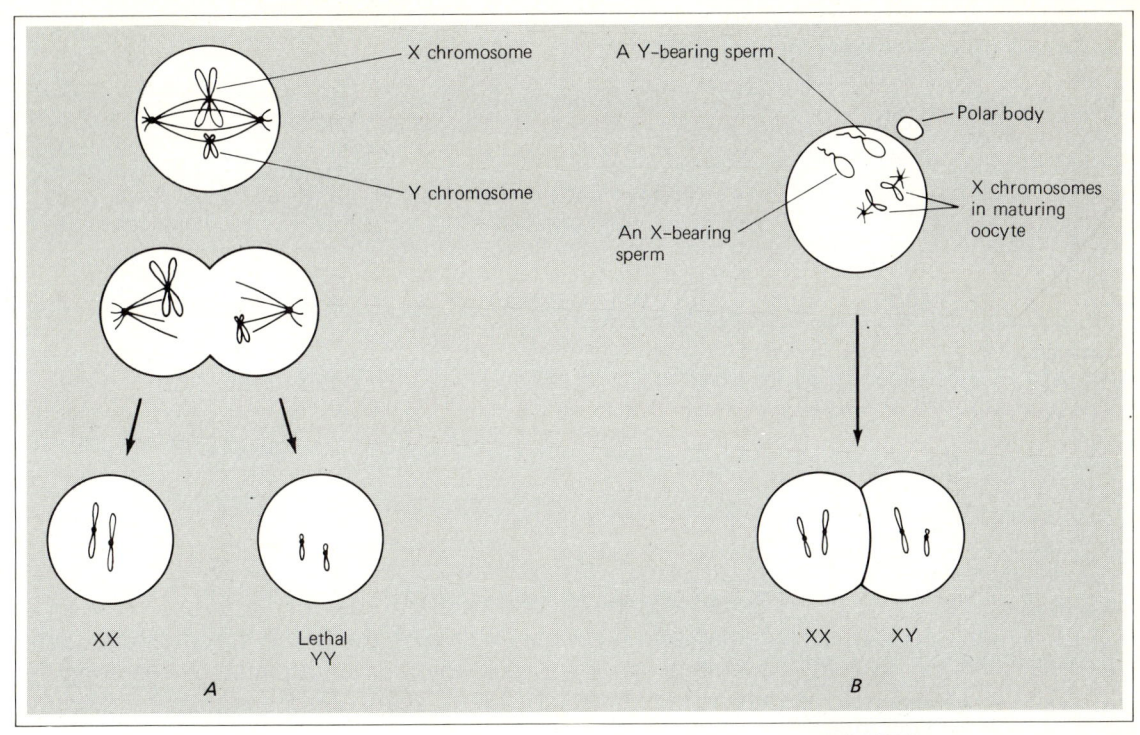

Many chromosomally abnormal zygotes never develop to term. Studies of aborted human fetuses have shown that many of them exhibit either aneuploidy or polyploidy. In fact, no living human has ever been known to be polyploid, though this condition has been found in aborted fetuses (Table 11.2).

Although the extra set of chromosomes for Homo sapiens is not apparently conducive to life, plant geneticists have pro-

FIGURE 11.9
(A) Mitotic nondisjunction of sex chromosomes in a cell of a male, resulting in a viable XX cell and a YY cell which is lethal. (B) Dispermic fertilization of a secondary oocyte by an X-bearing sperm and a Y-bearing sperm, resulting in two diploid cells, one XX and one XY. Both (A) and (B) would lead to mosaicism.

POLYPLOIDY

Chromosomal, or Gross, Mutations

	ABORTIONS KARYOTYPED	ABNORMAL KARYOTYPES	45,X	AUTOSOMAL TRISOMICS							TRIPLOIDS	TETRAPLOIDS	MIXOPLOID TRISOMICS	OTHERS
				A*	B	C	D	E	F	G				
No.	1291	322	64	7	3	20	30	48	4	42	53	16	16	19
						All trisomics = 154						Polyploids = 69		
Proportion total		0.25	0.05	A,B,F 0.001			0.015	0.023	0.037	0.___	0.032	0.041	0.012	0.027
Proportion abnormal		1.00	0.199	0.043			0.062	0.093	0.149		0.130	0.165	0.050	0.109
						All trisomics 0.478						Polyploids = 0.214		

TABLE 11.2
Frequency of different types of chromosome abnormalities found in spontaneous abortions

* Letters A to G refer to chromosome groups.
Source: J. L. Hamerton, *Human Cytogenetics*, vol. 2, Academic Press, Inc., New York, 1971, p. 382.

duced a viable hybrid from an interspecies cross between a radish and a cabbage in which the entire chromosome complement of the diploid hybrid is duplicated, resulting in tetraploidy and a new species, *Raphanobrassica* (Figure 11.10). Whereas the diploid hybrid is sterile, the tetraploid

FIGURE 11.10
Seed pods and somatic chromosome complements of (A) radish (*Raphanus*), (B) the diploid hybrid, (C) the tetraploid hybrid *Raphano brassica*, (D) cabbage (*Brassica*). [After E. M. Sinnott, L. C. Dunn, and T. Dobzhansky, *Principles of Genetics*, 5th ed. Copyright © 1958 by McGraw-Hill, Inc., used with permission of McGraw-Hill Book Company, p. 299.]

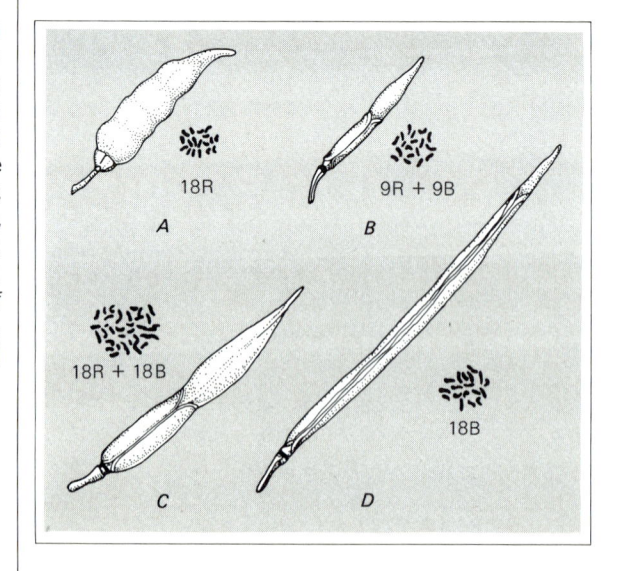

FIGURE 11.11
Diagrammatic illustration of the process of translocation involving chromosomes 15 and 21, in which a portion of the 21 chromosome becomes attached to a 15 chromosome.

plants are almost completely fertile. There do exist animal and plant species with extra sets of chromosomes that geneticists feel may have occurred naturally through nondisjunction in the same manner as was caused experimentally in the case of *Raphanobrassica*.

STRUCTURAL ABNORMALITIES OF CHROMOSOMES

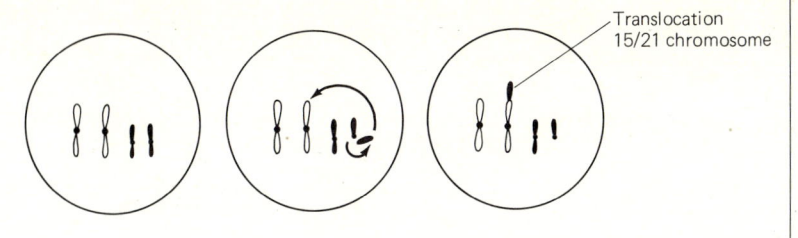

TRANSLOCATION

A structural abnormality of chromosomes which results in an imbalance of chromosomal material very like the trisomy discussed above is known as *translocation*. In this condition, a piece of one chromosome breaks off and attaches itself to another chromosome. During meiosis, therefore, one of the gametes may receive an extra piece of a particular chromosome. Should this gamete be involved in fertilization, the zygote will have an extra piece of the chromosome, as in trisomy.

One would expect the effect on the individual to be the same, and there is definite evidence that this is true. One of the well-known studies in humans involves translocation of the G chromosome, designated chromosome 21 (which results in Down's syndrome when the individual is trisomic), to chromosome 15. Figure 11.11 illustrates the effect of translocation.

It is possible that one of the gametes resulting from meiosis of the cell in Figure 11.11 could be viable, even though it contains the translocation chromosome. This is the gamete labeled *C* in Figure 11.12. Note that although the chromosome structure may be abnormal, the *total amount of genetic information* in the gamete is normal. A person who develops from such a gamete is known as a *translocation carrier*. The gametes that a carrier could produce are shown in Figure 11.12.

230
Chromosomal, or Gross, Mutations

FIGURE 11.12
Meiosis of an individual with a 15/21 translocation results in (A) a lethal gamete, (B) a gamete with extra 21 chromosome material which will produce a Down's-syndrome child upon fertilization by a normal gamete, (C) a 15/21 translocation carrier with an abnormal chromosome structure but normal genetic complement, and (D) a normal gamete. The last two gametes will produce phenotypically normal individuals if fertilized by normal gametes.

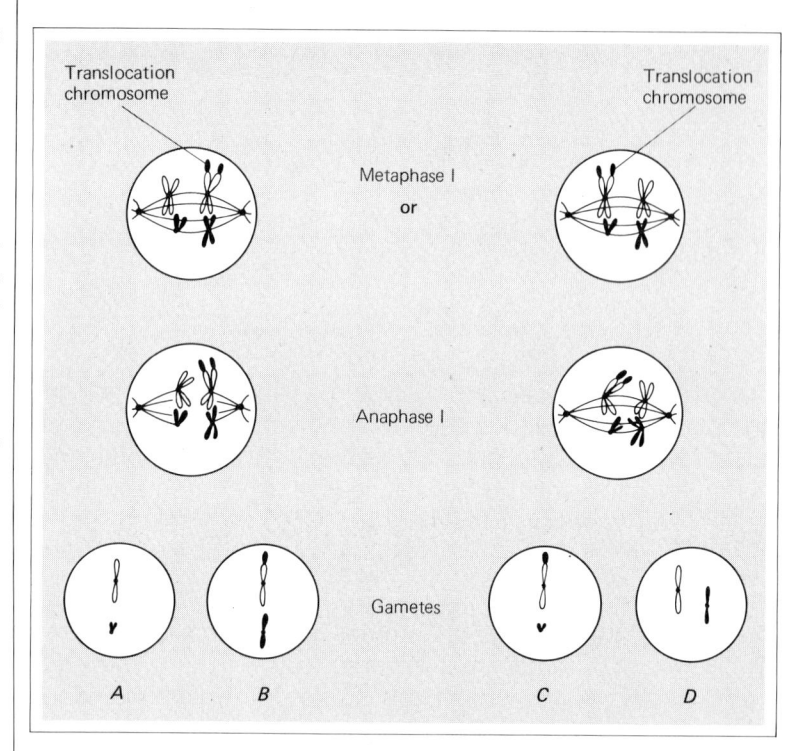

DUPLICATION AND DELETION

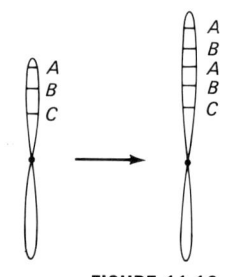

FIGURE 11.13
Schematic illustration of duplication, resulting in extra genetic material.

One of the obstacles to detailed examination of structural abnormalities in humans which does not arise in studying simpler forms of life, such as *Drosophila,* is that until recently it has been very difficult to distinguish visually between the chromosomes within a particular group. For example, we cannot tell by looking at the karyotypes in Figures 2.1 and 2.2 which of the C-group chromosomes is actually the sex chromosome, X. Consequently, it has been very difficult to detect smaller changes in chromosome structure.

In *Drosophila,* one structural abnormality in chromosomes which has been clearly identified is known as *duplication.* This is the condition in which, for some unknown reason, a piece of genetic material is doubled and added to the chromosome (Figure 11.13). The banding pattern on *Drosophila* chromosomes produces clear structural evidence to support genetic data that indicate a duplication of the Bar locus,

FIGURE 11.14
(A) Altered phenotypes of eye shape in *Drosophila* due to duplication at the Bar locus. [*After T. H. Morgan*, The Theory of the Gene, *Yale University Press, New Haven, 1926, p. 87.*] (B) Duplications of bands in *Drosophila* chromosome at the Bar locus. [*After C. B. Bridges*, "The Bar Gene a Duplication," *Science*, **83:**210, 1936.]

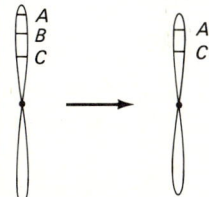

FIGURE 11.15
Schematic illustration of deletion or loss of genetic material from chromosomes.

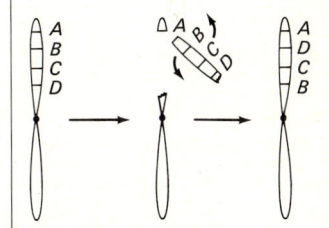

FIGURE 11.16
Schematic illustration of inversion, causing changes in gene sequence.

INVERSION

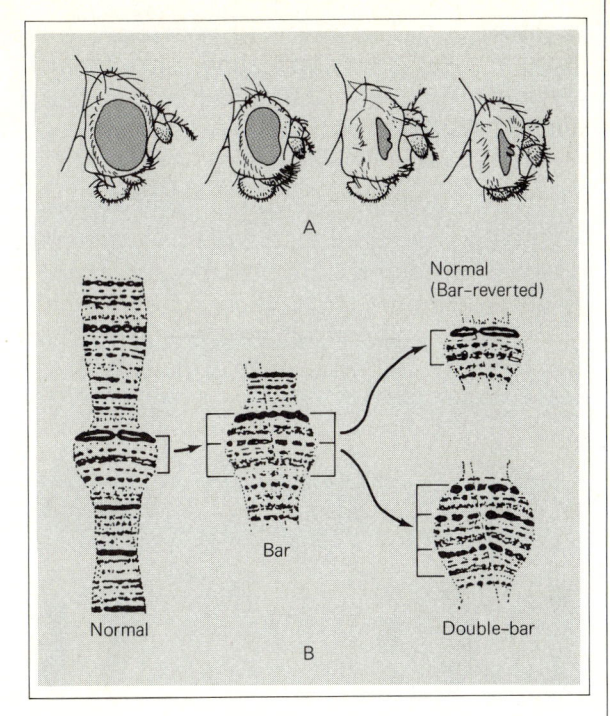

which results in abnormal structure of the eyes (Figure 11.14). Duplication has not been detected in human chromosomes.

The opposite of duplication is *deletion,* or the loss of chromosomal material, which has also been found in *Drosophila* (Figure 11.15). At least one lethal syndrome in man, cri du chat, is believed to be due to a deletion in chromosome 5. We have already mentioned the Philadelphia chromosome and its relation to cancer (see page 208).

Another type of structural abnormality which is even more difficult to identify visually is *inversion,* in which the order of genes is reversed. This occurs when a chromosome breaks in two places and the middle piece becomes reversed as it is reincorporated into the chromosome (Figure 11.16). As you might suspect, the effects of inversion are less severe than those of deletion or duplication, since there is no imbalance in genetic material. One effect of inversion is the suppression of crossing-over.

NEW TECHNIQUES FOR STUDYING CHROMOSOMES

A recently developed technique has modified the traditional methods of staining chromosomes for composing karyotypes. This may allow us to study chromosomal defects in humans in the same manner as cytogeneticists have been able to study them in *Drosophila*. Scientists have developed

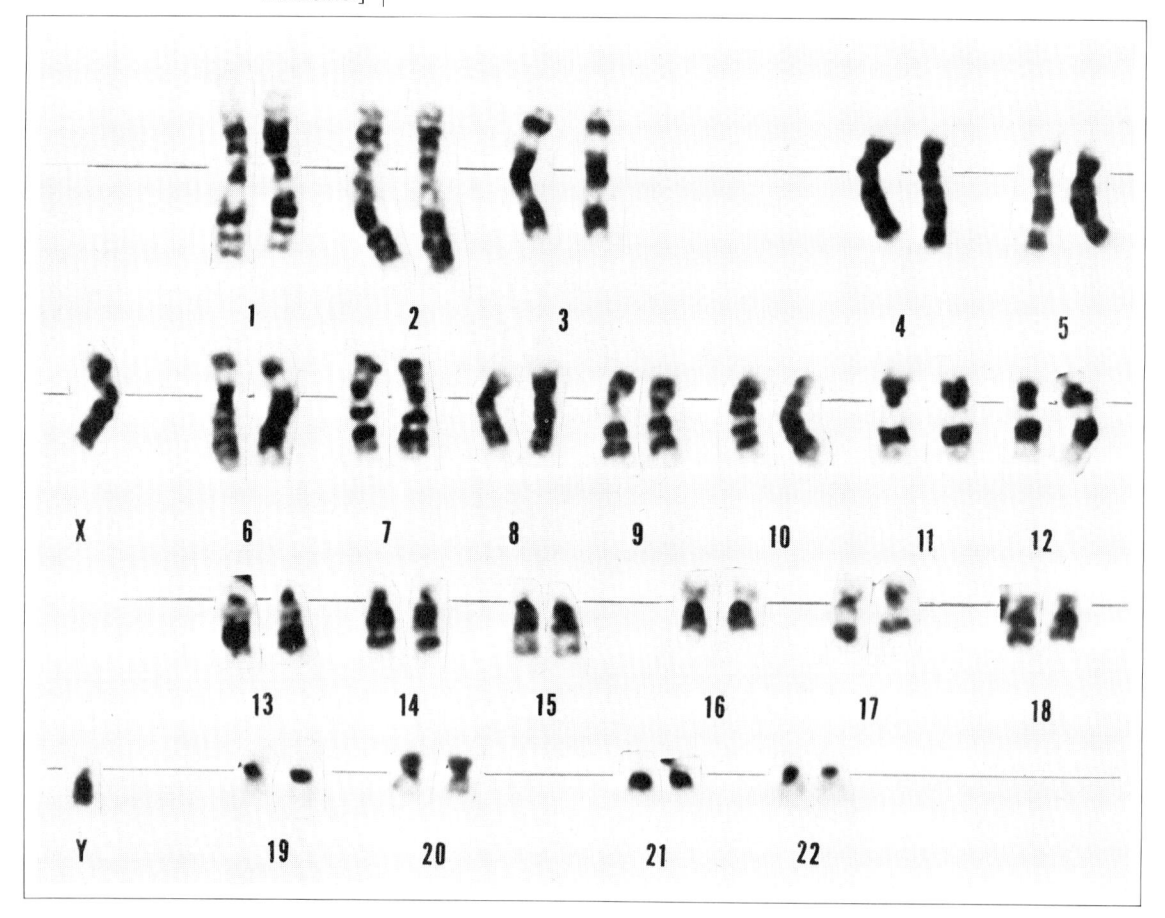

FIGURE 11.17
Banding patterns of human chromosomes revealed by giemsa staining. [*Courtesy of Dr. Harold Klinger, Albert Einstein College of Medicine.*]

methods using either quinicrine mustard dyes or giemsa stain, to show that human chromosomes also form banding patterns. It appears that every pair of homologs has its own banding pattern (Figure 11.17). With these new techniques, scientists can now expect to find smaller chromosomal structural alterations, such as deletions or duplications, and to identify specific chromosomes in man.

REFERENCES

The texts on human genetics referred to at the end of previous chapters discuss chromosomal mutations. The most detailed discussion can be found in J. L. Hamerton, *Human Cytogenetics,* vol. 2, Academic Press, Inc., New York, 1971.

Review articles on the XYY syndrome:

Hook, E. B., 1973: "Behavioral Implications of the Human XYY Genotype," *Science,* **179**:139–149.

Kessler, S., and R. H. Moos, 1971: "The XYY Karyotype and Criminality: A Review," *Journal of Psychiatric Research,* **7**:153–170.

A technical paper on Turner's syndrome patients with unusual sex chromosomes:

Ferguson-Smith, M. A., 1965: "Karyotype-Phenotype Correlations in Gonadal Dysgenesis and Their Bearing on the Pathogenesis of Malformations," *Journal of Medical Genetics,* **2**:142–155.

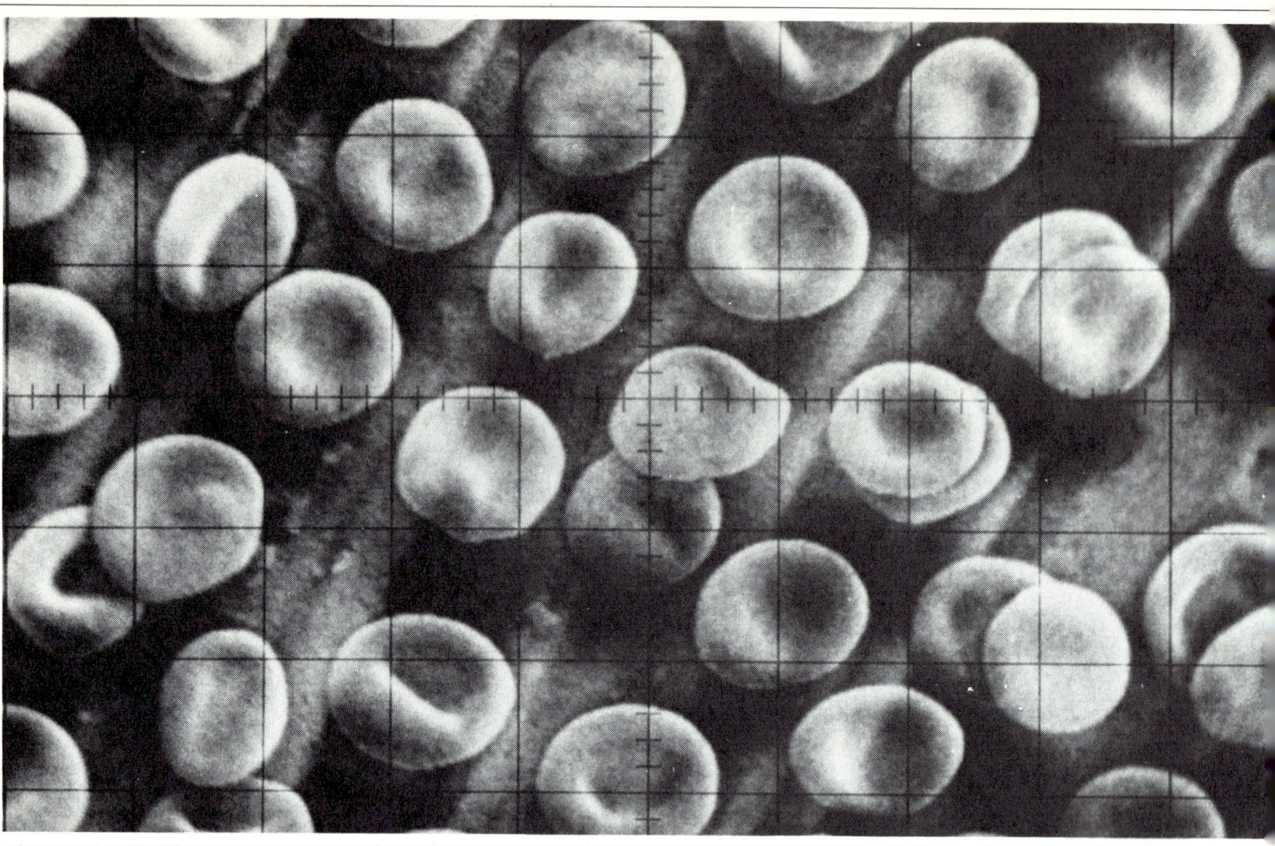
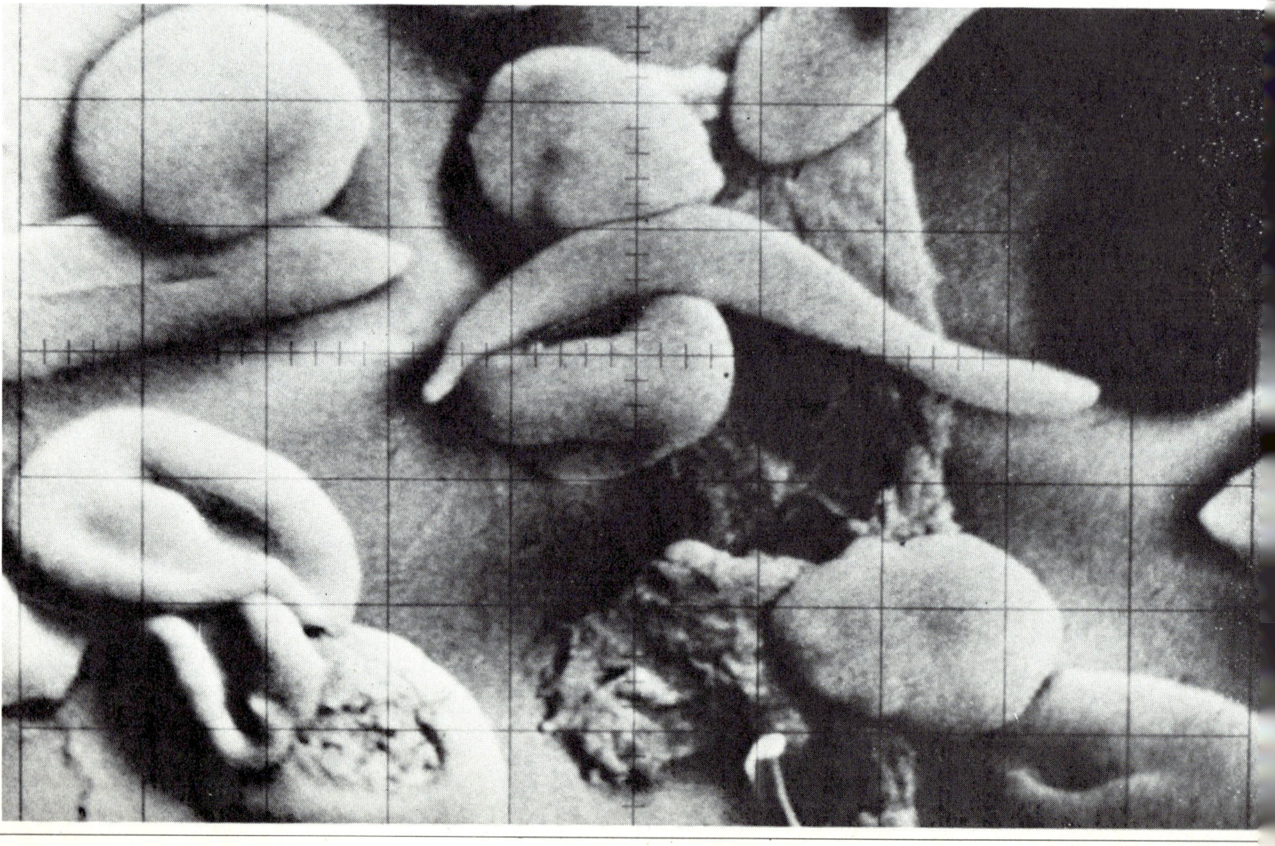

Chapter 12

Point Mutations and Population Genetics

The structural alterations of genetic material on the gross, or chromosomal, level also exist on the molecular level. They do not involve many genes but rather occur within a single gene. The term used for such mutations is *point mutation*. These mutations involve the deletion, addition, or substitution of bases within a gene. Because we know that the genetic message is determined by the sequence of bases in the polynucleotide, and the code is a triplet code, then it stands to reason that any change in the sequence of bases within a triplet or the sequence of triplets in a cistron would result in mutation.

One can make the analogy that if words are changed in some way, the meaning of an entire sentence can be altered. Figure 12.1 is a compilation of typographical errors illustrating five different kinds of intragenic alterations that result in mutation. In addition to deletion and inversion, note the change

POINT MUTATIONS

DIFFERENT KINDS OF MUTATIONS

(A) Normal and (B) sickled red blood cells. This condition is the result of a substitution of one amino acid for another in the mutant hemoglobin. [*Courtesy of Irene Piscopo, Phillips Electronic Instruments.*]

Point Mutations and Population Genetics

FIGURE 12.1
Typographical errors analogous to intragenic or point mutations. [*From Seymour Benzer, Harvey Lectures*, 1961, no. 56.]

> away, and already the doomsday warnings are arriving, the foreboding accounts of a Russian horde that will come sweeping out of the East like Attila and his Nuns.
> —*Red Smith in the Boston Globe*

SUBSTITUTION

> "I can speak just as good nglish as you," Gorbulove corrected in a merry voice.
> —*Seattle Times.*

DELETION

> "I have no fears that Mr. Khrushchev can contaminate the American people," he said. "We can take in stride the best brain-washington he can offer."—*Hartford Courant.*

INSERTION

> He charged the bus door opened into a snowbank, casuing him to slip as he stepped out and fall beneath the bus, which ran over him.—*St. Paul Pioneer Press.*

INVERSION

> Tomorrow: "Give Baby Time to Learn to Swallow Solid Food."
> etaoin-oshrdluemfwypvbgkq
> —*Youngstown (Ohio) Vindicator.*

NONSENSE

* If you are interested in chemical aspects of mutation, turn to Appendix D for a basic discussion of chemical changes in genes which can lead to some of the above-mentioned mutations.

called *insertion*, which is the addition of an extra base or bases, and changes in DNA structure involving substitution and nonsense sequences. Later in the chapter we will discuss some new discoveries showing an enormous amount of DNA duplication in chromosomes of higher organisms.*

As with so many aspects of genetics, much of what we know about mutation on the molecular level comes from studies of haploid microorganisms. Each type of point mutation illustrated in Figure 12.1 has been identified and studied in great detail in phage, bacteria, and fungi. As in Beadle and Tatum's experiments on *Neurospora,* all mutations would be reflected immediately in the altered phenotype of the microorganism. This is, of course, not true in higher organisms due to diploidy. Because most mutations which involve single genes are recessive, we become aware of their existence in diploid organisms only when two carriers for the same mutation produce homozygous young (with the exception of X-linked genes, which would be expressed in males).

In identifying such mutations in humans, we can usually describe only their effects. Because of the complexity of events occurring between primary gene action and the phenotype of higher organisms (see Chapter 8), we cannot as yet trace these effects to the gene that has mutated, much less discern the molecular change which has occurred to render the gene mutant.

One exception to this statement can be found in the studies of hemoglobin abnormalities in humans resulting from point mutations. We shall discuss some of these in detail, since they provide the best evidence we have that molecular changes comparable to those found in microorganisms also occur in man.

HEMOGLOBIN ABNORMALITIES IN MAN

One reason why the genetic determination of hemoglobin proteins in red blood tissue is known in such detail is that this tissue can be removed for study with relatively little effect on the donor, and within limits can be removed time and again, since the body is capable of replenishing the blood supply within a week or so. It is, in short, the most accessible human tissue for experimental study.

SICKLE CELL ANEMIA

Perhaps the most well-known "molecular disease" of hemoglobin is one determined by an autosomal recessive gene which in the homozygous condition leads to a fatal blood disorder known as *sickle cell anemia*. This disease, found primarily in black populations, derives its name from the fact that the recessive homozygotes possess abnormal blood cells which, under certain conditions, such as low oxygen tension, lose their normal shape and take on the shape of a sickle. The introductory illustration shows this change in blood cell appearance; the pleiotropic effects of the disease are listed in Figure 12.2.

In the late 1940s, Linus Pauling and his co-workers discovered that when hemoglobin from normal cells is compared with that from sickle cells, there is a discernible molecular

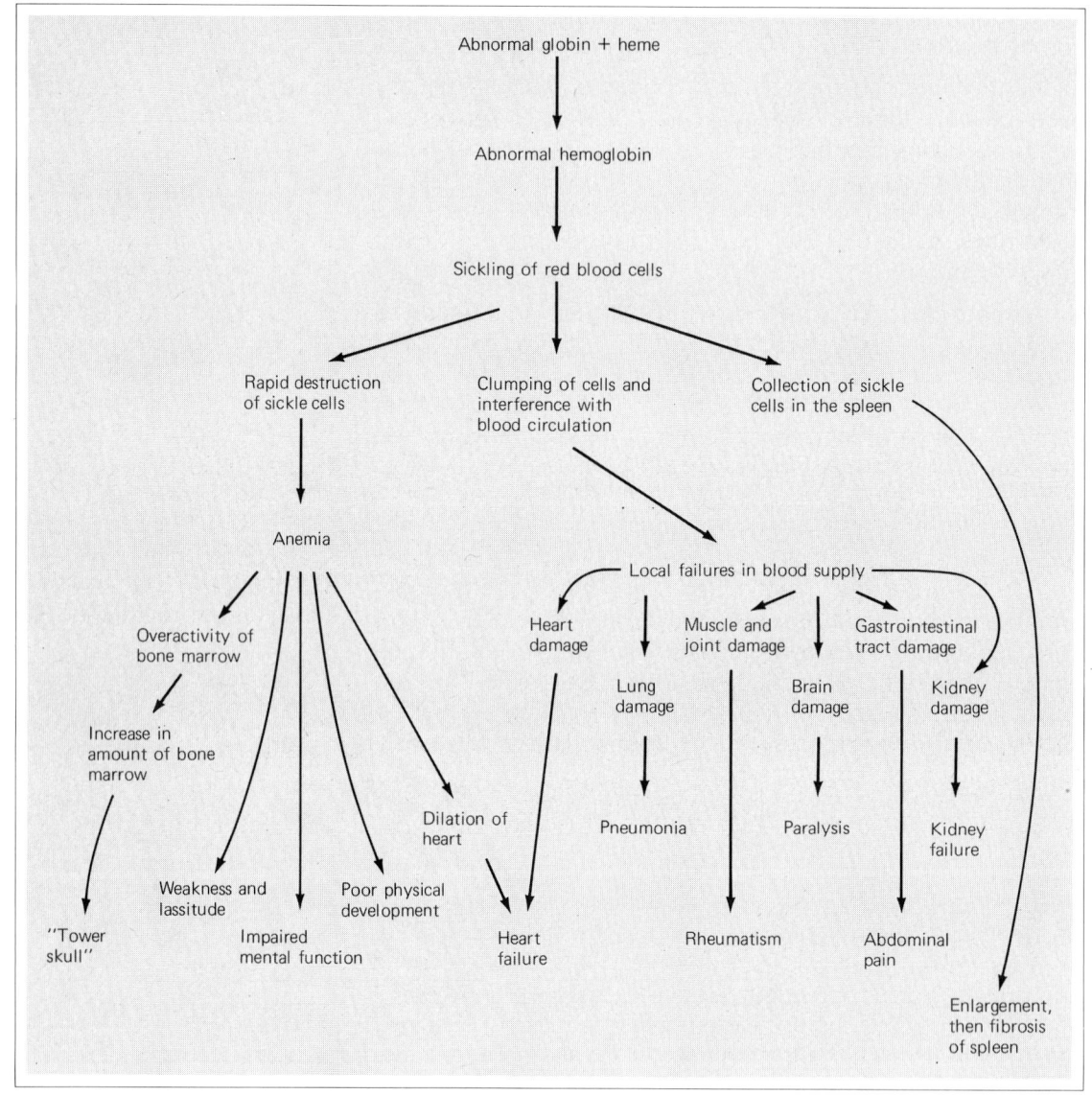

FIGURE 12.2
Disease conditions that have been associated with sickle cell anemia. [After J. V. Neel and W. J. Schull. Reprinted with permission of Macmillan Publishing Co., Inc., from Genetics by Monroe W. Strickberger. Copyright © Monroe W. Strickberger, 1968.]

difference when they are exposed to an electrical field. A technique known as *electrophoresis* measures molecular differences between substances when the proteins migrate at different rates toward the poles (Figure 12.3).

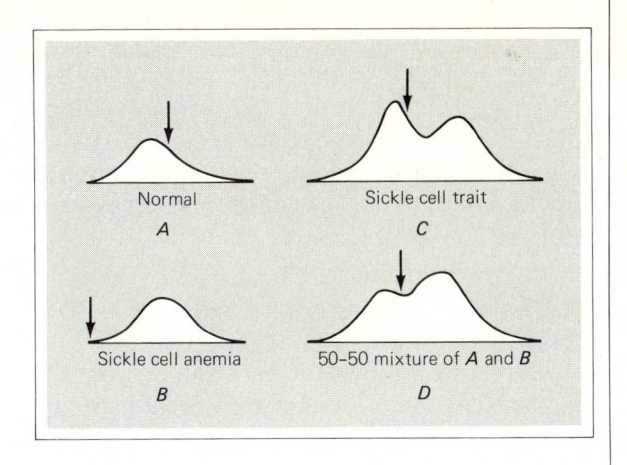

FIGURE 12.3
Movement of hemoglobin from normal and sickle cell homozygotes and heterozygotes in an electrical field. (A) Position of normal hemoglobin in an electrical field. (B) Position of sickle cell anemia hemoglobin. (C) Positions of sickle trait hemoglobin (from heterozygotes). (D) Positions of hemoglobins when normal and sickle cell anemia proteins are mixed. [After Pauling, L., H. A. Itano, S. J. Singer, and I. C. Wells, 1949. "Sickle Cell Anemia, A Molecular Disease." Science, 110:543–548.]

Since then, Vernon Ingram and others, using an elaboration of the electrophoresis technique, have found that the molecular difference between normal hemoglobin, referred to as hemoglobin A, and sickle cell hemoglobin, called hemoglobin S, is caused by the substitution of one amino acid, valine, for the normal amino acid, glutamic acid, at one specific position in the chain of amino acids that make up the beta-chain polypeptide in the hemoglobin protein (Figure 12.4).

It is this single substitution in a chain composed of 146 amino acids that leads to all the physical problems of the homozygote in Figure 12.2. As you can see from Figure 12.3, the heterozygote for the sickle gene shows the presence of both hemoglobin A and hemoglobin S. Such individuals are said to have the sickling trait.

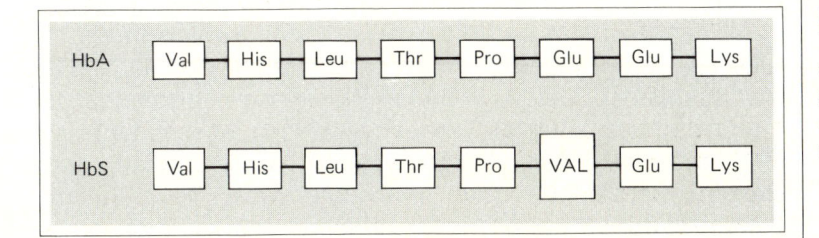

FIGURE 12.4
The substitution of one amino acid, valine, for the normal amino acid, glutamic acid, in the beta chain of hemoglobin results in sickle cell anemia in the homozygotes.

LEPORE HEMOGLOBIN

Hemoglobin Lepore (Hb Lepore) is a hemoglobin variant that appears to result from crossing-over between the genes determining beta-chain and delta-chain synthesis. Part of the chain has the beta sequence, and part has the delta sequence. Heterozygotes have a mild anemia, referred to as Lepore trait, and homozygotes have a severe anemia which is lethal.

Hb Lepore appears to be caused by abnormal synthesis of the aberrant beta chain. Recent studies have indicated that alpha-chain synthesis is at a normal rate but that the ability to synthesize the mutant beta-chain protein in maturing red blood cells rapidly decreases. Scientists studying this phenomenon conjecture that this may be due to a slower rate of synthesis at the delta locus than at the beta locus. The recombinant gene determining Lepore hemoglobin may have synthesis beginning at the delta portion, which would reduce the amount of mutant hemoglobin formation.

OTHER HEMOGLOBIN VARIANTS

Actually, about 100 hemoglobin variants have been identified and found to be caused by substitutions of amino acids at various positions on either the alpha or the beta chain. Table 12.1 shows a partial list of those that have been found to date.

Most of these variants are not as familiar to the general public as is sickle cell anemia, because they are, for the most part, rare mutations found in only a few individuals, whereas the sickle cell trait has been found in frequencies as high as 10 or 20 percent of the population in some parts of Africa. In addition, not all the variants of the hemoglobin molecule produce the same degree of disease, and some affect the functioning of the red blood cells in a negligible way.

COOLEY'S ANEMIA

Another anemia which, like sickle cell anemia, has been characterized as an "ethnic disease" because it is found in relatively high frequency among certain ethnic populations is *thalassemia,* or *Cooley's anemia.* This disease, also lethal, is found in relatively high frequency among descendants of Mediterranean populations such as the Spanish, Italians, and Greeks. Determined by an autosomal recessive mutation, the defect appears due to an insufficient production of hemo-

TABLE 12.1
Variant human hemoglobins caused by point mutations substituting single amino acids at certain positions

		AMINO ACID	
VARIANT	POSITION	OLD	NEW
ALPHA CHAIN			
J Toronto	5	Ala	Asp
J Oxford, I Interlaken	15	Gly	Asp
I, I Texas	16	Lys	Glu
Shimonoseki	54	Gln	Arg
Russ	51	Gly	Arg
BETA CHAIN			
Tokuchi	2	His	Tyr
Porto Alegre	9	Ser	Cys
G Copenhagen	47	Asp	Asn
D Ibadan	87	Thr	Lys
Kansas	102	Asp	Thr

Source: T. Dobzhansky, *Genetics of the Evolutionary Process*, Columbia University Press, New York, 1972, p. 50.

globin in cells of homozygotes. The homozygotes suffer from a severe anemia referred to as *thalassemia major* and the heterozygotes from a mild anemia referred to as *thalassemia minor*. These anemias clinically resemble those found in homozygotes and heterozygotes with Lepore hemoglobin, respectively.

Cooley's anemia differs from sickle cell anemia and other hemoglobin variants in which the hemoglobin protein is abnormal. In thalassemia patients the hemoglobin is normal, but there is too little to sustain normal functioning. Scientists have conjectured that the mutation causes a reduction of the mRNA molecule for the beta chain of hemoglobin.

REPETITIVE DNA AND GENE DUPLICATION

Although duplications are usually considered mutant conditions (see the discussion of the Bar mutation in *Drosophila* in Chapter 11), one of the most intriguing discoveries of recent years involves the duplication of sequences of DNA which does not result in a mutant phenotype. In fact, large numbers of these identical short segments of DNA (involving fewer than 20 nucleotides; an average gene is estimated to be 1000 nucleotides long) have been identified in the chromosomes of almost all eukaryotes. It has been estimated that there are more than a million copies of these sequences, now known as *repetitive DNA,* in mammals.

In addition, there exists a repeated DNA, somewhat longer than repetitive DNA, which has been termed *intermediate repetitive DNA.* This material is believed to be duplicated up to 100,000 times. Geneticists believe that these sequences arise during evolution from the duplication of existing genes. The origin and function of this DNA remain a mystery, but most of it does not seem to be involved in transcription and the formation of proteins. This probably accounts for the fact that it does not affect the normal phenotype of the organism.* We shall mention it further in the chapter on evolution.

* A symposium on "Repetitive DNA, Chromosomal Defects and Neoplasia" was held in September 1972 at the University of Minnesota's Center for Continuing Education.

One specific example of functional gene duplication in man has come from studies by O. Smithies and his co-workers on a group of proteins called *haptoglobins,* found in blood serum. The haptoglobin molecule, like the hemoglobin molecule, is formed by four polypeptide chains. There are two a^{hp} and two β^{hp} chains. There exist three kinds of a^{hp} chains, designated a^{1hp}, a^{1Shp}, and a^{2hp}. A study of the sequence of amino acids in each of these chains has shown that the a^{2hp} polypeptide differs from a^{1hp} only in that it includes an incompletely duplicated section of the a^{1hp}. This indicates that the gene for a^{2hp} is probably one that includes a partial duplication.

VISUAL EVIDENCE FOR GENE DUPLICATION: GENE AMPLIFICATION

The only known process of repeated duplication of genes that can actually be observed is in the maturation of eggs in various species of animals, such as certain amphibians and beetles. During oogenesis, genes which determine ribosomal RNA formation appear to undergo great multiplication in a process which has been termed *gene amplification.*

During this process, the DNA involved is estimated to be duplicated several thousand times, presumably to supply the egg with abundant ribosomal particles that will be distributed to embryonic cells as division proceeds at its furious pace following fertilization. The extra copies of DNA are discarded after the ribosome particles are formed, since the cells of the embryo do not show the presence of extra DNA. Figure 12.5 shows gene amplification in the amphibian *Xenopus.*

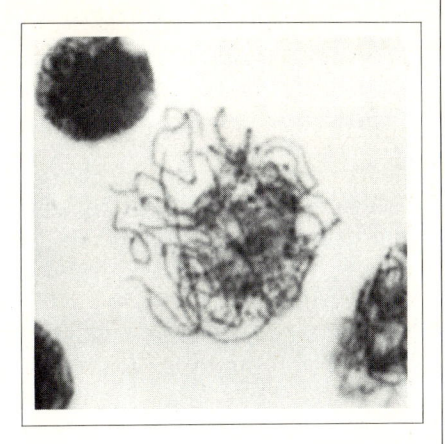

FIGURE 12.5
Gene amplification in amphibian oocytes. [*Courtesy of Dr. Joseph Gall, Yale University.*]

Although gene amplification results from duplication of genetic material, it is not permanent and does not result in an abnormal phenotype. Therefore, it is not considered a mutation. On the contrary, it appears to be important for normal development. Nonetheless, gene amplification does bear testimony to the potential for gene duplication in complex forms of life.

It is important not only for the individual, but also for future generations of any species, to determine how frequently mutation occurs.

With regard to dominant traits, determining the mutation rate is fairly straightforward. Let us take the study of *achondroplasia,* a form of dwarfism inherited as an autosomal dominant, as an example of how to estimate the rate at which a normal gene mutates to the dominant form causing dwarfism. A study was made of birth records in a babies' hospital in Denmark,* in which out of approximately 94,000 births, 10 were chondrodystrophic dwarfs. Of these, 2 were born to dwarf parents and thus could not be considered new mutations; however, 8 of the 10 were born to normal parents, so that they *are* new mutations. As 188,000

DETERMINATION OF MUTATION RATES

MUTATION TO A DOMINANT GENE

* E. T. Mørch, *Op. dom. Biol. hered. hum.,* (KBN)3, 1941.

gametes were involved in the procreation of these 94,000 babies, the natural or spontaneous mutation rate for this gene to the dominant state is approximately $\frac{8}{188,000}$ or 1 in 23,500 gametes.

ESTIMATION OF RECESSIVE MUTATIONS: DIFFICULTIES

Determining the rate of recessive mutations is a much different problem. Geneticists are limited to mathematical techniques which are beyond the scope of our discussion here. However, because of the importance of such mutations, not only for the unsuspecting carrier but also for the survival of the species (a theme we shall discuss in detail in Chapter 15), we shall mention here some of the problems geneticists encounter in attempting to estimate mutation rates of recessive genes.

DIPLOIDY The primary difficulty, of course, lies in the diploid condition of chromosomes in complex organisms which masks new recessive mutations by the normal, dominant allele. For a recessive mutation to be expressed, the organism would have to be a homozygote. This is unlikely to occur if the mutation is a rare one, and most mutations are (see Table 12.2). Many geneticists are now advocating the development of new screening methods to permit estimation of mutation rates in human populations by observing large numbers of individuals. Electrophoresis and other techniques to determine molecular changes discernible in heterozygotes, such as the sickling trait for sickle cell anemia, could prove useful.

BACK MUTATION If a recessive mutation is not expressed, there is always the possibility for *back mutation,* a process by which a gene that has mutated may mutate back to the original form. Evidence in studies of mice indicates, however, that there are more "forward" mutations—that is, mutations to a new allele—than back mutations.

If recessive mutations are not expressed and then mutate back, there is no way to tell that the original mutation occurred. Then why should it concern us? Because the frequency of mutations at particular loci gives us some idea of the susceptibility of these genes to change, either spontaneously or in response to man-made factors that induce mutation. We shall deal with this subject in Chapter 15.

NEUTRAL MUTATIONS Another factor that obscures the estimation of mutation rates is that changes in protein molecules can occur that do not result in a net change of electrical charge. Such *neutral mutations* are undetectable using the techniques applied to studies of hemoglobin variants.

REPAIR MECHANISMS A phenomenon in bacteria known as *photoreactivation* has led to a series of studies that revealed the ability of cells to repair defects in the DNA of the cell (Figure 12.6). It was found that if colonies of bacteria are irradiated with ultraviolet light, the resulting effects on the DNA would be lethal for the great majority of cells. However, if the irradiated plates are exposed to visible light, harmful effects are greatly reduced.

Further studies showed that the ultraviolet irradiation causes abnormal bonding: thymine molecules bond with adjacent thymine on the same strand, forming what is known as a *dimer.* The presence of dimers in a strand of DNA prevents normal replication; the cells would not be able to reproduce and would die. The visible light reverses this process by serving as a source of energy to activate enzymes in the cell which detect aberrant structures in DNA such as dimers. These enzymes are capable of removing the two thymines and replacing them with normal thymine nucleotides, which allow the cell to undergo replication and division.

Thus, it appears that there are *repair mechanisms* in cells which are able to detect some alterations of the DNA struc-

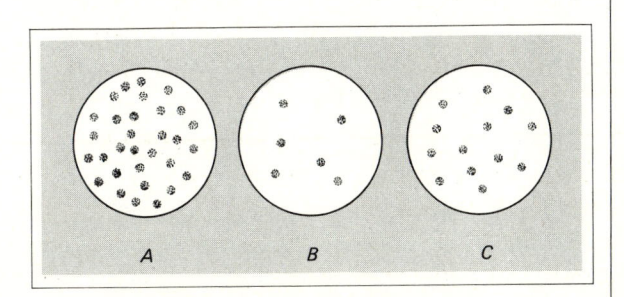

FIGURE 12.6
Schematic illustration of the effect of photoreactivation in bacteria cultures. Plate *A* shows a control plate of colonies. Plate *B* was irradiated with ultraviolet light, resulting in the deaths of many colonies. Plate *C* was also irradiated and then exposed to visible light. There are many more survivors of the radiation than in plate *B*.

FIGURE 12.7
Illustration of two theories of DNA repair. The "cut and patch" theory proposes removal of the abnormal portion of the DNA first, and then replacement with normal nucleotides. The "patch and cut" theory proposes that as the abnormal DNA is removed, there is simultaneous replacement with normal nucleotides. [*After "The Repair of DNA," by Philip C. Hanawalt and Robert H. Haynes. Copyright © Feb. 1967 by Scientific American, Inc. All right reserved.*]

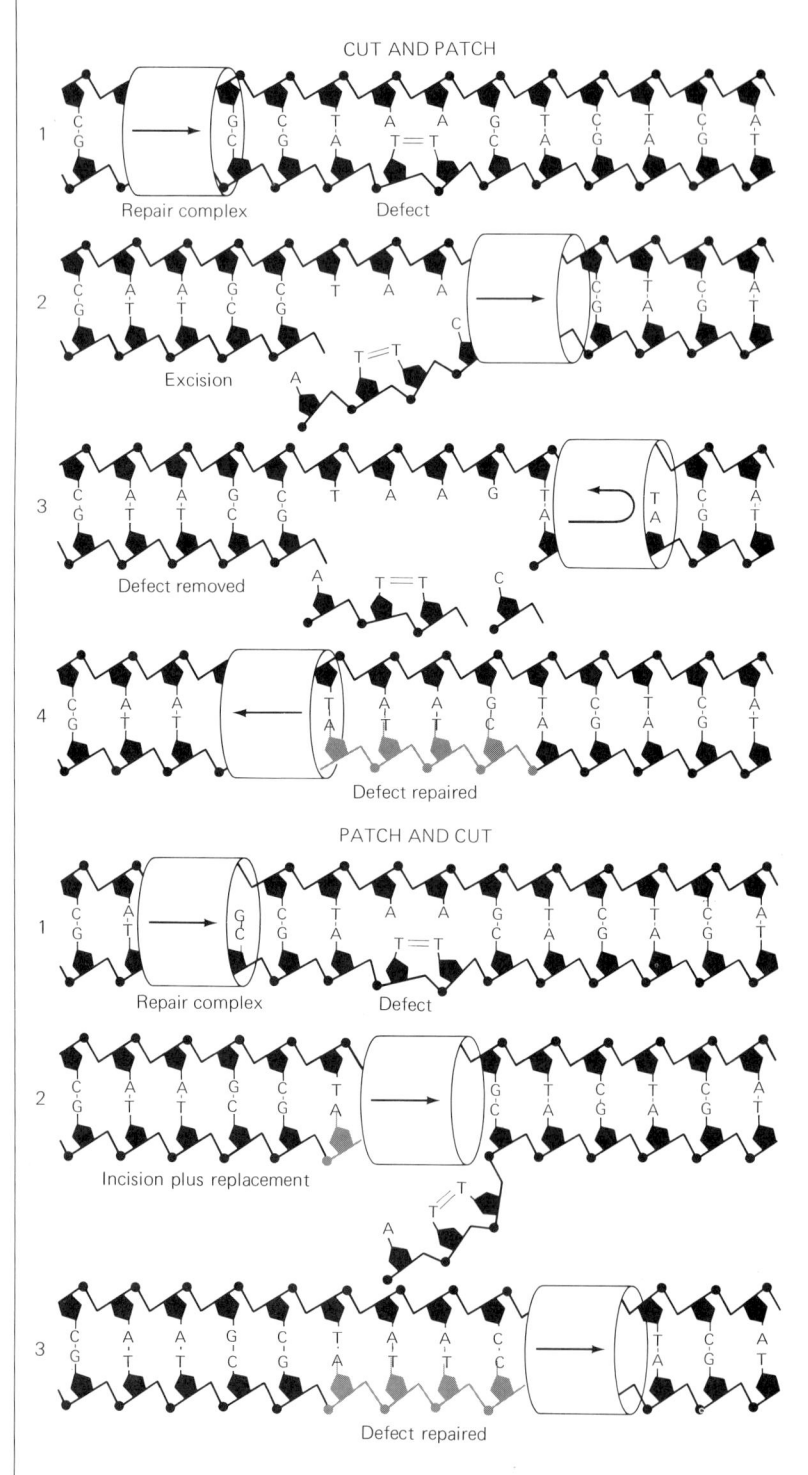

ture and restore the normal structure by excising the aberrant piece of DNA and replacing it with newly replicated, normal DNA, as in photoreactivation. The regulation of this mechanism is not well known, but Figure 12.7 illustrates two recent theories of the sequence involved in repair.

If data such as that from the study on dwarfism is so easy to obtain, you may wonder why we cannot determine the precise rate of all mutations in man, since obviously all genes are composed of DNA, and whatever causes a change in one portion of the DNA should affect other portions equally. Unfortunately, it has been well established in microorganisms, especially by S. Benzer and his studies on bacteriophage,* that this is not true. Certain sites mutate far more frequently than others, as illustrated in Figure 12.8.

DIFFERENTIAL MUTABILITY

* Benzer, S., "Fine Structure of a Genetic Region in Bacteriophage," *Proceedings of the National Academy of Science*, **41**: 344–354, 1955.

FIGURE 12.8
Map of two cistrons in a virus. Each square symbolizes one mutation at that locus. Note that some loci mutate much more frequently than others. These are known as "hot spots." [*Reprinted with permission of Macmillan Publishing Co., Inc., from* Genetics *by Monroe W. Strickberger. Copyright © Monroe W. Strickberger, 1968.*]

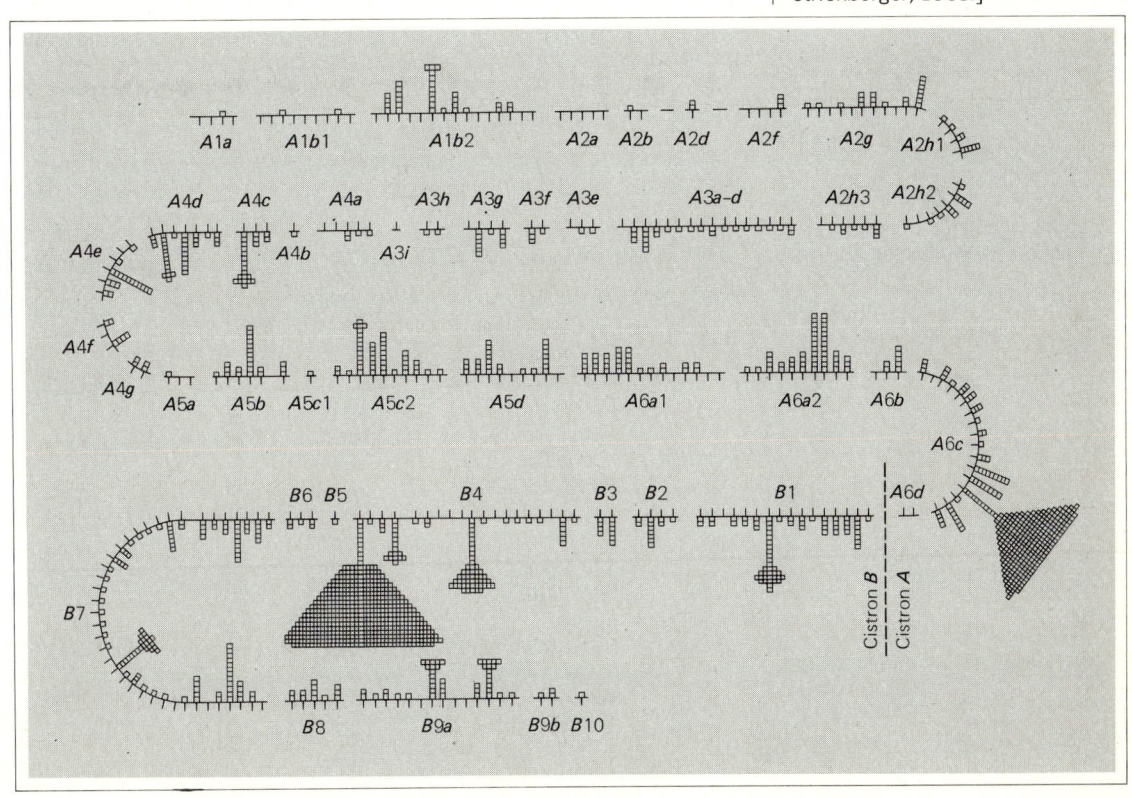

Table 12.2 lists some estimated rates of mutation at several loci in some of the higher plants and animals, and in man. Note that even in the case of achondroplasia, other studies have arrived at different figures than the study we described of babies in Denmark.

POPULATION GENETICS

You may have wondered why, if so many mutations are not expressed because of diploidy, neutral mutation, or repair, we should be concerned about them. The main reason is that because of these factors, we sustain many more mutations than are expressed. And for the most part, these mutations, if not repaired, are simply retained in our genome. Geneticists have coined a term for the number of unexpressed mutations we possess: our *genetic load*.

The true significance of our genetic load will not become apparent to you until we discuss evolution, but for now, it should be understood that the word *load* is used intentionally to imply some sort of burden. The burden in this case is that most mutations are harmful, and therefore, the greater the load of a population, the greater the probability of some con-

Table 12.2
Spontaneous mutation rates at specific loci for various organisms

ORGANISM	TRAIT	MUTATION PER 100,000 GAMETES
Corn	*Wx* to waxy	0.00
	Sh to shrunken	0.12
	C to colorless	0.23
	Pr to purple	1.10
	I to *i*	10.60
	R^r to r^r	49.20
Drosophila	*Y* to yellow	12
	Bw to brown	3
	Ey to eyeless	6
Mouse	*B* to brown	0.85
	S to piebald	1.7
	D to dilute	3.4
Man	Epiloia	0.4–0.8
	Huntington's chorea	0.5
	Pelger's anomaly	1.2–14.3
	Achondroplasia	4.2–14.3
	Retinoblastoma	1.2–2.3

Source: M. Strickberger, *Genetics*, Macmillan Company, New York, 1968, p. 524.

dition being expressed that is detrimental to the individual. The greater the number of abnormal individuals, the weaker the species.

The area of genetics that concerns itself with the estimation of mutation rates and the frequency of existing mutations in a population is called *population genetics*. The tools of population geneticists consist of fairly sophisticated mathematic formulas which we shall not go into here. Those who are interested should consult the references at the end of this chapter, which discuss these methods in detail. There is one simple formula that we *can* discuss, as an example of how one can estimate the frequency of particular genes within a population. This is the Hardy-Weinberg law, formulated independently in 1908 by the two gentlemen for whom it is named.

THE HARDY-WEINBERG LAW

Suppose we are interested in the extent to which a gene determining a particular form of albinism is present in a population. A study of the population has revealed that this condition is determined by a recessive autosomal gene, and that the frequency of homozygotes is about 1 in 20,000. What is the frequency of carriers of this mutation in the population?

If we assume that the frequency of the normal allele for pigmentation equals p and the frequency of the recessive mutant allele equals q, then $p + q = 1$, the total of all the genes in the population for that particular locus. If we assume that mating is random and that the transmission of the dominant gene is equal in probability to the transmission of the recessive gene, then the frequency of individuals homozygous for the dominant allele would be $p \times p$, or p^2. This is so because the probability that two genes would be found in one person at the same time is equal to the product of their individual probabilities, or in this case, their frequency in the population. (Remember our discussion in Chapter 1?) For the same reason, the proportion of albinos in the population would be q^2.

As for the heterozygotes, the probability of having a dominant and a recessive allele at the same time is $p \times q$, or pq. But since the dominant gene could have come from either parent, there are two ways of obtaining pq (just as there are two ways of obtaining a head and a tail when flipping two coins at once); therefore, the frequency of heterozygotes is considered to be *2pq*. The Hardy-Weinberg formula for the total number of genotypes in the population, then, is the sum of the homozygous dominants, the heterozygotes, the homozygous recessives, or $p^2 + 2pq + q^2$.

With this formula we can estimate the frequency of carriers of the recessive mutation for albinism in the problem mentioned above. If the frequency of albinism is 1 in 20,000, then $q^2 = \frac{1}{20,000}$; $q = \frac{1}{141.4}$. To simplify the problem, let us assume that $q = \frac{1}{141}$. Since $p + q = 1$, then $p = 1 - q$, or $1 - \frac{1}{141}$, or $\frac{140}{141}$. Since the frequency of heterozygotes is *2pq*, then the frequency of carriers for this mutant gene in this population is $2 \times \frac{140}{141} \times \frac{1}{141}$, or approximately $\frac{1}{69}$.

If one were studying codominant alleles, such as those in blood groups, one could modify the use of the Hardy-Weinberg formula by directly assessing the frequency of heterozygotes in a population since heterozygotes would be as apparent as homozygotes. Note that in the case of X-linked recessive genes, the frequency of affected males would not be q^2, as they are hemizygous and not homozygous for the gene; the frequency would be simply *q*. The frequency of affected females would, of course, be q^2 because of the two X chromosomes.

FACTORS AFFECTING THE HARDY-WEINBERG LAW

Note that the Hardy-Weinberg law is valid only if one assumes that the different alleles are transmitted equally in a freely breeding population. One could not use the formula if factors existed that might affect the genetic equilibrium of the alleles from generation to generation.

SELECTION One such unbalancing factor would be lethality, which would eliminate one genotype from the population and cause a proportional increase in the remaining genotypes. As defined by geneticists, the resulting "differential perpetuation of genes from generation to generation constitutes *selection*."* In other words, if we assume that all the individuals in a population are heterozygous for a particular lethal gene, *s*, and the homozygous condition is lethal, then one-fourth of the next generation would be *ss* and die. This results in selection against these genes or, in other words, their removal from the population.

* E. W. Sinnott, L. C. Dunn, and T. Dobzhansky, *Principles of Genetics,* McGraw-Hill Book Company, New York, 1958, p. 245.

The following diagram illustrates such a population with a lethal gene. You can see that where the first generation had equal numbers of the dominant and recessive alleles, the second generation, because of lethality, has only half as many recessive as normal alleles in the individuals who live to reproduce:

P_1 Ss × Ss   2S:2s
F_1 Ss Ss Ss   4S:2s

POLYMORPHISM Although selection against lethal mutations would be expected to cause a decrease in the frequency of the genes from one generation to another, it was of interest to geneticists studying populations in Africa that the gene determining sickle cell anemia seemed to be maintained almost at equilibrium, even though the disease is fatal. A certain number of the mutant genes could be expected to appear every generation as a result of spontaneous mutation, but mutation rates are very low and could not account for the high frequency of the sickle gene in these populations.

Eventually it was realized that the frequency of the recessive mutation was being maintained because the heterozygote was more resistant to falciparum malaria than the normal homozygote. Presumably alterations of blood proteins in the heterozygote rendered him more resistant to another fatal disease. Thus, while the recessive homozygotes died of the sickling disease, the dominant homozygotes died of falciparum malaria. The heterozygote thus transmitted as many recessive genes as dominant genes to the next generation. This situation is known as *balanced polymorphism.*

The existence of widespread polymorphism, as revealed by studies on sickle cell anemia and recent studies on protein variants, discussed earlier, is one aspect of population genetics that is receiving much attention. There is far more polymorphism than can be expected on the basis of selection forces versus new mutations. Two eminent human geneticists, James V. Neel and William J. Schull, state:

To put it very simply, *they* [polymorphisms] *have now reached such numbers, and new ones are being recognized at such a rate, that the understanding of their significance is unquestionably a focal problem of modern biomedicine. . . .* The fact of intraspecific diversity has long been recognized. . . . What is new is our recognition of its potential extent and our ability to reduce this to the primary manifestation of genetic differences, namely, variation in proteins and related substances. . . . The polymorphisms resulting from this variability are thought to be the genetic equivalent of biochemical "buffer" systems, in that the gene frequencies they reflect are not easily disturbed by changes in mutation rate or temporary fluctuations in selective pressure. Some of the functions of the polymorphisms undoubtedly relate to important human differences and disease susceptibilities.*

* J. V. Neel and W. J. Schull, "On Some Trends in Understanding the Genetics of Man," *Perspectives in Biology and Medicine,* 1968, pp. 565–602.

MUTATION As mentioned above, *mutation* tends to upset the equilibrium upon which the Hardy-Weinberg law depends, since new spontaneous mutations would increase the frequency of a particular gene in a population. The term used to refer to all the genes distributed among the members of a freely interbreeding population is *gene pool.*

MIGRATION Another factor that would obviously change the frequency of genes in the gene pool of a population is *migration* into or out of populations by individuals of certain genotypes. In this connection, we should note that a commonly held notion of the origin of the American Indian as being from the Orient has had to be reevaluated from the point of view of population geneticists. Studies on the frequency of different alleles determining the ABO blood groups in populations of the Orient show a high frequency of the I^B allele, which is practically nonexistent in the gene pool of the American Indian. We shall speak more on this when we discuss the genetic basis of race formation in Chapter 14.

GENETIC DRIFT One other factor that could affect genetic equilibrium is *genetic drift.* As the name implies, this is a random change in gene frequency from one generation to another caused by chance, for example, from genocide of a particular population in war. The smaller the population, the more effect genetic drift would have. Figure 12.9 illustrates this point.

REFERENCES

Benzer, S., 1955: "Fine structure of a Genetic Region in Bacteriophage," *Proceedings of the National Academy of Science,* **41**: 344–354. (Highly technical paper.)

Culliton, B., 1972: "Cooley's Anemia: Special Treatment for Another Ethnic Disease," *Science,* **178**: 590–593. (A good review article.)

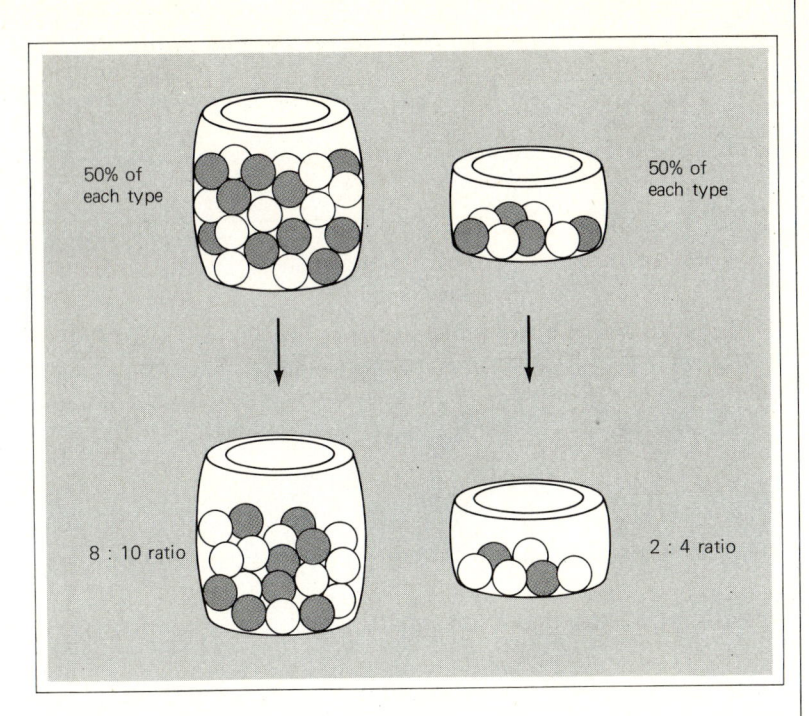

FIGURE 12.9
Schematic illustration of the effects of removing the same number of marbles from different-sized barrels. The smaller the number of marbles in the original "population," the greater the effect on the remaining ratio of different marbles. This is analogous to the effects of genetic drift on small populations.

Gill, F., J. Atwater, and E. Schwartz, 1972: "Hemoglobin Lepore Trait: Globin Synthesis in Bone Marrow and Peripheral Blood," *Science,* **178**: 623–625.

Hanawalt, P. C., and R. H. Haynes, 1967: "The Repair of DNA," in *Facets of Genetics,* W. H. Freeman and Company, San Francisco.

Ingram, V., 1961: *Hemoglobin and Its Abnormalities,* Charles C Thomas, Publisher, Springfield, Ill.

Mørch, E. T., 1941: *Op. dom. Biol. hered. hum.,* (KBN)3.

Neel, J. V., and W. J. Schull, 1968: "On Some Trends in Understanding the Genetics of Man," *Perspectives in Biology and Medicine,* pp. 565–602.

Those of you who are interested in mathematical aspects of genetics may wish to use the following:

Falconer, D. S., 1960: *Introduction to Quantitative Genetics,* Noland Press, New York.

Li, C. C., 1955: *Population Genetics,* University of Chicago Press, Chicago.

Wallace, B., 1970: *Genetic Load: Its Biological and Conceptual Aspects,* Prentice-Hall, Inc., Englewood Cliffs, N.J.

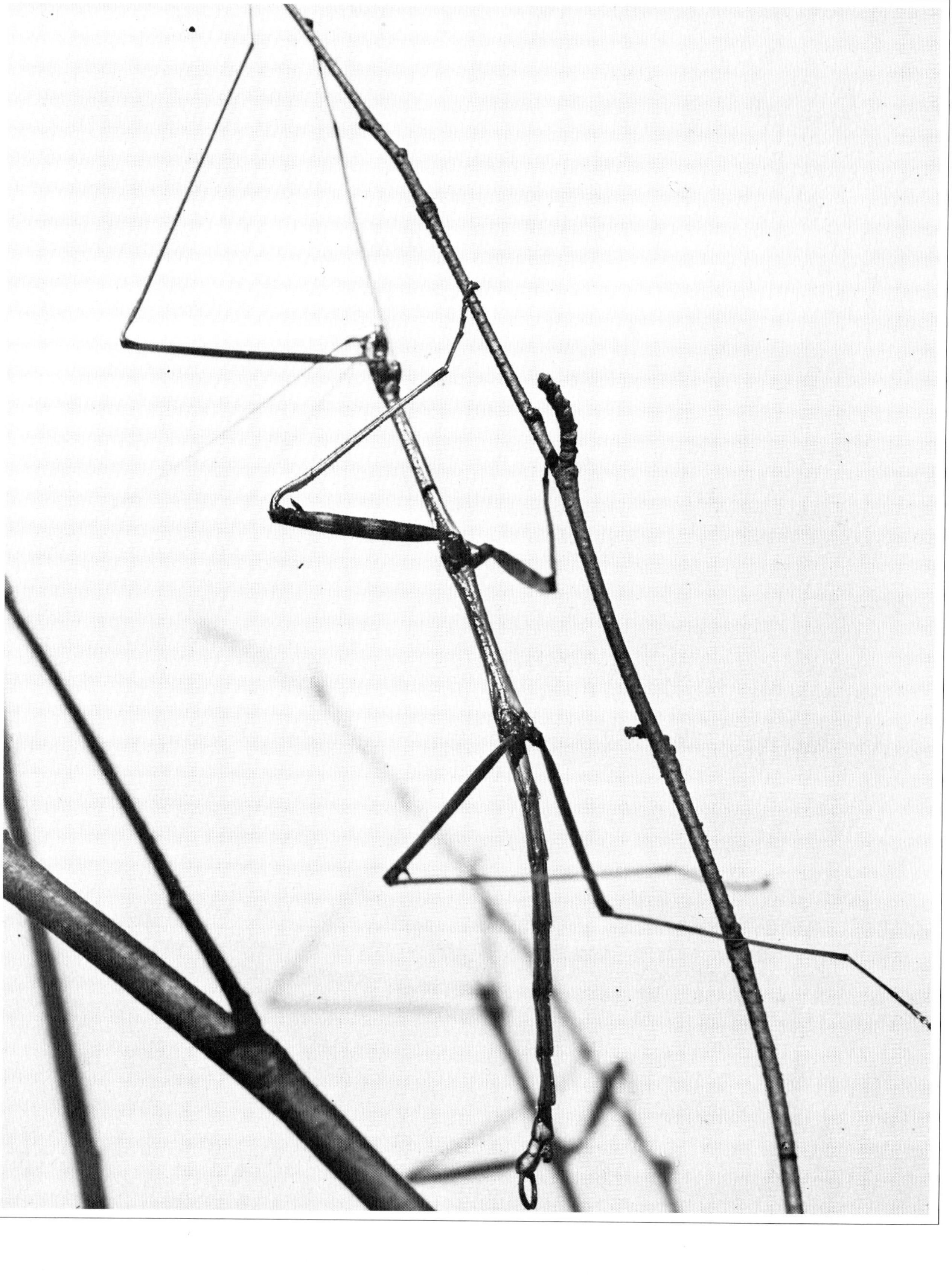

Chapter 13

The Genetic Basis of Evolution

From our discussion so far, it should be clear that cells and organisms are what they are and function as they do because of the genes they possess. Yet no geneticist would ever state that to know the genome of an individual is to know automatically his phenotype, or vice versa. Our *potential* is determined by our genome, but whether we fulfill our potential is largely determined by our *environment*.

For example, there is a genetic component to human intelligence and height, though it is far from clear. Yet whether a child ever fully develops his potential for these traits depends on such environmental factors as good health and nutrition. Thus, deprivation as a result of war and poverty takes its toll especially on young victims, whose cells desperately need sources of amino acids for protein synthesis and mitosis and, lacking these sources, cease to divide and develop. The resulting stunting of growth and mental capacity is mute testimony to the role of the environment in the development of an individual's full genetic potential.

A walking stick from Peru is a good example of protective adaptations that have allowed this species to persist in the struggle for survival. [*Courtesy of the American Museum of Natural History.*]

THE ROLE OF MUTATIONS

As our discussion of mutation in the last few chapters indicates, the gene pools of populations are constantly changing, due to mutation and other genetic phenomena such as drift and migration. Harmful mutations tend to be lost by selection, whereas other mutations which do not affect reproduction or are beneficial are incorporated into the gene pool. In this way, traits characterizing the individuals of the population correspondingly change.

We should be aware that mutations in and of themselves are not beneficial or detrimental. Whether they benefit the carrier is really determined by the environment. For example, galactosemia, one of the inborn errors of metabolism discussed earlier, would be considered a detrimental mutation if the homozygote for the recessive mutation continued to receive milk. Yet if the milk intake of such children is restricted, the mutation no longer causes severe mental retardation. This alteration of the child's environment thus renders the mutation neutral with respect to brain development.

In bacteria, this point can also be easily made. We know, for example, of mutations that render bacteria resistant to certain antibiotics, such as streptomycin. In the absence of streptomycin in the environment such mutations could not be considered beneficial. However, in media containing streptomycin, the mutation surely has to be considered beneficial (Figure 13.1).

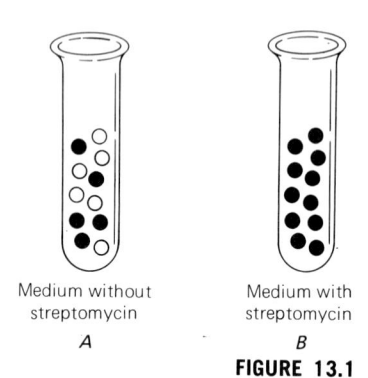

FIGURE 13.1
Illustration of the role of the environment in determining whether a mutation will benefit the organism or be detrimental. Bacteria carrying a mutation to streptomycin resistance are represented by dark spheres. In tube A, in the absence of the antibiotic, the mutants have no particular advantage. In tube B, they are the only survivors in media containing streptomycin.

THE SIGNIFICANCE OF VARIABILITY

To extend the thought further, any population of bacteria which contains mutants carrying the streptomycin-resistant gene would be far more likely to survive than populations which are homogeneously normal. In other words, the greater the variability of individuals in a population, the more likely the population is to survive changes in its environment. This idea is crucial to the concept of the interaction of genes and environment—or nature and nurture—in the gradual process of change in all populations of living organisms known as *evolution*. In this chapter we shall discuss the role of the environment in the formation of species, of all the wonderfully varied life forms which exist in our natural world.

Even before Mendel reported his results on experimentation with garden peas, a young English naturalist was persuaded to join a sea voyage on the H.M.S. *Beagle,* the purpose of which was to explore the Galápagos islands to the west of South America and observe the wildlife found there. His name was Charles Darwin, and the conclusions which he eventually reached, based on these observations, revolutionized not only biology but every field of study dealing with the existence of man and shook religion and philosophy to their core (Figure 13.2).

CHARLES DARWIN AND THE THEORY OF NATURAL SELECTION
THE VOYAGE OF THE H.M.S. *BEAGLE*

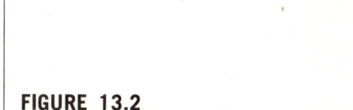

FIGURE 13.2
Charles Darwin. [*Courtesy of the Bettmann Archive Inc.*]

Of particular interest to Darwin were the variations he noticed in certain structures of related animals on the different islands. For example, various species of finches each had a different beak structure, and Darwin realized that these structures were adapted to those particular environments (Figure 13.3).

LYELL AND MALTHUS

Darwin mulled over his observations for many years. He related them to earlier concepts that had been formulated by Charles Lyell, a great geologist and his mentor and friend, and by the economist Thomas Malthus.

Lyell had concluded that the earth was very old, that geologic changes occurred gradually, rather than through cataclysmic events that overnight changed the earth and caused life to begin anew, as had been held previously and supported by religious dogma and creationism theories. Malthus, in the eighteenth century, had published a study contending that the human population tends to increase faster than the food supply, and consequently, that survival necessitates a struggle within the species.

NATURAL SELECTION

Darwin took these concepts and, adding his own observations on variation in nature, formulated the basis of modern concepts of evolutionary change: natural selection. As he wrote:

If under changing conditions of life organic beings present individual differences in almost every part of their structure, and this cannot be disputed; if there be, owing to their geometrical rate of increase, a severe struggle for life at some age, season, or year, and this certainly cannot be disputed; then, considering the infinite complexity of the relations of all organic beings to each other and to their conditions of life, causing an infinite diversity . . . it would be a most extraordinary fact if no variations had ever occurred useful to each being's own welfare. . . . But if variations useful to any organic being ever do occur, assuredly individuals thus characterized will have the best chance of being preserved in the struggle for life; and from the strong principle of inheritance, these will tend to produce offspring similarly characterized. This principle of preservation, or the survival of the fittest, I have called Natural Selection.*

* Charles Darwin, *The Origin of Species*, Modern Library, New York, p. 98.

FIGURE 13.3
Darwin's finches. Their differences reflect the natural selection of traits that are of adaptive advantage for the various environments of the different islands. For example, strong beaks would be of advantage for diets of hard seeds whereas slender beaks aid in dislodging insects from tree bark crevices. [*Courtesy of the American Museum of Natural History.*]

Natural selection, then, as Darwin envisioned it and as we hold it to be true today, results from favorable variations in populations; these variations cause the individuals so endowed to have an advantage over others in the population. Greater percentages of such individuals would tend to survive and produce offspring which would inherit the favorable variation. These, in turn, would reproduce, thus gradually causing the frequency of this variation to increase in a population. Eventually, this characteristic, which was rare when it first appeared, will become so common as to be a trait by which the population is identified.

Eventually differences become so great between populations that they are no longer able to mate with each other. They are then considered separate species. Thus are species formed: slowly, gradually, randomly.

THE ORIGIN OF SPECIES AND THE DESCENT OF MAN

In part because of Lyell's strong influence on Darwin and his opposition to organic evolution, Darwin spent the next 20 years gathering data to support his theory and did not release this revolutionary concept of evolution until by chance a young British naturalist, Alfred Russel Wallace (Figure 13.4), sent a manuscript to Darwin for review. Written as a result of studies in Indonesia, where Wallace had noticed structural variations in animals on different islands just as Darwin had in the Galápagos, the manuscript contained the same hypotheses about natural selection and survival of the fittest over which Darwin had labored for two decades.

On the advice of Lyell, who at this time had begun to accept the logic of Darwin's arguments on organic evolution, Darwin sent a summary of his own work, together with Wallace's paper (with the latter's approval), to a scientific society for presentation. Publication, finally, of *The Origin of Species,* in 1859, and of *The Descent of Man,* in 1871, firmly established Darwin's eminence in the history of scientific thought.

VARIABILITY IN POPULATIONS

The variations in populations which Darwin and Wallace noted are caused by spontaneous mutations. These mutations, as we discussed in Chapter 12, arise by chance alone.

FIGURE 13.4
Alfred Russel Wallace. [*Courtesy of Radio Times Hulton Picture Library.*]

There is no process in nature that can direct specific mutation events. However, if the mutation is advantageous to the individual, it will be transmitted to future generations rather than its less advantageous alleles.

It should be noted that Darwin, of course, did not know about mutations, since the science of genetics was about a half century in the future. For this reason, modern concepts of evolution that relate mutation to natural selection form what is referred to as *neo-Darwinism*.

Note that what the geneticist feels is an advantageous trait is not necessarily one that endows the mutant with exceptional strength or — fellow humans, take note — exceptional intelligence. It is simply a trait that allows the mutant to sus-

EVOLUTIONARY FITNESS

tain itself better in its environment and, more important, to be more fit to reproduce offspring (that, in turn, reproduce and have an advantage over others in the population).

The history of life on earth is replete with examples of species that have ceased to exist even though they were the dominant species in their time. The prime example of this is the dinosaurs, which terrorized other forms of life because of their great strength and size. Yet by the Cenozoic era, which began 65 to 70 million years ago, they had become extinct. We do not know the exact reason, but it surely must have involved changes in the environment to which the giant reptiles could not adapt.

It is a fact of evolution that changes in the environment occur much more rapidly than changes within populations of complex organisms caused by mutation which might allow individuals to adapt to the changes. It is also a fact that once a species becomes extinct, it never reappears.

RANDOMNESS OF EVOLUTIONARY CHANGE

This brings us to a very important point, and one misunderstood by many people. Because variation in populations results from random mutations, *there is no way that individuals in a particular population can intentionally acquire hereditary characteristics that would allow them and their offspring to adapt to environmental change.*

For example, one widely known phenomenon of recent years, with our widespread use of pesticides, is the common occurrence that a treated insect population will contain certain individuals who are resistant to the chemicals. For example, DDT has been used in massive amounts to combat malaria mosquitos all over the world, and about 10 percent of the populations have been found to be resistant.

The wrong interpretation, from the neo-Darwinian point of view, is that the 10 percent in question reacted in some way and adapted to the pesticide to become resistant. What actually happened was that, even before the use of DDT, mutation to DDT resistance arose by chance in certain individuals in the population. If DDT had not existed, the mutation would have been neutral and unexpressed. However, when DDT was applied, only the offspring carrying this mutation to resistance survived. There was no *intentional* change in the body chemistry of the organisms that survived. *This inability of an individual to change hereditarily in response to environmental factors holds true for all living organisms, including man.*

There are many lines of evidence supporting Darwin's theory that natural selection is the basis for evolutionary change.

EVIDENCE FOR DARWIN'S THEORY OF EVOLUTION

FOSSILS

Geology and paleontology have contributed one line of information which supports current concepts of evolution. Evidence embodied in fossils shows gradual changes associated with geological change. This has led to conclusions about the progression of life forms that have evolved during the history of our planet, and about the evolutionary relationships between various phyla and species (Table 13.1 and Figure 13.5).

SIMILARITIES BETWEEN EMBRYOS OF DIFFERENT SPECIES

Another line of evidence comes from an area of biology called *embryology*. There are stages in the embryonic development of related species, such as the various species of vertebrates, that reveal structures so similar among the embryos of these species that even trained scientists would find it difficult to distinguish one from another (Figure 13.6). Even terrestrial organisms, such as man, possess in the embryonic stages what appear to be gill-like structures and a tail. Because the human embryo manifests these similarities to lower forms of life during embryogenesis, these changes are believed to reflect man's evolutionary history leading to its present form.

This has led to the coining of the phrase "ontogeny recapitulates phylogeny," meaning that the progressive changes of the embryo sum up the progressive changes that the species underwent during its evolution. Although the statement is not altogether accurate (since, for example, man is not believed to be a direct descendant of fishes), there is no doubt that some evolutionary relationship between species is reflected in these embryonic changes.

TABLE 13.1
The geologic ages

NAME OF ERA	NAME OF PERIOD	APPROX. DATE OF BEGINNING (MILLIONS OF YEARS AGO)	RELATIVE TIME	BIOLOGICAL CHARACTERISTICS
(Present)		0	Dec. 31, midnight	
Cenozoic	Quaternary	1		Appearance and dominance (?) of man.
	Tertiary	58	Dec. 27	Dominance of birds, mammals, flowering plants, insects.
Mesozoic	Cretaceous	125	Dec. 21	Reptiles dominant; toothed birds, marsupial mammals appear. Insects and flowering plants increase in number and variety. Extinction of dinosaurs.
	Jurassic	150	Dec. 19	First flowering plants. Age of giant dinosaurs. Pterodactyls, bony fish, archaeopteryx. First hardwood forests.
	Triassic	180	Dec. 17	Reptiles, conifers dominant. First mammals.
Paleozoic	Permian	215	Dec. 14	First conifers. Diversification of reptiles, insects. Extinction of trilobites. Deserts common.
	Carboniferous	260		Extensive forests of giant ferns, clubmosses, horsetails (all spore-bearers). First appearance of reptiles, seed plants, insects. Trilobites almost extinct. Horseshoe crab (still living) appears. Swamps common.
	Devonian	310	Dec. 6	First land forests—of ferns, clubmosses, horsetails. Corals, armored fish abundant. First amphibians.
	Silurian	350	Dec. 3	Definite land plants. Eurypterids and primitive fishes dominant; corals, trilobites abundant. First cephalopods. Lungfishes. Scorpions.
	Ordovician	430	Nov. 27	Land plants (?). Algae. Corals abundant. Bivalve and coiled shell mollusks. Eurypterids. First vertebrates (fishlike forms).
	Cambrian	510	Nov. 20	Algae. Trilobites dominant. Many mollusks; *Lingula.* Sponges, jellyfish.
Proterozoic		900	Oct. 20	Evidences (some disputable) of sponges, radiolaria, bacteria, algae; all water-dwelling forms.
Archeozoic		2000	July 20	Presumptive origin of life.
Azoic		4500	Jan. 1	Origin of earth.

* This table is constructed in accord with geologic convention: the oldest era is placed at the bottom, with succeedingly younger ages above. This is the order in which the sedimentary rocks are deposited. The table should be read from the bottom upward.

Dates under relative time column represent the time converted to the scale of a solar year.

Source: G. Hardin, *Biology; Its Principles and Implications,* 2d ed., W. H. Freeman and Company, San Francisco, 1952.

265
Evidence for Darwin's
Theory of Evolution

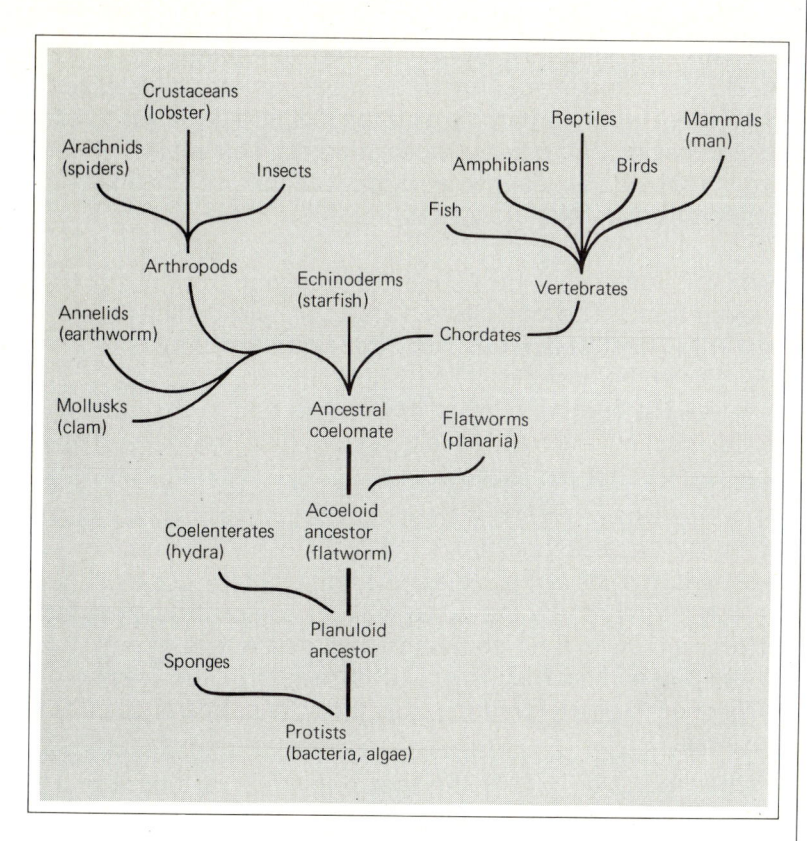

FIGURE 13.5
Evolutionary relationships in the animal kingdom. [*After Helena Curtis*, An Invitation to Biology, *Worth Publishers, New York, 1972, p. 228.*]

FIGURE 13.6
Stages in the development of different vertebrates showing similarities in early stages. [*After* Biology: Its Human Implications, *first edition, by Garret Hardin, W. H. Freeman and Company. Copyright © 1949.*]

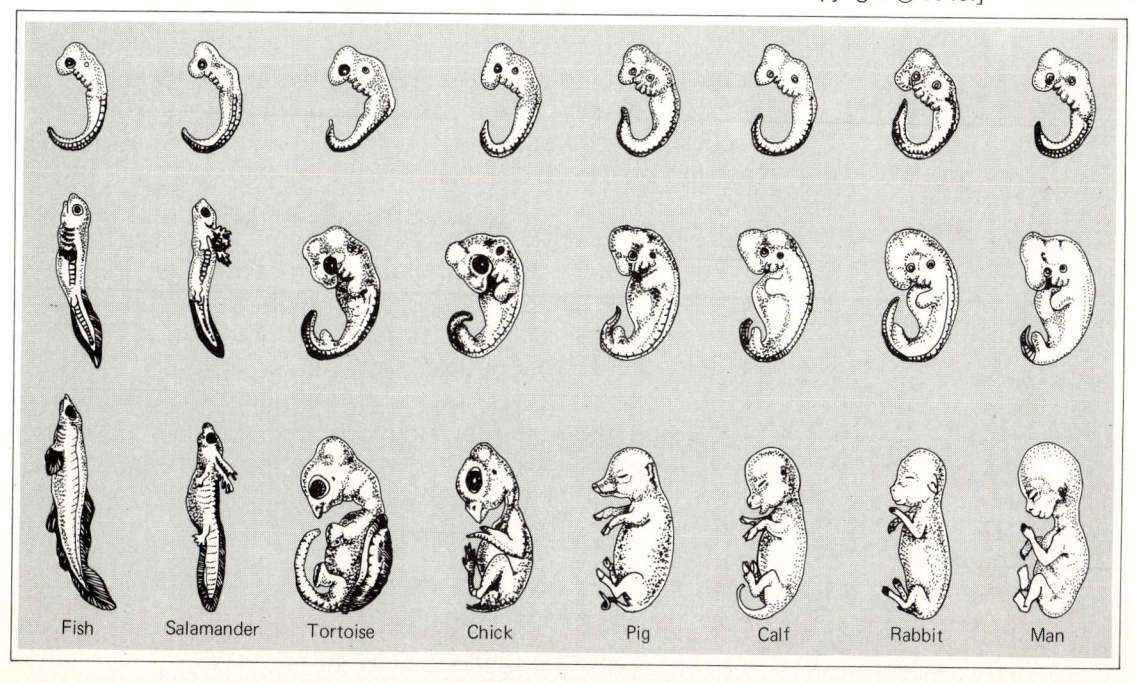

Fish   Salamander   Tortoise   Chick   Pig   Calf   Rabbit   Man

The Genetic Basis of Evolution

ADAPTATIONS IN LIVING ORGANISMS

Perhaps the strongest evidence, and certainly among the most interesting in support of the theory of natural selection, consists of the innumerable examples of structural variations and behavior patterns found in animals and plants, which are of particularly adaptive advantage to the lives, habits, and environments of the individual.

PARASITIC BIRDS One of the most interesting examples is the cuckoo bird. The cuckoo is a parasitic bird which lays its egg in the nest of another bird after first throwing out one of the eggs of the host bird. Thus, when the host bird returns to the nest, the number of eggs is the same, and further, the egg of the cuckoo frequently looks exactly like the eggs of the species which it parasitizes.

Of interest to scientists who study the behavior of parasitic birds, the baby cuckoo learns the songs of its foster parent as well as that of its own species. As adults, those cuckoos that have been reared by the same host species will be attracted to each other; in their courtship, they sing songs of both the foster parent and their own. This results in the mating of individuals that will produce eggs with the appearance of the host egg.

Another parasitic bird is the sparrow, which lays its eggs in the nest of African widow birds.* Although the adult sparrow does not look like the adult widow bird, the chicks are virtually indistinguishable from the chicks of the host, and thus will be accepted by the host as its own young (Figure 13.7).

How do we interpret these remarkable examples of mimicry in nature according to the theory of natural selection? The association between the parasite and the host species came about as a result of random mutations which resulted in eggs that looked like those of the host. Such eggs and chicks were accepted, and these then reproduced similar young. Those birds that attempted to parasitize species whose eggs and young were so dissimilar that the parasitic intruders were rejected simply did not survive to reproduce. Those that lived were, in other words, the more fit, and won the battle of the survival of the fittest.

ADAPTATION IN COLORATION Any lover of nature has marveled at the protective coloration of various species of animals and plants. One of the classic examples of coloration as an adap-

*W. Sullivan, "They're Not All in the Family," *The New York Times*, News of the Week in Review, July 30, 1972.

FIGURE 13.7
An adult widowbird (center) and its parasite (right). The chicks of the two (left) bear remarkable resemblance to each other. [*After J. Nicolai, Max Planck Institute, in* The New York Times, *News of the Week in Review, July 30, 1972.*]

tive characteristic was the experiment on moths in different environments in England, commonly known as *industrial melanism*. An Oxford biologist, H. B. Kettlewell, released equal numbers of light- and dark-colored moths in two different areas of England, one heavily industrial and the other a rural area with no industry. Over a period of time, the population of experimental moths in the industrial site was found to be predominantly dark, whereas the population in the rural area was predominantly light.

Studies of the moths revealed that the cause for this shift in the proportions of dark and light moths was the fact that in the heavily industrialized environment, most surfaces of trees were black from the soot and discharge of factories, and the dark moths blended well with the background, whereas the light-colored moths stood out and were victimized by predators. The opposite situation occurred in the

268
The Genetic Basis of Evolution

rural areas, where the dark moths were easily distinguishable against the light background of lichen-covered trees. Figure 13.8 illustrates this example of natural selection.

There are numerous other examples of protective coloration in nature which presumably arose in the same manner as seen in industrial melanism.

PHYSIOLOGICAL ADAPTATION Detailed studies of the body chemistry of animals that must adapt to extreme environments reveal physiological changes which allow the organism to survive where others cannot. For example, studies have been made of animals native to desert regions, such as the desert rat and the camel (Figure 13.9). It has long been

FIGURE 13.8
White and black moths on lichen-covered trees in rural areas (left) and soot-blackened trees in industrial regions (right). [*Photo on the left from Heather Angel, MSc. FRPS; photo on the right courtesy of the American Museum of Natural History.*]

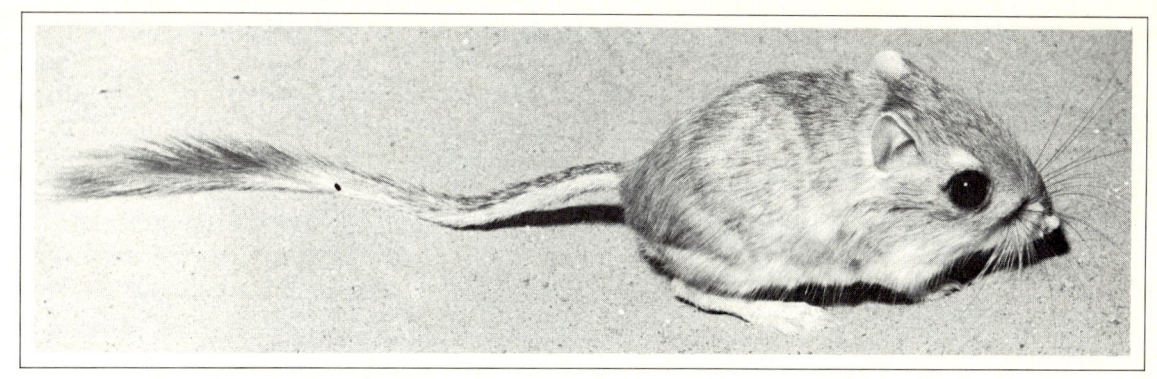

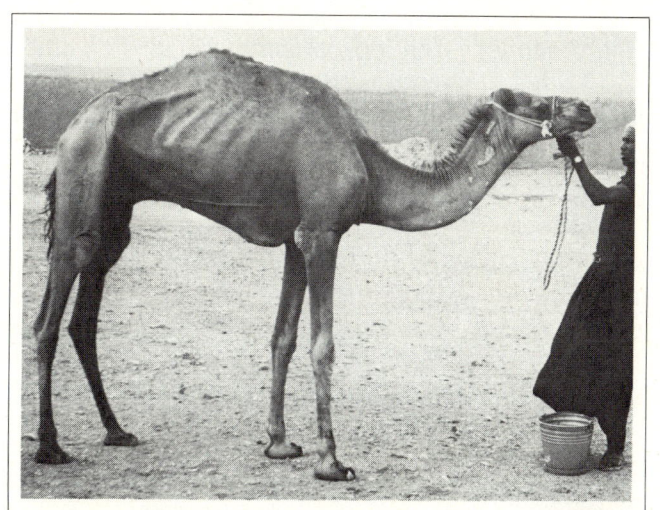

FIGURE 13.9
(A) The kangaroo rat. (B) Two photographs of a camel taken ten minutes apart, while it is in a state of extreme dehydration (top) that would be fatal to man, and after it has been watered (bottom). Both animals are physiologically well adapted to hot, arid environments. [*Photo of kangaroo rat from Russ Kinne Photo Researchers; photos of camel courtesy of Dr. Knut Schmidt-Nielsen, Duke University.*]

known that these animals subsist on far less water than other related species. Physiologist Knut Schmidt-Nielsen has found that the desert rat can exist in arid environments partly because its kidneys are far more efficient than man's; it is able to excrete urine containing such low amounts of water as to be almost solid. One of the ways the camel conserves water is by tolerating increases in body temperature up to 105° before it begins to sweat, whereas a man in the same environment would lose water by sweating continuously.

Again, such adaptations have arisen not in response to environmental conditions but through chance mutations which have allowed certain organisms to thrive in places where others cannot. The lack of competition with other animals also serves to maintain these species.

CHROMOSOMAL EVOLUTION

If evolution occurs as a result of gradual genetic change, it stands to reason that closely related species of living orga-

FIGURE 13.10
Illustration of the different combinations of chromosome arms in various species of *Drosophila* which prevent breeding between the different groups. [*After E. M. Sinnott, L. C. Dunn, and T. Dobzhansky, Principles of Genetics. Copyright 1958 by McGraw-Hill, Inc. Used with permission of McGraw-Hill Book Company.*]

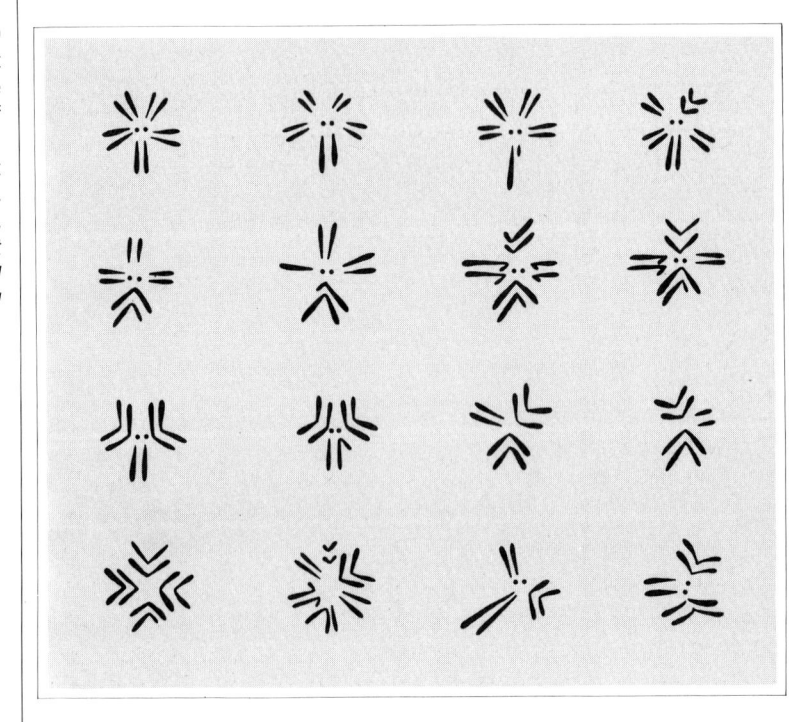

nisms should manifest some divergence in genetic constitution that would indicate their evolutionary relationship. This divergence has been found in the chromosomes of various species of the fruit fly Drosophila.

Drosophila karyotypes are, of course, especially well suited to studies of chromosomal differences because, as mentioned earlier, the giant salivary gland polytene chromosomes show banding patterns that allow their detailed characterization. Although the various species of fruit flies may look similar to the lay observer, they cannot interbreed, and a study of the chromosomes reveals gross differences among the species. Figure 13.10 shows how arms of chromosomes have combined in various ways in the different species.

INTRACHROMOSOMAL DIVERGENCE

Before two populations diverge to the point of not being able to interbreed as a result of gross chromosomal changes, one would expect that evidence for less extreme genetic divergence would exist; Drosophila geneticists have, in fact, found this evidence. Figure 13.11 shows chromosomes of different species in which inversions have been found, although the gross appearance of the chromosomes appears to be similar. Presumably these genetic changes caused the inability of the different populations of Drosophila to interbreed, or what is known as *reproductive isolation*. These species can be expected to continue to diverge through further mutations. Different gene arrangements do in fact have differential adaptive advantages, as has been found in populations which inhabit the same geographical territory. There are fluctuations in the percentages of populations carrying different gene arrangements, depending on the season (Figure 13.12).

With new staining techniques for chromosomes of vertebrates, similar studies on related species are expected to provide evidence for evolutionary relationships. For example, the location of areas of repetitive DNA in related species of vertebrates has been found to be quite similar (Figure 13.13). Recently, K. W. Jones of Edinburgh University, using techniques which cause homologous DNA to bind together, found that the position of DNA in chimpanzee and orangutan which binds to human satellite DNA is remarkably

FIGURE 13.11
Various chromosome complements of species of the *virilis* group of *Drosophila* showing inversions in some of the chromosomes. [After E. M. Sinnott, L. C. Dunn, and T. Dobzhansky, Principles of Genetics. Copyright 1958 by McGraw-Hill, Inc. Used with permission of McGraw-Hill Book Company.]

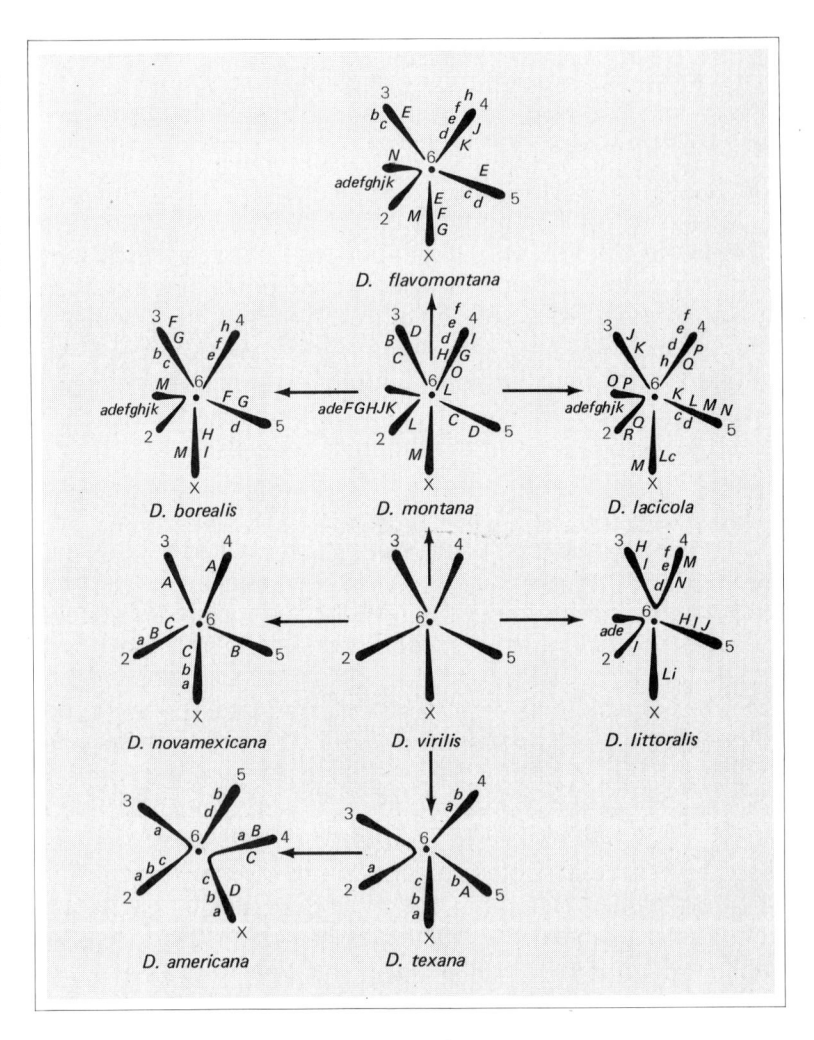

FIGURE 13.12
Seasonal changes in the frequencies of third chromosomes with two different gene arrangements (*ST* and *CH*) in a population of *Drosophila pseudoobscura* from a certain locality in California. [After T. Dobzhansky, Genetics of the Evolutionary Process, Columbia University Press, New York, 1972, p. 135.]

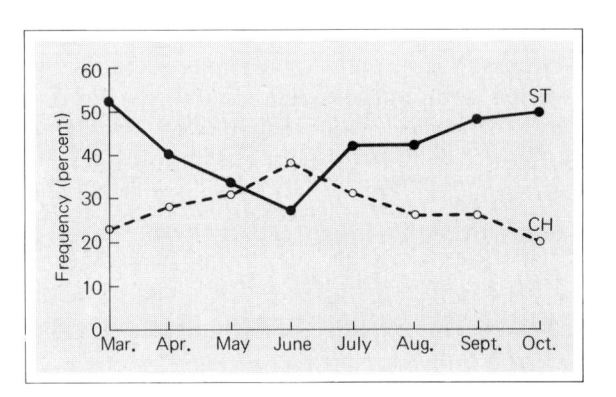

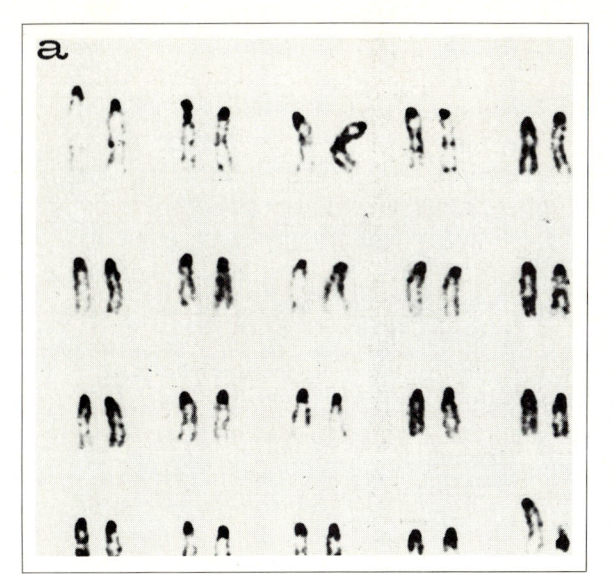

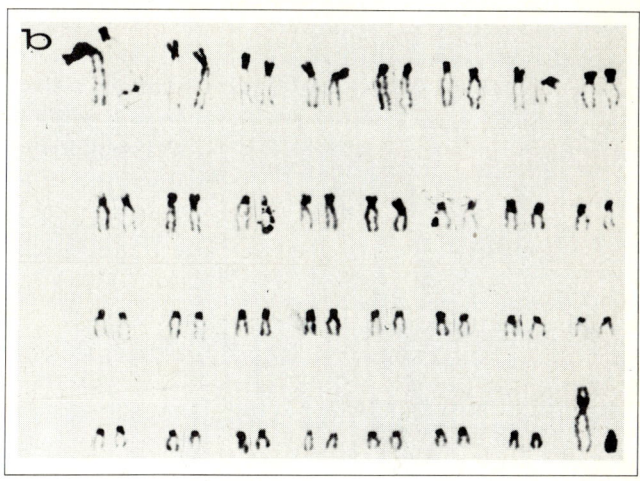

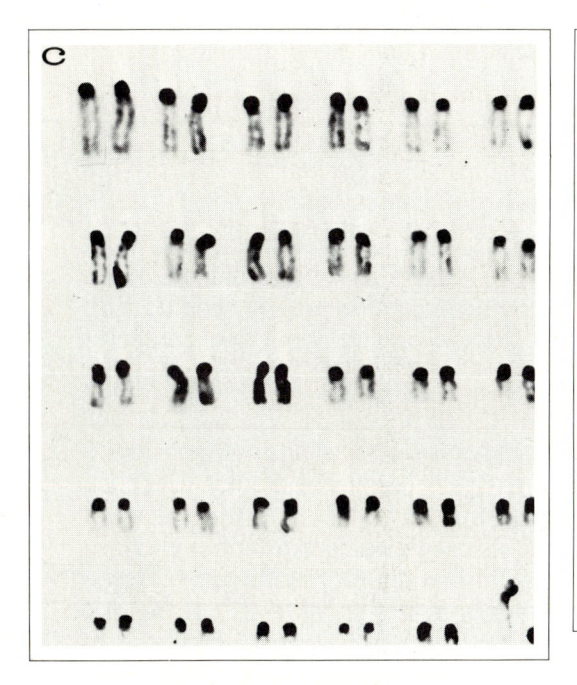

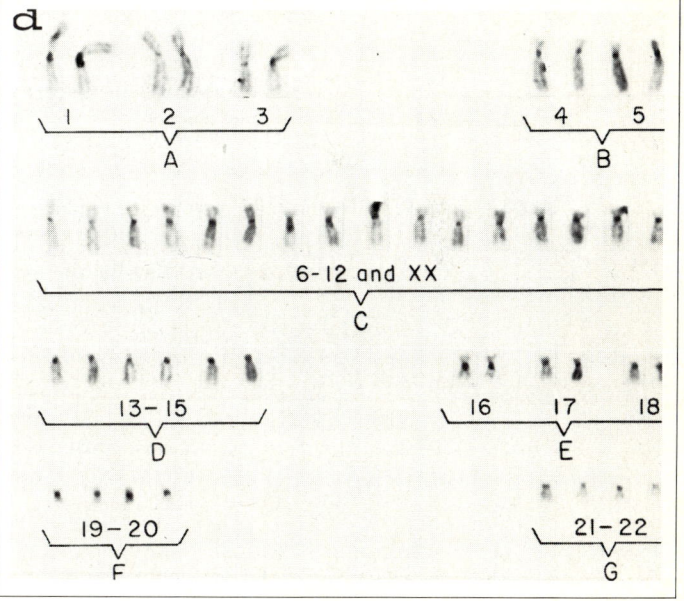

FIGURE 13.13
Position of repetitive DNA in chromosomes of (A) mouse, (B) guinea pig, (C) calf, and (D) man, showing similarities of position. [From J. J. Yunis and W. G. Yasmineh, "Heterochromatin, Satellite DNA, and Cell Function," Science, **174**:1203, 1971. Copyright 1971 by the American Association for the Advancement of Science upon publication.]

similar to the position of these DNA molecules in human chromosomes (Figure 13.14).

MOLECULAR EVOLUTION

Although detecting gene rearrangements in higher organisms is still a technological problem, present methods allow the breakdown of protein chains into component amino acids. These breakdowns reveal evolutionary divergence on the molecular level. The more alike species are, generally, the more similar is the molecular structure of certain proteins. Conversely, as expected, the more dissimilar organisms are, the greater the differences in molecular structure of these proteins. Hemoglobin in man and the chimpanzee, for example, are identical. In gorilla hemoglobin there is only one site that differs from that of man. The rate of evolution of different proteins, including hemoglobin, has been estimated as shown in Figure 13.15.

Actually, the universality of the genetic code and the existence of biochemical pathways, or identical sequences of biochemical reactions that exist in many diverse species, all are new lines of evidence pointing to the evolutionary relationships between all forms of living organisms.

FACTORS INFLUENCING THE COURSE OF EVOLUTION

Because of these different lines of supporting evidence, Darwin's theory of evolution based on natural selection has been widely accepted in the scientific world. Let us turn briefly to some of the factors which influence the course of evolution.

MUTATION

Since it is the ultimate source of variability, mutation is the basic mechanism by which natural selection acts. Because of the importance of mutation to neo-Darwinism, we must once again emphasize that mutation is a random process and occurs rarely. In addition, most mutations appear to be recessive and detrimental. They are considered detrimental mainly because the characteristics of present organisms are presumed to be advantageous, since they have been retained through natural selection. Therefore, any change from the "norm" would tend to be a change for the worse, that is, toward some selective disadvantage.

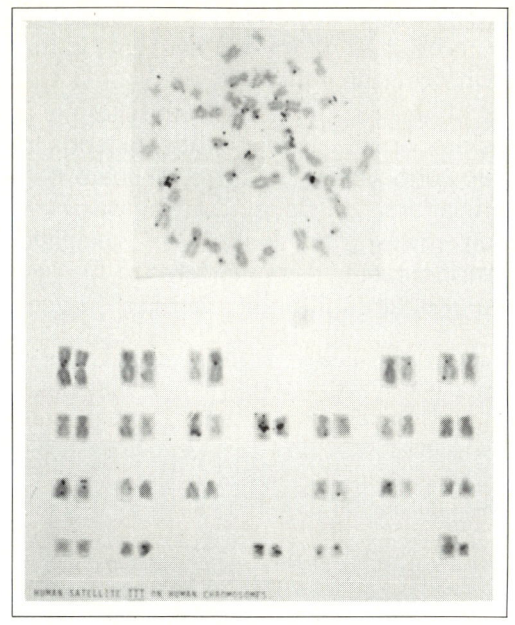

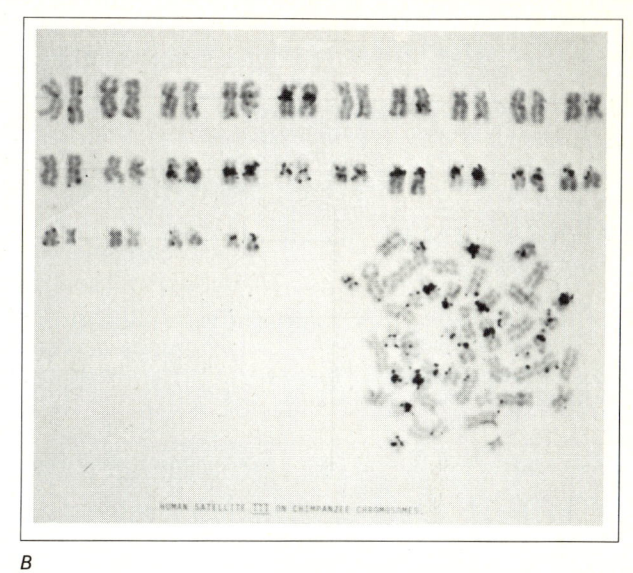

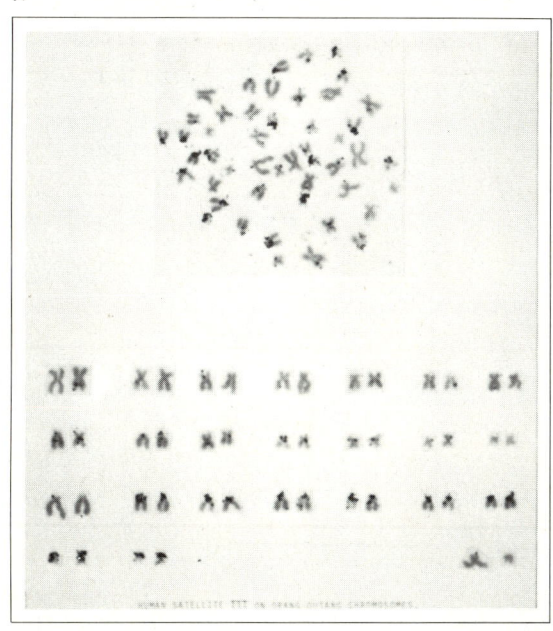

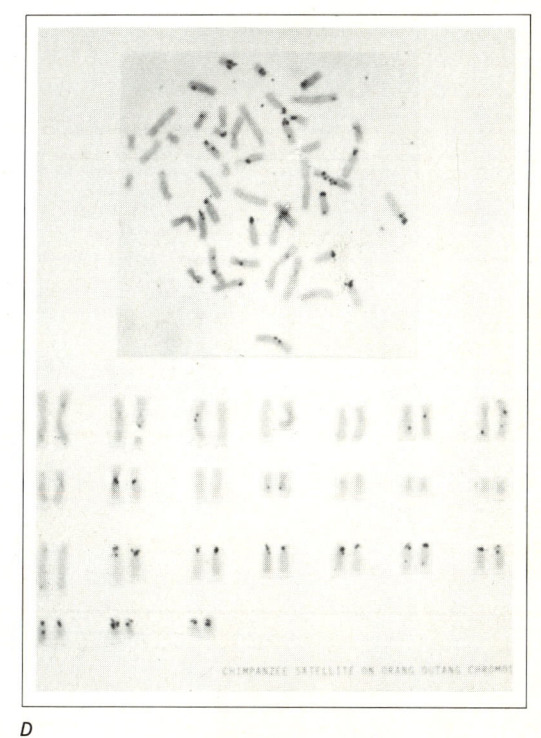

275
Factors Influencing
the Course of Evolution

FIGURE 13.14
(A) Satellite DNA on human chromosomes; (B) human satellite DNA on chimpanzee chromosomes; (C) human satellite DNA on orangutan chromosomes; (D) chimpanzee satellite DNA on orangutan chromosomes. The existence of homologous DNA and their similar positions illustrates evolutionary relationships between these species. [*Courtesy of Dr. Kenneth W. Jones, Edinburgh University.*]

ENVIRONMENT

Environment affects the course of evolution because, regardless of what mutation occurs, it is the environment that in the long run determines whether the mutation will benefit the organism. The effect of the environment necessitating a struggle for survival (as first proposed by Malthus) extends to all levels of the natural world and results in a delicate balance of life in geographical areas. The interplay between living organisms and their environment forms what biologists call *ecosystems*, and an increasingly popular area of biology which studies factors affecting different aspects of ecosystems is *ecology*.

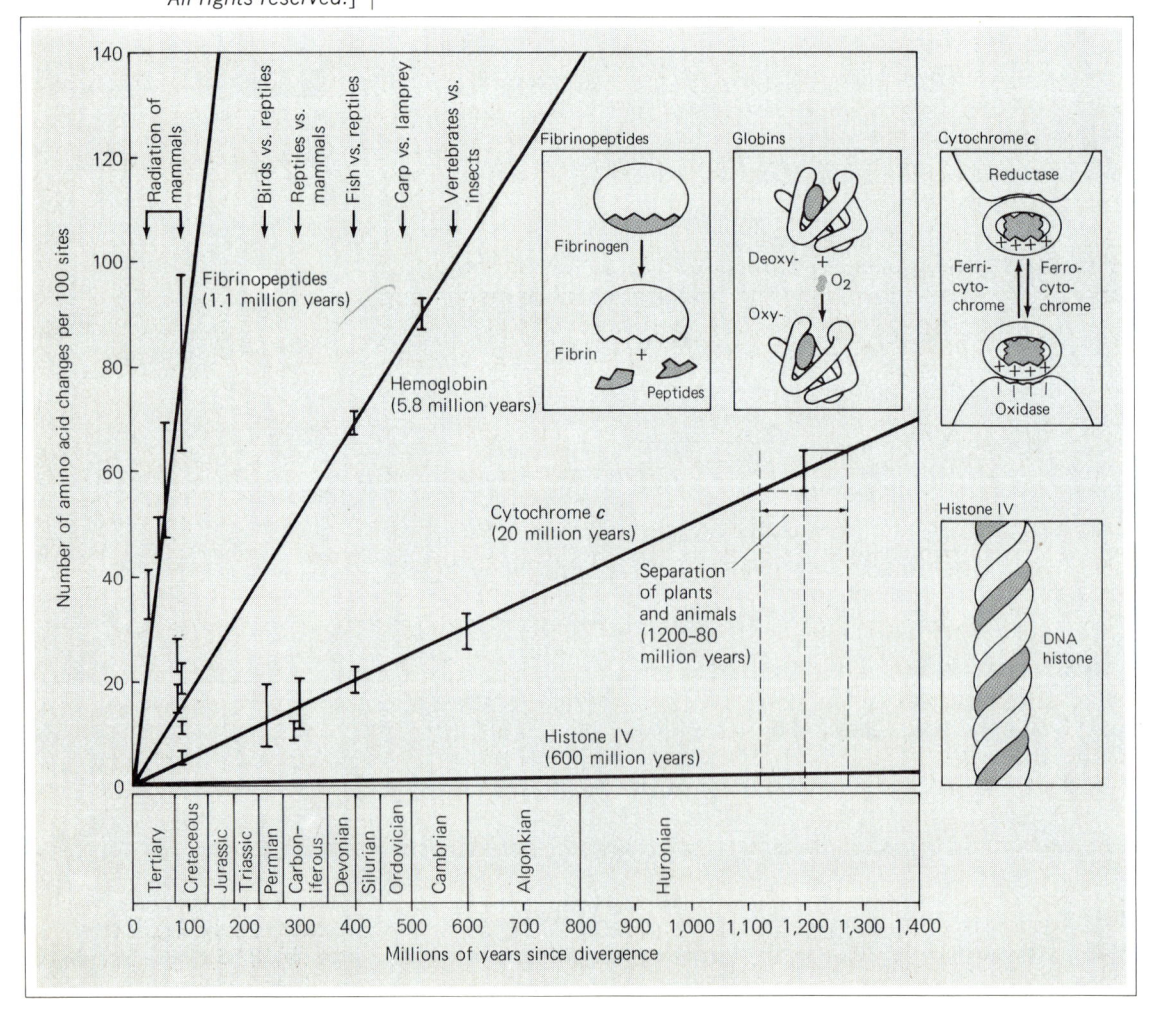

FIGURE 13.15
The rates of evolution of protein as determined by plotting the average amino acid difference between two species on two sides of an evolutionary branch point. (Drawings illustrate the biological function of the proteins.) The rate of change is proportional to the steepness of the curve. [After "The Structure and History of an Ancient Protein," by Richard Dickerson. Copyright © 1972 by Scientific American, Inc., **226**:68. All rights reserved.]

Occasionally a species will develop traits that allow it to survive in an environment uninhabitable to other forms of life; this environment is known as its *ecological niche*. This adaptive process is usually followed by a burst of reproduction, which is generally referred to as a *population explosion* because of the absence of competition and predation.

One modern example of this process, which has caused considerable distress to many areas of the United States, is the infestation of the gypsy moth. Originating in Europe, Asia, and North Africa, the gypsy moth was brought to the United States in 1869 for experimentation as a source of silk. Some moths escaped, and because they are not indigenous to the United States and therefore had no predators against them here, within 20 years they had become unmanageable in parts of New England, destroying forests by defoliation (Figure 13.16). Today they are found as far south as Florida and as far west as Wisconsin. In 1970, they defoliated 800,000 acres and in 1971 destroyed 1.9 million acres more. Although pesticides have been widely used against the moths, they continue to increase in number and to spread. With the banning of DDT, current efforts center on biological means of control, including the importation of parasites and predators of the moth from their native environments, a practical application of natural selection and the principle of survival of the fittest.

SEXUAL REPRODUCTION

One other factor which can affect the course of evolution is sexual reproduction. Because adaptability of a species depends on variability, any biological process leading to variability is important for evolutionary change. Sexual reproduction is a source of variability since meiosis ensures random assortments of genes in the gametes. In addition, crossing-over contributes to variability, and with random fertilization of one gamete by another, the great numbers of gene combinations in offspring of sexually reproducing organisms is of great advantage from the standpoint of survival.

SOME UNANSWERED QUESTIONS

Even though natural selection is recognized as the paramount factor in the diversification of organisms, some questions still remain to be answered. One problem difficult to

278
The Genetic Basis of Evolution

FIGURE 13.16
(A) Aerial view of a forest in northwestern New Jersey in which various species of oak trees have been totally defoliated by gypsy moths. Other less preferred trees and shrubs which have not been defoliated can be seen interspersed through the forest (B). A ground-level picture of the same forest, which resembles a winter landscape; in fact, the picture was taken in the month of July. (C) A cluster of female gypsy moths laying masses of eggs. The absence of natural predators allowed the moth population to undergo an explosion that has proved devastating for forests in the United States. [*Photographs courtesy of Dr. George Luedemann.*]

A

B

C

resolve on the basis of natural selection is the existence of complex structures, such as organs found in highly developed organisms. Let us take the eye of a vertebrate for an example.

If evolutionary change is as gradual as we assume it must be, based on fossil evidence and our knowledge of the rarity of mutational events, it becomes difficult to envision how the eye as a whole evolved. We know that the eye does not work properly unless all of its many parts are present and functioning. Yet these many essential components could not have appeared simultaneously. If they did not, how functional could the eye have been? And if not functional, then by our theory of evolution it should not have been retained in the population.

Indeed we have examples of structures which have been rendered less important because of decreasing need for them as the species evolves. They are known as *vestigial organs.* In man, wisdom teeth and the appendix are vestigial structures.

Along a different line, those who support creationism point to discontinuities in fossil records. For example, no adequate transitional forms between fish and amphibians have been found that would reflect the gradual change from aquatic to terrestrial environments.*

Another problem arises from the existence of traits which do not appear to have any selective advantage in nature. A good example of this is the existence of the different blood groups, which seem to have biological significance only if there is a transfusion and mixing of blood from different individuals. Since transfusions obviously never occur in nature, it is curious that not only does polymorphism exist, but also the ABO alleles differ in frequency in different parts of the world (Figure 13.17). The existence of so much genetic polymorphism is perplexing from the point of view of natural selection (see again the comment by Neel and Schull, page 251).

These difficulties, however, have not caused scientists to abandon the neo-Darwinian concept of evolution. Much is not understood about natural phenomena, but this is not sufficient reason to render invalid concepts that have much support in other biological phenomena. It is more likely that

* D. T. Gish, "Creation, Evolution, and the Historical Evidence," *The American Biology Teacher,* **35**:132–140, 1973.

FIGURE 13.17
The frequencies of (A) blood group A and (B) blood group B in different parts of the world. [*After E. M. Sinnott, L. C. Dunn, and T. Dobzhansky,* Principles of Genetics. *Copyright 1958 by McGraw-Hill, Inc. Used with permission of McGraw-Hill Book Company.*]

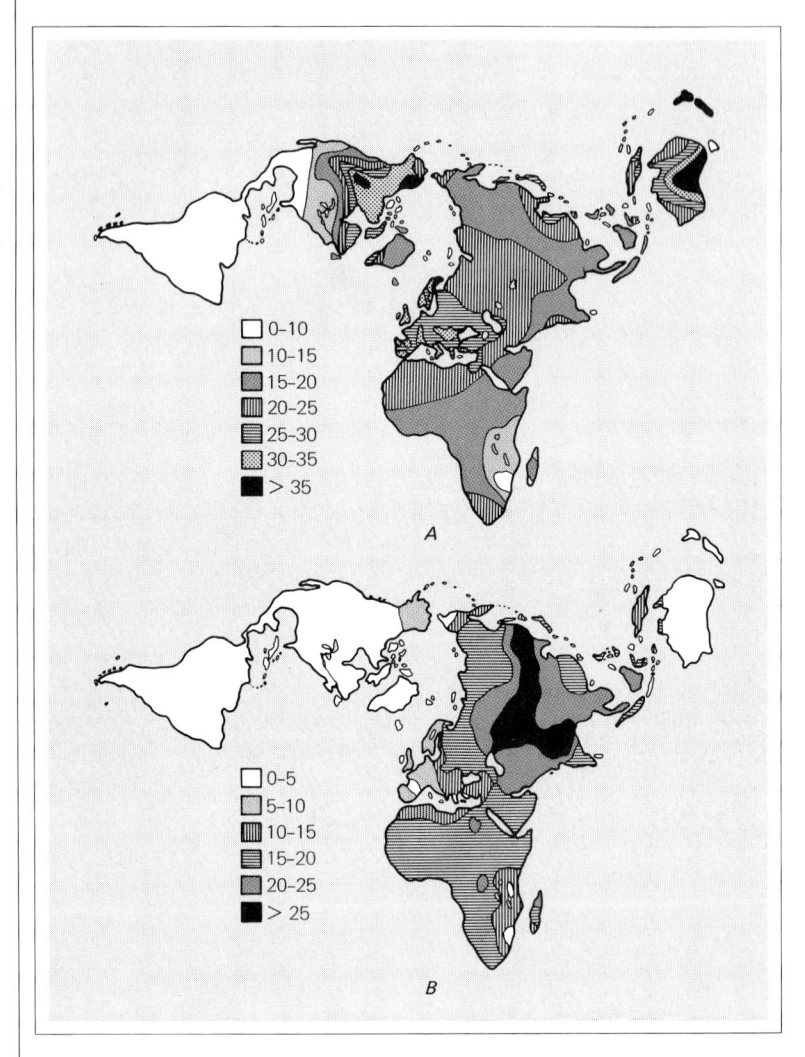

these problems will eventually be found to have some basis in natural selection. For example, there is evidence that type O blood is related to a higher incidence of ulcers, as shown in Table 13.2.

Knowing the genetic basis of evolution, it is now appropriate to discuss the application of these concepts to man, who, as a biological organism, is as subject to the pressures of selec-

TABLE 13.2
Relationship between ABO blood groups and ulcer

	PERCENTAGE OF PATIENTS IN BLOOD GROUPS			
	O	A	B	AB
Duodenal ulcer	58.7	30.6	7.2	3.5
Gastric ulcer	49.1	40.0	8.2	2.7

Source: T. Dobzhansky, *Genetics of the Evolutionary Process*, Columbia University Press, New York, 1972, p. 279.

tion and other evolutionary factors as is any other organism on earth. We shall devote the next chapter to the implications of evolutionary change for man as a living species.

REFERENCES

Darwin's works have been published as one volume:
Darwin, C.: *The Origin of Species* and *The Descent of Man,* Modern Library, Inc., New York.

The fundamental concepts of evolution are discussed in every general biology text and many general genetics texts. One that can be recommended for the layman:
Lerner, I. M., 1968: *Heredity, Evolution and Society,* W. H. Freeman and Company, San Francisco.

Some interesting references on adaptations in nature:
Rothschild, M., 1967: "Mimicry," *Natural History,* **76**:44–51.
Sullivan, W., July 30, 1972: "They're Not All in the Family," *The New York Times.*
Vertebrate Adaptations, 1968: (A collection of articles from *Scientific American*), W. H. Freeman and Company, San Francisco. This includes articles on the desert rat and the camel: Schmidt-Nielsen, K., and B. Schmidt-Nielsen, 1953: "The Desert Rat," pp. 189–191; Schmidt-Nielsen, K., 1959: "The Physiology of the Camel," pp. 193–198.

A technical discussion on various genetic aspects of evolution can be found in:
Dobzhansky, T., 1972: *Genetics of the Evolutionary Process,* Columbia University Press, New York.

An article presenting the creationist theory of evolution:
Gish, D. T., 1973: "Creation, Evolution, and the Historical Evidence," *The American Biology Teacher,* **35**:132–140.

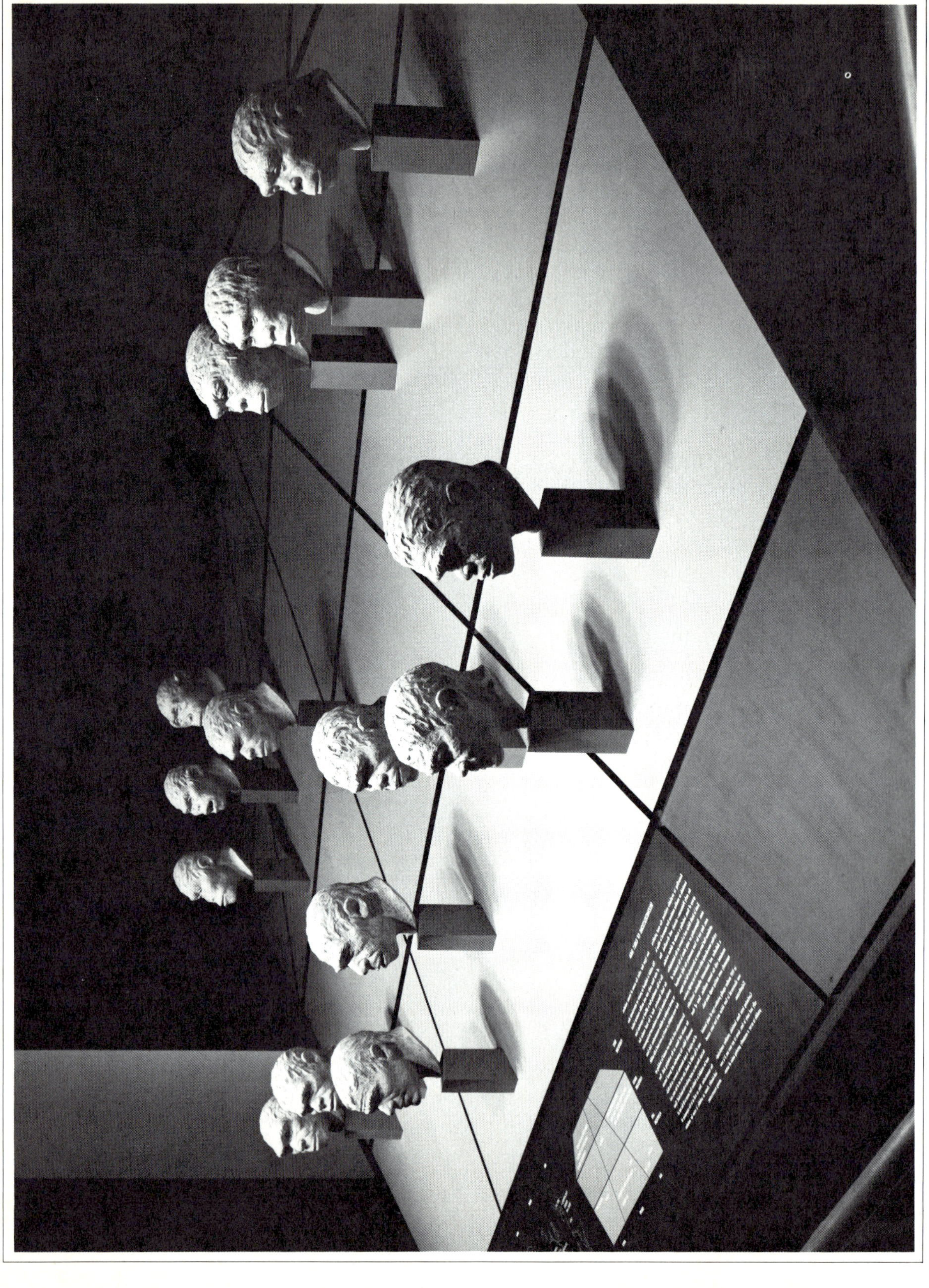

Chapter 14

Man and Evolution

We may infer from the theory of natural selection that man, as a biological organism, arose in the same manner through evolution as any of those species which we still refer to in an antiquated anthropocentric sense as "lower": namely, by chance. Although man's intelligence renders him unique in many ways, every living species is unique in its own way, having survived through adaptation to its environment. As the Nobel Prize winner P. B. Medawar has written, " . . . far from being one of his higher or nobler qualities, his individuality shows man nearer kin to mice and goldfish than the angels; it is not his individuality but only his awareness of it that sets man apart."*

* P. B. Medawar, *The Uniqueness of the Individual,* Basic Books, Inc., New York, 1957, p. 185.

RESISTANCE TO DARWINISM

The concept of man evolving by chance was and is, to many, difficult to accept, especially for those who espouse fundamentalist religions that declare man to be a special creation of a supernatural being. Ernst Mayr† has noted six major elements in the Darwinian revolution, several of which directly contradict religious dogma:

† Ernst Mayr, "The Nature of the Darwinian Revolution," *Science,* **176**:981–998, 1972.

The evolution of man. The groups of evolutionary antecedents depicted are from: Asia (upper left), Australia (upper right), Africa (lower right), and Europe (lower left). [*Courtesy of the American Museum of Natural History.*]

1 Discovery that the earth is millions of years old.

2 Refutation of catastrophism, and replacement of the idea that different life forms arose after catastrophic upheavals (such as the Flood) that destroyed all existing life by the concept of a gradual, steadily evolving world.

3 Refutation of the concept of automatic evolution to "higher" forms and the notion that the later-evolving, more complex species are more perfect than earlier, simpler forms, man of course being the epitome of evolution.

4 Rejection of creationism, requiring a "new concept of God and a new basis for religion."

5 Refutation of essentialism, the philosophy that beneath the variability of the world, the only things that are real and permanent are the "fixed, unchangeable 'ideas' underlying the observed variability," which implies an immutability of living organisms.

6 Abolition of anthropocentrism and acceptance of the non-exceptional nature of the origin of man through evolution.

THE "MONKEY TRIAL"

Many of these new concepts were unacceptable to large numbers of people. Their resistance took the form of laws forbidding the teaching of Darwinism in public schools in many states in the United States. One of the most famous controversies that arose as a result of this attempt to legislate a scientific theory out of existence was the so-called Monkey Trial in Tennessee in 1925 of John Scopes, a biology teacher who discussed Darwin's theories in his biology classes (Figure 14.1). Although ably defended by Clarence Darrow against the prosecutor William Jennings Bryan, Scopes was convicted for violating the state law, a conviction which was nullified by higher courts in 1927. In many states, however, anti-Darwinian laws have persisted. As recently as 1971 Mississippi still prohibited the teaching of modern concepts of evolution in its schools, although the law is no longer in the state code.

THE SCIENCE FRAMEWORK FOR CALIFORNIA PUBLIC SCHOOLS

In 1967, a "Science Framework for California Public Schools" was developed in which guidelines were set for the teaching of science in primary and secondary schools. When the final

FIGURE 14.1
John Scopes (seated in center with arms folded) and his attorney, Clarence Darrow, listening during a moment of the famed "monkey trial" in 1925. Scopes was convicted for teaching the Darwinian theory of evolution in Tennessee. [*Courtesy of the Bettmann Archives.*]

draft was submitted to the California State Board of Education in 1969, the proposed guidelines were not immediately approved.

The Board of Education made some changes over the strenuous objections of the educators who had drawn up the original framework, and the revised copy was adopted. The revision insisted on the use of textbooks which include a creationism theory of evolution as well as the theory of natural selection and survival of the fittest.

As some biologists have noted,* the impact of this change could be significant, since California buys 10 percent of all textbooks published in the United States. Publishers will no doubt revise their texts according to the Science Framework, and these texts will be used throughout the country.

Many laymen have also criticized the Board's action. A resolution against the revised framework was submitted on behalf of 45,000 Presbyterians, stating, "The Presbytery of San

* J. P. Lightner, "Evolutionary Theory and the California Science Framework," *NABT News and Views,* **XVI**(5):2, 1972.

Francisco regrets the action of the State Board of Education. ... As Christians we find no conflict between the emphasis in Genesis that God is the creator and the objective findings of science that creation has occurred by means of an unimaginably long and complex evolution process."*

Early in 1973, the California State Board of Education compromised by requiring the inclusion of creationism theories as well as natural selection in social science textbooks but not in biology textbooks. However, a seven-man editorial committee was appointed to oversee the manner of presentation of subjects, including evolution, in the biology texts. Not one member of this committee is a biologist.†

* Ibid.

† J. P. Lightner, "Update on California Science Textbook Controversy," *NABT News and Views,* **XVI**(7):7, 1973.

REFUTATION BY RELIGIONS

Many organized religions still resist the theory of natural selection as the prime force in evolutionary change and the formation of species. For example, the Watchtower Bible and Tract Society, which is affiliated with Jehovah's Witnesses has published a book attacking modern evolutionary concepts. Some 1½ million people are reported to be actively preaching the tenets of this religion, and more than twice this number attended their meetings all over the world in 1971.

LAMARCKISM

One alternative theory which has been seized upon by some who doubt the chance nature of evolutionary change is the *theory of acquired characteristics,* sometimes referred to as *Lamarckism.* Named after its originator, the naturalist Jean Baptiste Lamarck, the theory, which was proposed in 1800, states that organisms can react to environmental changes in such a way that the adaptations can be transmitted to future generations.

The classic example used to illustrate this theory is the manner in which giraffes might have developed their long necks. Because of their eating habits, which include taking leaves from trees, giraffes were thought to extend their necks to reach higher and higher, especially in times of drought when the leaves were scarce. This stretched condition was transmitted to the next generation, which would have slightly longer necks that would, in turn, stretch more. Over many years, the increase in neck length eventually culminated in the unusual anatomy of that species (Figure 14.2).

Lamarck's hypothesis was formulated before the modern era of biology. The cell theory was not in existence, and it was to be another century before the significance of genes and chromosomes was recognized. So it is not surprising that a theory which is clearly untenable to modern science was drawn up at the time. This theory has been discarded because, if it were true, then people who have had limbs amputated would produce offspring with abnormally short limbs.

FIGURE 14.2
Theories of how the giraffe evolved its long neck, from the Lamarckian and neo-Darwinian points of view.

LYSENKO AND LAMARCKISM What *is* surprising is that as late as the mid-twentieth century, one of the world's great powers espoused Lamarckism as a national scientific policy. This was the Soviet Union. In the first three decades of this century, Russian scientists were contributing strongly to advances in the field of genetics under the direction of a botanist, N. I. Vavilov. However, through political finagling, Trofim D. Lysenko, described by I. M. Lerner as a "fanatical charlatan,"* rose to power as the scientific dictator, and Vavilov fell into disfavor with the Stalinist regime.

Lysenko openly favored Lamarckism as being the true basis for evolution. Unacceptable to Lysenko and others was the chance aspect of natural selection, which they felt was a racist concept and degrading to man. Their displeasure extended to Mendel as well, and the Mendelian laws were declared to be untrue. To the Soviet authoritarian regime, the idea of being able to control the development of plants and animals was much more acceptable. An entire area of science was suppressed for 20 years. Vavilov and other geneticists were arrested, imprisoned, and finally perished. All existing genetic experimentation was destroyed, and references to Mendel and Darwin were removed from textbooks. Fraudulent experiments were published, such as those claiming to have transformed one species into another at will.

It was not until Khrushchev was deposed in 1965 that Lysenko and his policies were overturned, and Russia joined the rest of the world in twentieth-century genetics. By this time, Russian agriculture was in chaos because of the peculiar experiments which had been carried out in an attempt to implement Lamarckism and influence the development of crops and animals through the manipulation of the environment rather than breeding experiments based on Mendelian genetics.† It is ironic that one scientific concept should be so vehemently opposed by both organized religions and an atheistic system.

* I. M. Lerner, *Heredity, Evolution and Society*, W. H. Freeman Company, San Francisco, 1968, p. 277.

† Z. A. Medvedev, *The Rise and Fall of T. D. Lysenko*, Columbia University Press, New York, 1969.

ORIGINS OF MAN

Information obtained by paleontologists and anthropologists, who have studied fossils and remains of ancient communi-

FIGURE 14.3
Proconsul, believed by some to be the common ancestor of man and the apes. [*Courtesy of the British Museum of Natural History.*]

ties, has allowed scientists to trace what is believed to be the pedigree of the human species as it evolved. It includes the idea that man is related to apes which exist today, though not directly. It is believed that a common ancestor known as proconsul gave rise to the two divergent lines of descent (Figure 14.3). Through the forces affecting natural selection which were discussed in Chapter 13, man subsequently evolved into the present species, Homo sapiens, with its own distinguishing traits.

Man is a biological organism subject to natural selection and other evolutionary pressures, like any other living organism. However, our intelligence has allowed us to compensate for

MAN AND HIS ENVIRONMENT

certain disadvantages that we have when compared to many other species—such as relative physical weakness, lack of natural protective covering, and slowness of foot—by erecting shelters and developing weapons. In short, we can do what no other species has been able to do: alter our environment to suit our purposes. The ability to do so has led to our supreme position in the natural world and to our somewhat premature and smug attitude of superiority over the forces of nature. This has proven to be a mistake from two main points of view.

MAN'S EFFECT ON ECOSYSTEMS

First of all, only recently have we begun to realize that we may not have the knowledge necessary to control the balance of life in the delicate manner that occurs in nature. Belatedly we are trying, for example, to save many species of animals threatened with extinction, having previously altered their environment to the point where it is no longer fit to support them. Wildlife management programs have been set up in many parts of the world, and it is a fact that wherever man has tried to create a balance of life, an ecosystem, he has usually caused an imbalance.

The altered balance of nature has created a favorable environment for species which have undergone a population explosion, threatening the existence of other species. Only when man has allowed the forces of natural selection and survival of the fittest to work without interference have ecosystems achieved new balances, flourished, and remained in equilibrium.

MAN AS A BIOLOGICAL ORGANISM

Our second mistake has been in assuming that man is exempt from the problems of survival facing other living organisms. To geneticists, our environmental problems pose the question of the survival of not only various species of wildlife but of our own species as well. Simply put, our manipulation of our own environment (discussed in detail in Chapter 15), such as changing the chemical composition of the air we breathe, could lead to the same end as that of other species which we have shot or poisoned out of existence: namely, our own extinction.

In the past, we have abused natural resources as if there were no end to their supply. We have taken for granted that fresh air will always be there for us to breathe and that pure streams and an abundance of fuel for our comfort would never be exhausted. And now our air is filled with acrid particles (Figure 14.4), our streams are fouled by detergents, and we are faced with an energy crisis.

Whether our abuse of the environment is a reversible process remains to be seen. What *is* definite is that if the evolutionary destiny of a species is extinction, once reached, there is no recourse. Evolution simply takes too long, and the environment constantly changes so that the conditions leading to the appearance of a particular species are unlikely ever to exist again.

Should the day come when our environment has changed so drastically that even our intelligence will not allow us to compensate, then we as a species face extinction just as surely as any of the species we have destroyed. In fact, because of our long generation span and relatively small numbers of progeny, man is actually at a disadvantage in terms of natu-

THE EXTINCTION OF MAN?

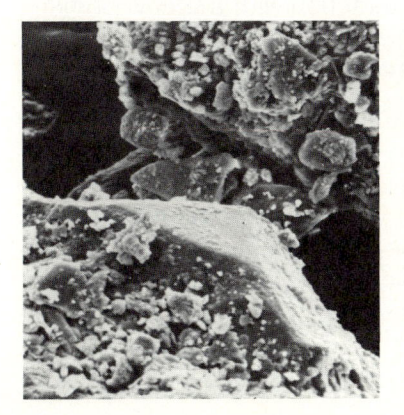

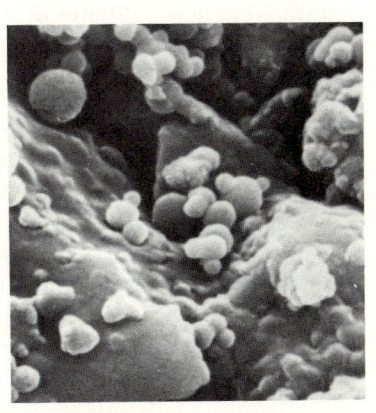

FIGURE 14.4
Electron micrographs of highly magnified particles of mineral flyash, concrete dust, glass, paper fibers, and other components found in city air. [*Courtesy of* The Particles Atlas.]

ral selection. It is far more likely that populations of insects, which breed in days and by the hundreds, would have the variability in their populations to carry chance mutations causing them to adapt to any change in environment. And of course, this is even more true of microorganisms which breed in minutes and exist by the millions in the space of a test tube.

Because of man's uniqueness, not only in being the only species to intentionally affect the course of evolution but also in having developed different modes of living which we call cultures, there is an aspect of man's evolution which has played an extremely important role not found in any other species. Evolutionary changes occur gradually in all living organisms, leading first to the formation of subgroups within the species and finally to the formation of the species with reproductive isolation. However, the subgroups in man are the only ones which have intentionally used these differences as reason for abuse and violence against each other. The subgroups in man are generally referred to as *races*.

THE FORMATION OF HUMAN RACES

How did different races appear in man? The major factors leading to all evolutionary changes are the same ones involved in the formation of human races: (1) variation in populations as a result of chance mutation, (2) selection of advantageous traits as a result of interaction with the environment, and (3) isolation, resulting in traits being transmitted within groups, causing them to diverge from other groups.

If we concentrate on pigmentation of the skin as a criterion to divide humans into races, we can illustrate the process by which different races, such as the black and the white, may have arisen. As discussed by Goldsby,* the pigmentation present in skin determines the amount of sunlight that is absorbed. This, in turn, determines the amount of vitamin D to be synthesized by cells just under the skin, since the synthesis of this substance depends on ultraviolet rays of the sun.

Vitamin D, as you may know, is important for the use of calcium by the body. A deficiency of vitamin D results in a

* R. A. Goldsby, *Race and Races,* The Macmillan Company, New York, 1971.

deficiency of calcium deposits in bones, which then soften; the disease that results is known as *rickets*. On the other hand, too much vitamin D causes an excess of calcium deposits in bones, which then become brittle and subject to breakage. Either condition, if severe enough, could be fatal.

Because of its role in vitamin D production, the amount of sunlight in the environment is believed to have played an important role in the formation of darkly pigmented races in tropical areas of the world, such as Africa, where sunlight is most intense, and of lightly pigmented groups where sunlight is less intense, such as in northern Europe. In other words, these differences in degree of pigmentation arose as a result of chance mutation. Individuals more darkly pigmented were better adapted to the tropics than lighter ones with regard to normal bone formation, and the situation was reversed in northern Europe. In both areas, the better adapted individuals were the ones most likely to survive and reproduce.

The isolation of human populations resulted in greater and greater divergence, in the same manner that isolation led to the divergence of Darwin's finches on the different islands of the Galápagos (see Chapter 13). The result is the existence of different races in different areas of the world. Today, with the breaking down of geographical barriers as a result of better transportation methods, an increasing amount of interbreeding between races will probably take place in the future. The trend will then be toward amalgamation rather than divergence.

THE GENETIC BASIS OF RACE

An important question to pose here, however, is whether pigmentation of the skin is the only legitimate criterion for determining racial differences. The answer, from the geneticist's point of view, is no. Geneticists define race as simply a group of individuals who share certain genes with higher frequency than other groups. Note that there is no specification of *which* genes are involved, which is the difference between the scientific concept of race and the social concept that has traditionally emphasized differences between breeding groups determined by skin pigmentation (Figure 14.5).

294
Man and Evolution

FIGURE 14.5
Racial varieties and areas of the world in which they are concentrated. [*After John Dawson, from* Biology Today, © 1972 by Communications Research Machines, Inc., reprinted by permission.]

① North American Colored   ② American Indian   ③ Polynesian

⑨ Bantu   ⑩ Hamite   ⑪ Hindu

⑰ North Chinese   ⑱ Ainu   ⑲ Classic Mongoloid

④ Ladino   ⑤ Sudanese   ⑥ Negrito   ⑦ Bushman   ⑧ South African Colored

⑫ Tibeto-Indonesian Mongoloid   ⑬ Murrayian   ⑭ Carpentarian   ⑮ Melanesian   ⑯ Southeast Asiatic

⑳ Turkic   ㉑ Northeast European   ㉒ Northwest European   ㉓ Nordic   ㉔ Mediterranean

By our definition, races can certainly be categorized on the basis of skin color. But this is not the only possible criterion. If one were to use pigmentation as the criterion for grouping people into races, the individuals who may resemble each other in this particular trait are by no means similar in other traits. Table 14.1, for example, shows different populations which are similar with regard to the frequency of different alleles for the ABO blood groups.

If races can be discerned on the basis of one trait alone, it is as valid to group the Chinese, Japanese, African pygmies, and Hottentots as one race, based on the frequencies of the ABO alleles in these populations, as it is to separate them into different races based on the color of their skin. And it makes as much sense for these people to abuse white Americans, Swiss, and Eskimos who might be living in their communities because of different genes for blood types as it does to discriminate because of skin color.

RACES AND RELIGIONS

An even cruder application of the concept of race is that which causes social and political isolation of groups as races based on their religious beliefs. Nazi Germany passed laws forbidding marriage between Jews and Germans "to ensure the purity of the German blood that is the basis for the permanence of the German people. . . ."* This invoking of the ancient blood theory of heredity was not limited to the Reich-

* Goldsby, op. cit., p. 7.

TABLE 14.1
Frequencies of the different alleles determining the ABO blood groups in various populations

POPULATION	NUMBER TESTED	i	I^A	I^B
Japanese	29,799	0.55	0.28	0.17
Chinese (Huang Ho)	2,127	0.59	0.22	0.20
African pygmies	1,032	0.55	0.23	0.22
Hottentots	506	0.59	0.20	0.19
Americans (white)	20,000	0.67	0.26	0.07
Swiss	275,644	0.65	0.29	0.06
Eskimos	484	0.64	0.33	0.03
Navajo Indians	359	0.87	0.13	0.00
Blackfoot Indians	115	0.49	0.51	0.00

Source: Adapted from E. W. Sinnott, L. C. Dunn, and T. Dobzhansky, *Principles of Genetics*, McGraw-Hill Book Company, New York, 1958, p. 279.

stag. Until the mid-1960s, laws forbidding marriage between whites and persons with "Negro blood" existed in some states of the United States.

Although it is true that religious beliefs do sometimes cause populations to interbreed so that there are higher frequencies of some genes among the individuals of that population that would tend toward the formation of a race, obviously religious groups on the whole are not formed on the basis of shared biological traits.

RACES AND IQ

Because races do differ in the frequency of certain genes, there is always the temptation to compare one race with another with regard to their relative superiority or inferiority in certain characteristics. Because we have long been conscious of our intellectual superiority over other living species, the level of intelligence of an individual or a race is a source of interest and pride to those involved.

In 1969, A. R. Jensen discussed in the *Harvard Educational Review* a study conducted on black and white children which seemed to indicate that black children had a mean of 15 IQ points below that of white children. He conjectured that this difference was determined by different genes in the two populations. Although he himself did not state as such, the implication of superiority of whites over blacks was inescapable. Many professionals, however, would dispute the hypothesis of a gene-determined superiority of one race over another. The problem lies in determining how much of an individual's intelligence is influenced by his genotype and how much by his environment, the old problem of nature versus nurture.

Few scientists would dispute the statement that the intelligence level of an individual is inherited, probably to a very large extent. However, intelligence is believed to be a polygenic trait and cannot be traced in a Mendelian fashion. On the other hand, that the environment is definitely a factor is also indisputable. Since man is not an experimental animal to be handled at will, scientists are extremely limited in methods of ascertaining the effect of environment on the development of the individual.

TABLE 14.2
Intrapair correlation coefficients of twins

	MONOZYGOTIC TWINS		DIZYGOTIC TWINS
	REARED TOGETHER	REARED APART	
Binet IQ test	0.97	0.67	0.64
Otis intelligence test	0.92	0.73	0.62
Stanford achievement test	0.96	0.51	0.88
Intelligence quotient	0.87	0.75	0.53
Composite intelligence score	0.87		0.62

Source: I. M. Lerner, *Heredity, Evolution, and Society*, W. H. Freeman and Company, San Francisco, 1968, p. 156.

TWIN STUDIES The closest we can come to the ideal of studying genetically identical individuals under set environmental conditions is to study identical twins who have been reared in the same home environment and contrast their development with those who have been raised separately. Table 14.2 gives some data of studies done on the IQ of twins.

The degree of similarity (or *correlation coefficient*) is given as a percentage of *concordance,* which refers to whether the twins both manifest or do not manifest a trait. The high degree of concordance of identical twins reared together, as compared with that of dizygotic twins, points to the inheritable component of intelligence mentioned before. The decrease of concordance seen in twins reared separately is presumably due to the difference in environment.

NATURE VERSUS NURTURE These and other data all indicate that the phenotype of an individual with regard to intelligence has both a genetic and an environmental component. Jensen and a few others believe it is primarily genetic. Apart from the emotions this hypothesis evokes, the social implication is important: namely, that since genotype determines potential, it must follow that the black race does not have the genetic potential for improvement. Therefore, according to this argument, since data (Table 14.3) have been gathered that relate lower IQ to lower economic classes, blacks as a race will be forever consigned to an underprivileged status.

However, there are two additional factors which must be taken into account when reviewing data such as that in Table

14.3. One is the validity of IQ tests which are used to determine levels of intelligence. This is disputed by practitioners in many fields, including psychology, who feel that some of the tests contain a built-in bias against persons with lower socioeconomic backgrounds. The second point is that the IQ scores listed in the table are already the result of environmental effects, and therefore do not indicate the genetic potential of the children for intelligence. The same can be said of Jensen's work.

As Scarr-Salapatek has pointed out,* an experiment was carried out in 1969 by R. Heber on a group of ghetto children, beginning shortly after their birth, whose mothers had IQs of less than 70. For four years the children were tutored for several hours every day. At the end of this period, the tutored group tested at a mean IQ of 127, while the mean IQ of a control group not given the same attention was 90.

The feeling of many scientists about the controversy raised by Jensen's work is summarized by Scarr-Salapatek as follows:

For individual differences within populations, and for social-class differences, a genetic hypothesis is almost a necessity to explain some of the variance in IQ. . . . But what Jensen . . . and others propose is that genetic racial differences are necessary to account for the current phenotypic differences in mean I.Q. between populations. That may be so, but it would be extremely difficult, given current methodological limitations, to gather evidence that would dislodge an environmental hypothesis to account for the same data.

* S. Scarr-Salapatek, "Unknowns in the I. Q. Equation," *Science*, **174**:1223–1228, 1971.

OCCUPATIONAL STATUS OF PARENTS	AVERAGE IQ OF CHILDREN U.S.	U.S.S.R.	ENGLAND	AVERAGE SIZE OF ENGLISH FAMILY
Professional	116	117	115	1.73
Semiprofessional	112	109	113	1.60
Clerical and retail Business	107	105	106	1.54
Skilled	105	101	102	1.85
Semiskilled	98	97	97	2.03
Unskilled	96	92	95	2.12

TABLE 14.3
Correlation of IQ, occupation, and family size

Source: I. M. Lerner, *Heredity, Evolution, and Society,* W. H. Freeman and Company, San Francisco, 1968, p. 161.

And to assert, despite the absence of evidence, and in the present social climate, that a particular race is genetically disfavored in intelligence is to scream "FIRE! . . . I think" in a crowded theater. Given that so little is known, further scientific study seems far more justifiable than public speculations.*

* Ibid., p. 1228.

GENETICS AND BEHAVIOR

There have also been attempts to analyze the inheritability of behavior traits. (We have already discussed the controversial relationship of the XYY chromosomal constitution to criminality.) Twins studies have been carried out on personality and interests, but the data have so far been inconclusive. Again, there is a genetic component, and certainly environmental factors affect our behavior, but no tests at present can distinguish between the contributions of nature and nurture. A modern philosopher has approached the question in the manner illustrated in Figure 14.6.

One of the most interesting aspects of animal behavior is the existence of inborn patterns of reactions to environmental stimuli, which we call _instinct_. There are far too many fascinating examples of instinctive behavior for us to deal with in this text. Many of them are well known, such as the intercontinental migration of birds (Figure 14.7) and the suicidal drive of salmon to their breeding ponds. The parasitic birds mentioned in Chapter 13 are good examples of what must be genetically determined patterns of behavior. The cuckoo lays its eggs in other nests by instinct, and by instinct destroys an egg of the host so that the number of eggs remains the same.

FIGURE 14.6
A modern reaction to behavior problems. [© 1966 United Feature Syndicate, Inc.]

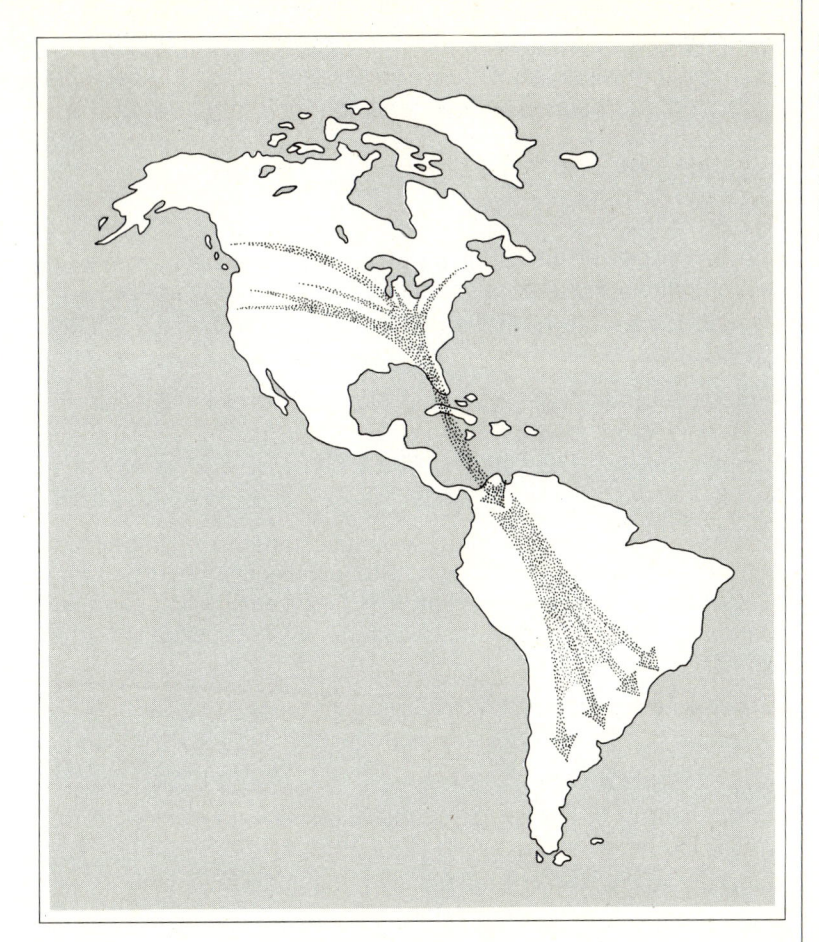

FIGURE 14.7
Migration routes of the North American songbird, the bobolink. [*After W. J. Hamilton, III, "Bobolink Migratory Pathways and Their Experimental Analysis under Night Skies,"* Auk, **79**:208–233, 1962.]

Since presumably all stimuli are received through the nervous system, and since, like any system, the nervous system consists of cells whose function depends on the activity of their genes, it is fascinating to speculate how instinct may be coded into the genes of nerve cells. The nervous system is one of the most poorly understood of all structures, especially in man, because of its unbelievable complexity. There are 100 million cells in a cubic inch of brain matter, each connected to tens of thousands of others in the penultimate example of a communications network. It may be expected that as our knowledge of the genetic control over cell activities increases, so will our understanding of the nervous systems of animals and their behavior, including that of man.

REFERENCES

A cogent discussion of the history of Darwinian evolution and its impact on human thought may be found in the following article:

Mayr, E. 1972: "The Nature of the Darwinian Revolution," *Science,* **176**:981–988.

A memorial article giving a brief biography of John Scopes, the defendant in the "Monkey Trial":

Tompkins, J. R., 1972: "Memoir of a Belated Hero," *The American Biology Teacher,* October 1972, p. 283.

A number of good articles relating evolution and man can be found in *Human Variations and Origins,* 1967, a collection of articles from *Scientific American,* published by W. H. Freeman and Company, San Francisco, including the following: L. Eisely, 1956, "Charles Darwin."

A book on the amazing career of Lysenko:

Medvedev, Z. A., 1969: *The Rise and Fall of T. D. Lysenko,* Columbia University Press, New York.

On the subject of human races, the following paperbacks are well written for the layman:

Dobzhansky, T., 1960: *Heredity and the Nature of Man,* Signet Books, New American Library, Inc., New York.

Goldsby, R. A., 1971: *Race and Races,* The Macmillan Company, New York.

Two articles by the same author on race and IQ. The first is a review of Jensen's publication and two others, and is easier for the layman to read than the second:

Scarr-Salapatek, S., 1971: "Unknowns in the IQ Equation," *Science,* **174**:1223–1228.

———, 1971: "Race, Social Class, and IQ," *Science,* **174**:1285–1295.

The controversial article by Jensen:

Jensen, A. R., 1969: "Environment, Heredity, and Intelligence," *Harvard Educational Review,* reprint series no. 2, 1–123.

Two short articles on the California science textbook controversy are:

Lightner, J. P., 1972: "Evolutionary Theory and the California Science Framework," *NABT News and Views,* **XVI**(5):2.

———, 1973: "Update on California Science Textbook Controversy," *NABT News and Views,* **XVI**(7):7.

A list of works dealing with creationist theories and evolution can be found in *BioScience,* **21**(4), 1971, 201. The list was compiled by Elwood B. Ehrle and H. James Birx.

Two books recommended for discussions on human evolution:

Simons, E. L., 1972: *Primate Evolution,* The Macmillan Company, New York. (part of the Macmillan Series in Physical Anthropology).

Young, L. B., 1970: *Evolution of Man,* Oxford University Press, New York.

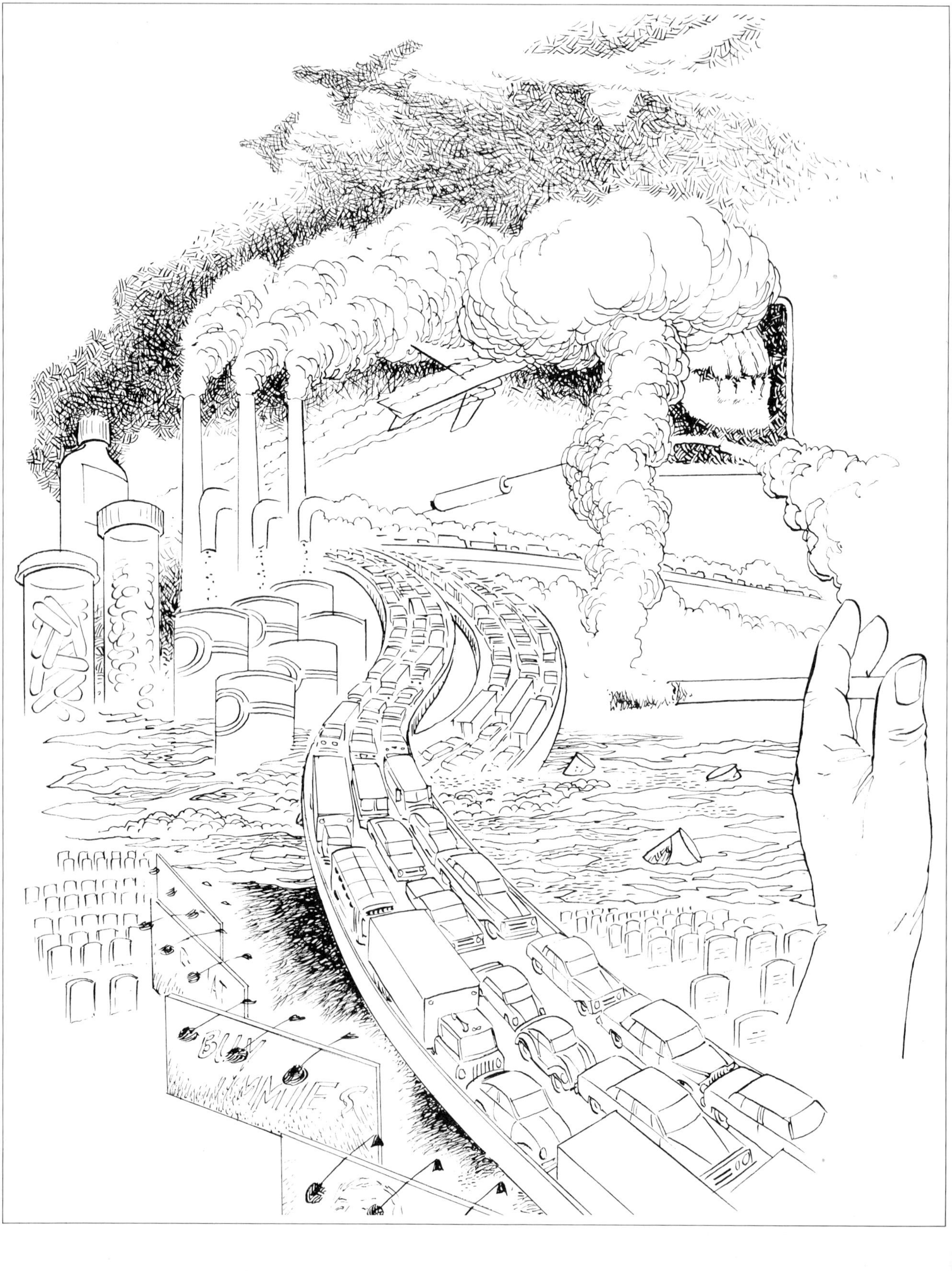
BUY

Chapter 15

Radiation and Chemical Mutagenesis

The role of the environment not only in the development of the individual but also in the evolution of the population (or race and species) to which he belongs has led biologists and many outside the life sciences to treat the environment with new respect. No longer can our environment be viewed as merely a space to be occupied, used, and abused at will; we now know that environmental factors play an active role in many fundamental biological phenomena.

One of these biologically essential roles is, of course, the production of mutations. To a geneticist, it is essential to analyze and determine the extent to which environmental factors, known as *mutagens,* can cause genetic changes. It is essential because all evolutionary change is determined ultimately by mutation, and with the increase of possible mutagens in our environment, our potential increase in genetic load and weakening of our species are of great concern.

There are two main classes of mutagens in our environment. One is radiation—both natural, from cosmic rays and radio-

Future generations may face serious increases in inherited diseases because of the mutagenic agents in our environment today.

active elements, and man-made, from x-ray machines and other technological advances in nuclear energy studies. The other class includes the increasing numbers of chemicals which we are inspiring and ingesting through our air, water, food, drugs, and medicine, and which geneticists feel may eventually prove to be a greater genetic hazard than radiation.

RADIATION MUTAGENESIS
COSMIC RAYS

Through the ages the main source of *spontaneous mutations*, those due to natural causes, has presumably been the small amount of radiation found in our environment known as *background radiation*. To a large extent, this comes from cosmic rays of the sun. Table 15.1 shows the amount of exposure to cosmic rays in various countries over a 30-year period. The differences stem from variations in elevation above sea level, which increases at 10,000 feet to about three times the amount found at sea level.

RADIOACTIVE ELEMENTS

Another source of natural radiation is the earth's crust, which contains various radioactive elements, such as radium. Kerala Province in India records some of the highest intensity of radiation in the world, due to relatively high concentrations of a radioactive element, thorium, in the soil. Kerala's inhabitants are reportedly exposed to as much as 84,000 Mrad over a 30-year period! Interestingly, no increase in mutation rate has been detected there.

THE NATURE OF RADIATION

It might be well to discuss briefly the nature of radiation and its effects on living systems. There are many different kinds

TABLE 15.1
Exposure to cosmic rays in different countries

* A *rad* is a unit of measurement of the amount of radiation absorbed. The absorption of one rad delivers 100 ergs of energy per gram of matter; a *millirad* (mrad) is one-thousandth of a rad.

Source: B. Wallace and T. Dobzhansky, *Radiation, Genes, and Man*, Holt, Rinehart and Winston, Inc., New York, 1959, p. 68.

COUNTRY	*MRADS PER 30 YEARS	COUNTRY	*MRADS PER 30 YEARS
Great Britain	840	Japan	1,020
United States	870	Northern Argentina	600
Austria	840	Southern Argentina	840
France	720		

FIGURE 15.1
The relationships of some types of radiation in terms of wavelengths. [*After* Radiation, Genes, and Man, by B. Wallace and T. Dobzhansky. Copyright © 1959 by Holt, Rinehart and Winston, Inc. Adapted by permission of Holt, Rinehart and Winston, Inc.]

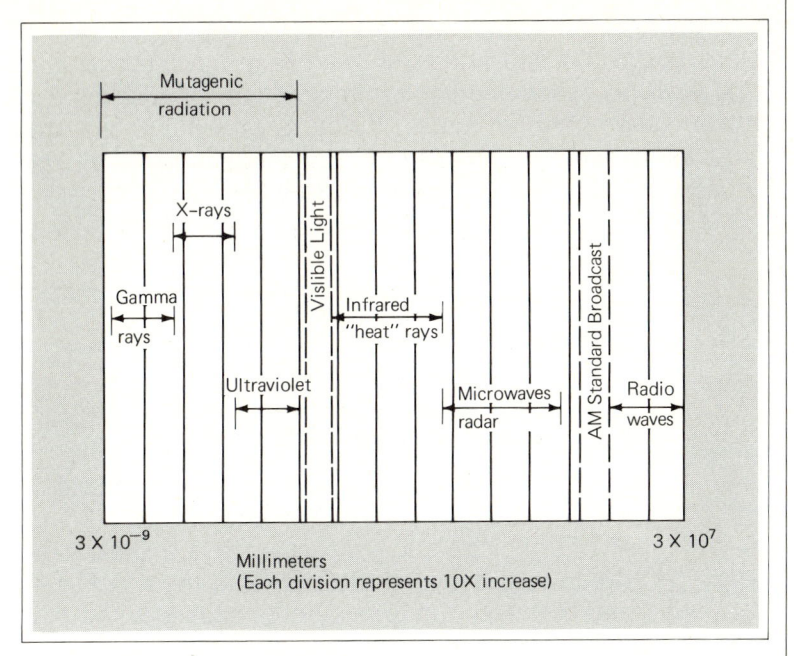

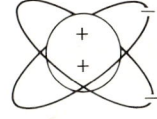

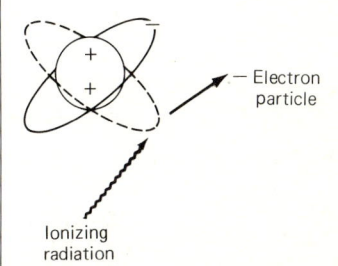

of rays, such as x-rays, gamma rays, and alpha rays (Figure 15.1). Not all rays are equally harmful. Alpha and beta rays do not penetrate beyond the skin of humans, and therefore would not affect internal cells. Others, such as gamma and x-rays, are high-energy particles that do penetrate living tissue easily. The biological effects of such high-energy rays on cells are believed to result from the rays colliding with atoms and molecules of the cells at high speed, breaking them up into electrically charged particles called *ions* (Figure 15.2). The result is abnormal function or perhaps cell death. Irradiated cells have been shown to possess abnormal chromosomes and to undergo abnormal mitosis.

Often various cancers develop as a result of radiation, especially cancers of the white blood cells, known as leukemia. Whether the cancers are a direct result of the radiation or the radiation activates viral DNA to transform normal cells into malignant ones is not known. Although radiation frequently leads to the development of neoplasias, it is also the tool used in attempts to arrest the course of cancers, especially in terminal stages.

FIGURE 15.2
Schematic illustration of the effects of ionizing radiation. The loss of an electron causes the atom to become electrically charged. Such an atom is known as an ion.

The increased incidence of various cancers associated with overexposure to x-rays has shown these rays to be *carcinogens,* or cancer-inducing. In addition, studies on exposure of embryos to x-rays during development have revealed the potential of x-rays to cause deformities in the fetuses. Factors which are fetus-deforming are called *teratogens.*

MAN-MADE RADIOACTIVITY
X-RAYS

The undesirable biological effects of radiation were not fully appreciated until technological advances brought widespread usage of radioactive substances and x-ray machines. In 1927 H. J. Muller showed that exposing *Drosophila* to x-rays (and later, also, to gamma rays) caused an increase in the number of mutations among the offspring of the irradiated flies. Further, he showed that there is a linear relationship between the dosage of radiation and the number of mutations, and that even the lowest doses caused some mutations to occur. This has since been confirmed by further experimentation (Figure 15.3).

Although x-rays can cause great harm if misused, they are nevertheless important as a diagnostic and therapeutic tool for medicine and research. The tragedies which have re-

FIGURE 15.3
Graph of numbers of sex-linked lethal mutations plotted against the amount of x-rays which the X chromosomes received. The solid line corresponds to actual observations; the dotted line represents the direct proportion between the frequency of induced lethals and the dosage. [*After Radiation, Genes, and Man,* by B. Wallace and T. Dobzhansky. Copyright © 1959 by Holt, Rinehart and Winston, Inc. Adapted by permission of Holt, Rinehart and Winston, Inc.]

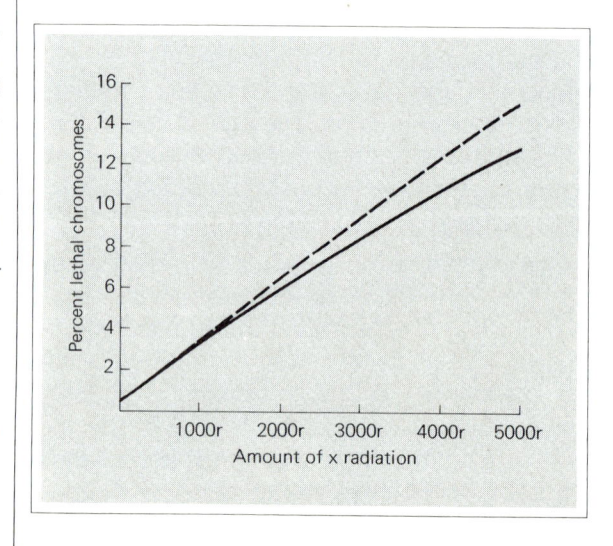

sulted in the past from overdoses of radiation simply reflect the fact that when a useful instrument is developed, man tends to overlook potential hazards of side effects in his eagerness to exploit it. Some of you may remember x-ray machines that were once widely used in shoe stores to aid in fitting shoes. It is disturbing to speculate how many young toes wiggled happily while being exposed to radiation minutes on end before the machines were removed.

THE SPECIFIC LOCUS TEST W. L. Russell, of the Oak Ridge National Laboratory, carried out an extensive experiment on mice to determine the effects of x-rays on higher organisms. The experiment, called the *specific locus test,* involved the mating of mice that were homozygous recessive for seven coat color loci to homozygous dominant animals that had been exposed to x-ray treatment (Figure 15.4).

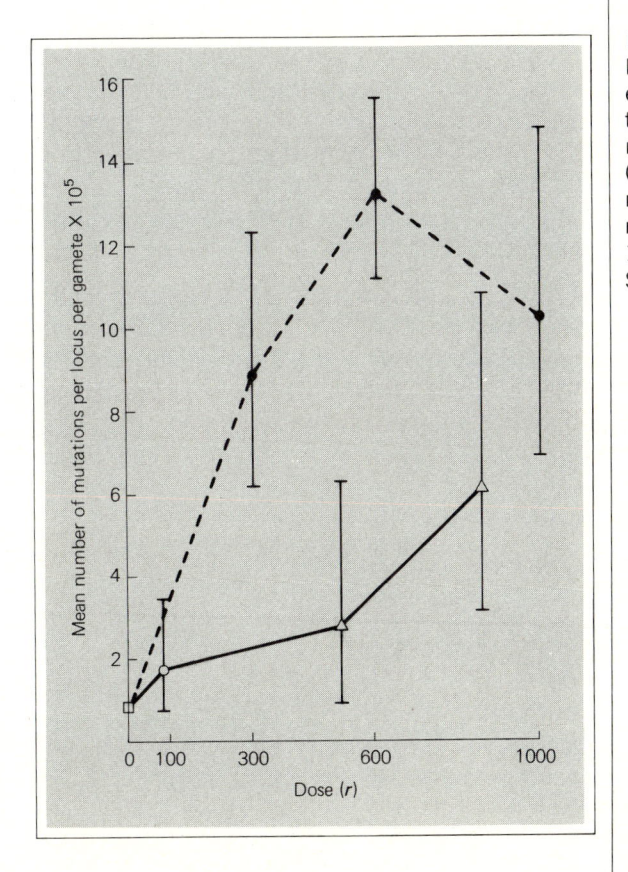

FIGURE 15.4
Results of W. L. Russell's radiation experiments on mice, known as the *specific loci test*. Solid circles represent results with acute x-rays (80–90 rpm). Open points represent chronic gamma ray results (triangles 90 rpwk); (circle 10 rpwk). [*After W. L. Russell,* Science, **128:***1547, 1958.*]

Any mutation to the recessive condition of any of the color loci that results from the radiation will appear in the offspring of this cross:

Control

aabbccddeeffgg × AABBCCDDEEFFGG
↓
AaBbCcDdEeFfGg

Experimental — X-rayed

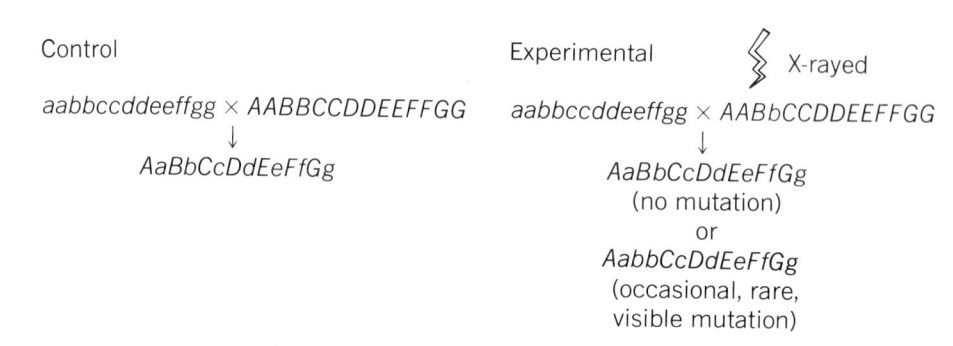

aabbccddeeffgg × AABbCCDDEEFFGG
↓
AaBbCcDdEeFfGg
(no mutation)
or
AabbCcDdEeFfGg
(occasional, rare, visible mutation)

Table 15.2 gives some of the results of this experiment.

Just as in the tests on *Drosophila*, a linear relationship was found between the amount of radiation and the number of induced mutations; even the lowest doses caused some mutations. Further, *acute* doses were found to be more mutagenic than *chronic* doses. (Acute doses are those given at one time to the individual, whereas chronic dosages are smaller amounts of radiation given at different times. In other words, one dose of 600 *r causes* more mutation than three doses at 200 *r*.)

It must be pointed out here that the specific locus test does not reveal all mutations induced by the treatment with x-rays, since it studies only the seven color loci. In Chapter 12 we had discussed the fact that genes mutate at different rates; therefore, the results here indicate only the mutation rates of

TABLE 15.2
Effect of x-ray treatment of mice in the specific locus test used by W. L. Russell at Oak Ridge National Laboratory

*r = roentgen, a unit of measure of the amount of radiation. One roentgen produces 2×10^9 ion pairs per cubic centimeter of air. See page 306 for an explanation of ionizing radiation.

DOSE OF IRRADIATION (r)*	NUMBER OF OFFSPRING	NUMBER OF MUTATIONS
0	42,833	1
300	40,408	25
0	106,408	6
600	119,326	111
0	33,972	2
1000	31,815	23

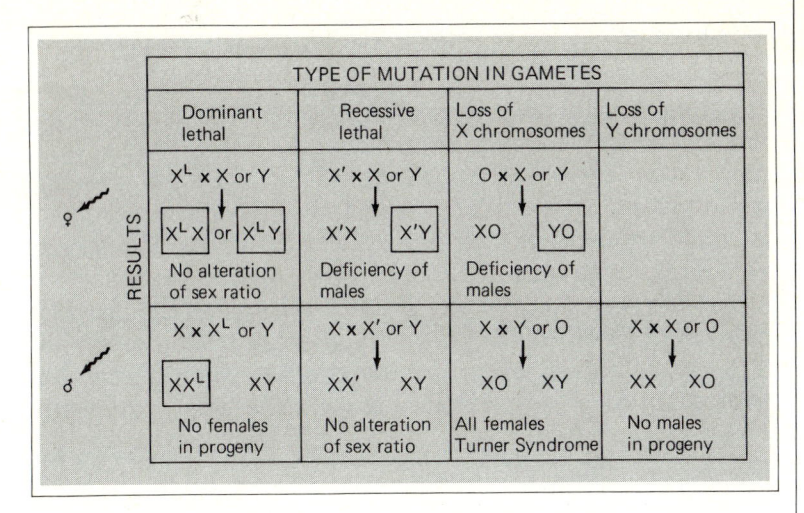

	TYPE OF MUTATION IN GAMETES			
	Dominant lethal	Recessive lethal	Loss of X chromosomes	Loss of Y chromosomes
♀ RESULTS	X^L x X or Y ↓ $X^L X$ or $X^L Y$ No alteration of sex ratio	X' x X or Y ↓ X'X X'Y Deficiency of males	O x X or Y ↓ XO YO Deficiency of males	
♂	X x X^L or Y ↓ XX^L XY No females in progeny	X x X' or Y ↓ XX' XY No alteration of sex ratio	X x Y or O ↓ XO XY All females Turner Syndrome	X x X or O ↓ XX XO No males in progeny

FIGURE 15.5
Possible effects of radiation in the alteration of sex ratios among progeny of radiated animals, as a result of sex-linked lethal mutations induced by the radiation.

these particular genes. In addition, only mutations to the specific recessive allele of the nonirradiated animals would be picked up. Neutral mutations would not be observed. Also, since different species are involved, we cannot apply the results directly to man and conclude that the same amount of radiation would produce the same results. Nonetheless, such data are the most extensive available to indicate the potential mutagenicity of x-rays in mammals.

POSSIBLE EFFECTS ON THE SEX RATIO OF OFFSPRING   Because of the genetic uniqueness of the Y chromosome in man and other sexually reproducing organisms, two types of mutations resulting from radiation involving the sex chromosomes may be expected to alter the sex ratio of offspring of males and females exposed to x-rays. First, a lethal mutation may be induced in the X chromosome of a gamete. Second, the sex ratio may be altered if, due to breakage or other severe damage, either the X or Y chromosome is lost. The results of these mutations are illustrated in Figure 15.5. Thus the alteration of sex ratios is another line of evidence for the genetic effects of radiation.

Although the potentially devastating effects of radiation on biological systems have been well documented, man persists in activities which do not reflect adequate concern over the

RADIATION FROM NUCLEAR ENERGY

long-term effects of radiation. Perhaps the most glaring example is the use of nuclear energy for weapons of war.

The dropping of two atomic bombs on heavily populated areas of Japan, Hiroshima and Nagasaki, during World War II has shaken the moral conscience of scientists all over the world. The culpability of scientists who, in their search for an understanding of their natural world, develop techniques potentially destructive of that world has been the basis for much debate in and out of science. We shall discuss this point in Chapter 16. For now, the effect of these bombs has afforded scientists the opportunity to study the reaction of the human organism to irradiation on a massive level (Figure 15.6).

Immediately after World War II, an Atomic Bomb Casualty Commission was formed to study the effects of the exposure on thousands of civilians who had survived enormous acute dosages of radiation. Table 15.3 summarizes some of the cytogenetic findings. Of the survivors who were thirty years old or younger when they received at least 200 rads of radia-

FIGURE 15.6
Result of the atomic bombing of Hiroshima in 1945. [*Wide World Photos.*]

TABLE 15.3
Frequencies of cytogenetic abnormalities among Japanese survivors of the atomic bombs

AGE AT TIME OF BOMB	DOSE (RADS)	EXPOSED NO. EXAMINED	% AFFECTED	CONTROL NO. EXAMINED	% AFFECTED
To 30 years	200	94	34	94	1
Over 30 years	200	77	61	80	16
In utero	100*	38	39	48	4
Not yet conceived	150	103	0		
Not yet conceived	100†	25	0		

* Maternal dose.
† Dose received by at least one parent.
Source: R. W. Miller, "Delayed Radiation Effects in Atomic Bomb Survivors," *Science*, **166**:570, 1969.

tion, 34 percent have been found to have chromosomal abnormalities in their cells. Those who were over thirty years of age have shown almost twice the number of visible chromosome abnormalities found in the younger age group. These abnormalities include translocation, deletion, and breakage. Of persons who were being carried in utero at the time that the mother was exposed, 39 percent display chromosomal abnormalities. Interestingly, children conceived after the bombings showed no chromosomal abnormalities at all, indicating that the gametes of exposed individuals were not affected to the extent that body cells were, although experiments indicated that spermatogonia are quite sensitive to radiation.

As for the induction of cancer, 1 out of 60 heavily exposed survivors of Hiroshima developed leukemia within 12 years. Other experimenters have found a decided increase in the incidence of thyroid cancer. In the years between 1950 and 1960, even excluding deaths caused by leukemia, the mortality rates for exposed persons compared to nonexposed controls increased by 15 percent.

The effects of detonating nuclear devices last far beyond the initial trauma. Fifteen years of atmospheric testing by Russia and the United States following the Second World War have further increased the amount of radiation in our environment because of the variety of unstable radioactive isotopes emitted. These isotopes decay at different rates. For example, strontium 90 is said to have a half-life of 28 years. This means that every 28 years, the radioactivity of a given quantity of strontium 90 decreases by half. In the mean-

FIGURE 15.7
Radioactive elements resulting from fallout from the detonation of nuclear weapons may affect man either by direct exposure or by the ingestion of plant and animal tissues which have incorporated the radioactive substance.

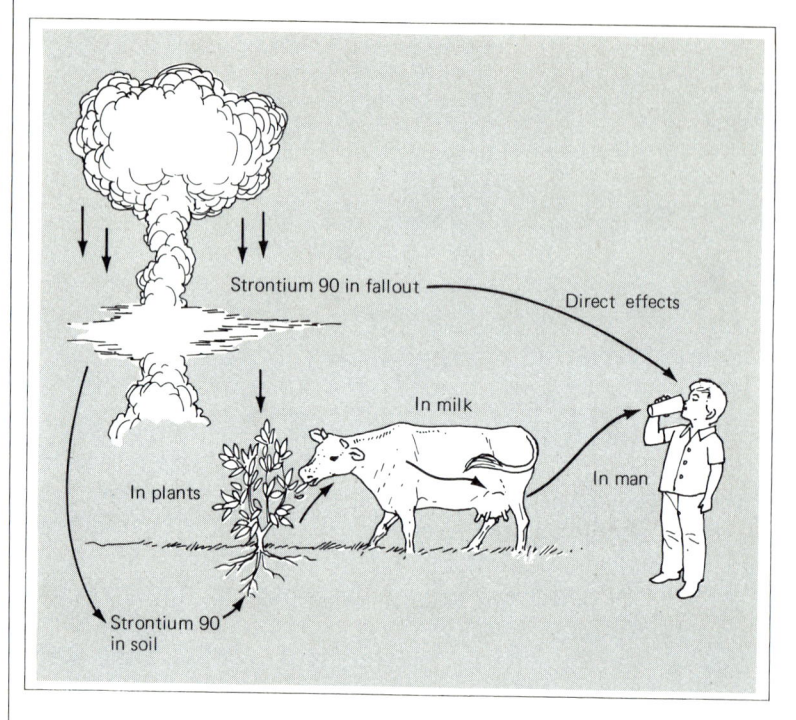

* B. Commoner, *Science and Survival*, The Viking Press, New York, 1966.

time, it is in our air and soil and will be taken up by animals and plants, which will, in turn, be ingested by other living organisms, including man (Figure 15.7).

In a scathing indictment of the U. S. Atomic Energy Commission and its management of the testing in those days, Barry Commoner,* a noted biologist, has pointed out the following disturbing facts:

1. Nuclear testing between 1948 and 1962 equaled 200 times the amount of all bombs dropped on Germany during World War II, or *500 million tons* of TNT.

2. In 1953, the AEC declared that fallout from the tests would be evenly distributed so that damage would be minimal. Figure 15.8 shows the actual distribution of the fallout. Most of it was concentrated in the north temperate zone, where most people of the world live.

3. In 1953 the AEC declared that fallout would be below the amount that could cause a "detectable" increase in muta-

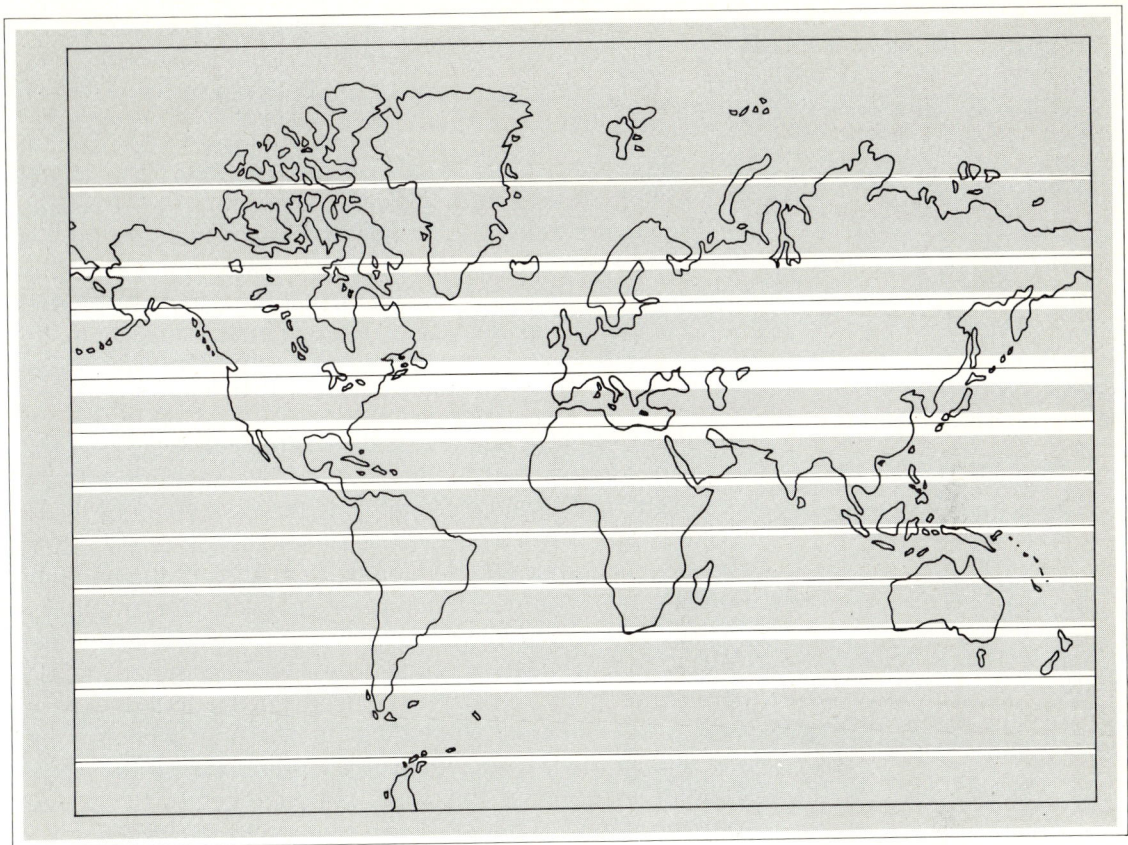

FIGURE 15.8
Schematic representation of worldwide fallout patterns. [*After Radiation, Genes, and Man*, by B. Wallace and T. Dobzhansky. Copyright © 1959 by Holt, Rinehart and Winston, Inc. Adapted by permission of Holt, Rinehart and Winston, Inc.]

tion. In 1957 it admitted that the fallout to date would lead to from 2500 to 13,000 cases of genetic defects all over the world.

Although most nations have prohibited atomic testing in the atmosphere, some still persist in doing so, such as the People's Republic of China and France. One can only hope that knowledge of the potential consequences of this action on their own as well as other people will lead to a voluntary cessation of a policy which increases radiation in our world.

The devastation wrought by atomic bombs at Hiroshima and Nagasaki is small compared to the potential destructiveness

NUCLEAR ENERGY FOR PEACE

of more sophisticated nuclear devices. The two atomic bombs detonated in Japan were equivalent to 20,000 tons of TNT. Today's hydrogen bombs are believed to equal at least 10 *million* tons!

With the easing of political tensions between the United States and other atomic powers, and with the increasing need for fuel energy, we are now turning to the possible use of nuclear energy for peacetime purposes. Here we must be careful not to overlook potential hazards in the use of what may be our major source of energy in the future, in the manner that x-rays were misused when they first were developed.

One of the inherent weaknesses of the present system of control in the United States is that the AEC, which undertakes to control the use of atomic energy, is the same agency which develops and advocates the use of new techniques. There have been instances of inadequate regulation of the handling of radioactive materials. One widely publicized example was the AEC's decision to allow real estate developers to use 300,000 tons of waste material from uranium mines in Colorado. The sandy material was used for the foundations of some 3000 buildings, including homes and schools. Not long after, radon, a radioactive gas produced by the decay of radium present in the waste material, began to seep up from the sand at measurable levels, in some cases up to 200 times the level of radioactivity considered safe for man. The AEC claimed that it could not be held accountable because it has jurisdiction only over man-made isotopes, and not radium, which is a natural element.*

* H. Peter Metzger, "Project Gasbuggy and Catch-85," *The New York Times Magazine*, 26–27, Feb. 22, 1970

These past mistakes made by the AEC are not meant to illustrate that the commission has total disregard for the environment or the well-being of people. Still, mistakes have occurred, and other errors in judgment may be made. The trouble with such errors is that once made they cannot be corrected. The radiation has been released, and there is no way it can be recovered.

It is to be hoped that lessons learned from past mistakes will prevent future ones from occurring. There is no question that the harnessing of nuclear energy for peaceful uses holds potential solutions for some of our more pressing problems, such as the exhaustion of our natural fuel supplies. Geneticists have proved the hazards of exposure to radiation; nuclear scientists must proceed with this knowledge in mind. Along these lines, it is reassuring that new systems being developed hold the promise of providing needed power with little danger of radioactivity leakage and less heat rejection to the environment than present systems. Those interested in a review of such a system should see the article by R. J. Creagan mentioned in the references at the end of this chapter.

CHEMICAL MUTAGENESIS
SCOPE OF THE PROBLEM

Although radiation has been proved to be damaging to biological systems if misused, many geneticists feel that chemical mutagenesis may prove to be a far greater hazard to man than radiation. There are a number of reasons for this. For one thing, the source of radiation is usually well documented, and if it is known to be excessive, man can be protected from it by removing the source of radiation or by installing protective shields that are impenetrable to the damaging rays.

Although it has been proved experimentally that chemicals can cause mutations, there is no comparable protection against chemicals that would shield the individual from possible mutagens. Also, whereas radiation can be measured, man is exposed to many chemicals in minute doses over a long period of time, and so a quantitative analysis of our exposure is most difficult. Further, the possible sources of chemical mutagenesis are unlimited, since the rate at which chemicals of all kinds are being introduced into our environment, and which we willingly ingest as medication and drugs, is constantly increasing.

As far back as the Second World War, Charlotte Auerbach and her co-workers were investigating the possible mutagenic effect of various chemicals. One of the first to be established as being definitely mutagenic was mustard gas. Work in laboratories all over the world has since pointed to many other chemicals that give evidence of increasing mutation rates in experimental organisms.

A recent survey of food additives* found that as many as 2500 chemical substances are being added to food for col-

---

* G. Kermode, "Food Additives," *Scientific American*, **226**, 15–22, 1972.

oring, preservative, and other purposes. Pesticides retained in foods add to the numbers of chemicals we eat. So many drugs are taken that we have been characterized as the "pill-popping" society, and the establishment of scientific organizations such as the Environmental Mutagen Society and the Registry of Tissue Reactions to Drugs, a part of the Armed Forces Institute of Pathology, reflects scientists' concern for the potential long-term effect of this kind of "internal pollution."

MUTAGENICITY OF DRUGS

The increased use of drugs and hallucinogens has caused concern that detrimental effects on germ cells may be transmitted to future generations, especially since so many young people of reproductive age are involved. The controversy that surrounds experimentation on the possible genetic effects of LSD (lysergic acid diethylamide) can serve as a good example of the problems facing geneticists attempting to arrive at some definitive answer to the mutagenicity of such substances.

Studies applying LSD to human cells in tissue culture have occasionally resulted in chromosomal abnormalities. However, other similar experiments have drawn negative results. A recent review of the plethora of studies on this problem concluded that, based on experimental results, LSD when "ingested in moderate doses does not damage chromosomes in vivo, does not cause detectable genetic damage, and is not a teratogen or a carcinogen in man."*

* N. I. Dishotsky, W. D. Loughman, R. E. Mogar, and W. R. Lipscomb, "LSD and Genetic Damage," Science, 172:439, 1971.

However, these conclusions, though scientifically sound, must be tempered with caution for a number of reasons that apply to the use of any drug. For one thing, the experiments done in laboratories under controlled circumstances cannot duplicate the actual situation of an individual who is illicitly buying, using, or perhaps making the drug. The substances used in laboratories are usually pure and the quantities controlled, whereas the chemicals used by addicts may be highly contaminated with many other substances, and there is no certainty of the amount ingested.

POTENTIATION

One important result of recent studies on mutagenesis is the discovery of a phenomenon known as *potentiation*. This means that while some chemicals tested individually are not mutagenic, when applied together they do increase the frequency of mutations. Some geneticists call such substances *comutagens*. For example, BHT (butylated hydroxytoluene), a food additive, has been found to be radio-potentiating, which means that organisms treated with BHT and radiation incur increased numbers of mutations. Individuals who are prone to hallucinogens and drugs are known to take an assortment of chemicals, the effects of which in combination we have no way of knowing. (The same can be said of the large numbers of food additives we ingest daily.)

THE DETERMINATION OF MUTAGENICITY

Furthermore, there is always the problem of whether we can apply experimental situations involving animals or small groups of cells in tissue culture to man. The effects of the substance on experimental models may not necessarily be duplicated in the living human body.

At this point, let us discuss some of the techniques being used at present to test the mutagenicity of various chemicals (and point out the deficiencies of each with regard to the application of results to man).

IN VITRO STUDIES OF HUMAN CELLS

This is the technique by which human cells of various types are grown in tissue culture and treated with chemical agents, and the resulting mutants counted and isolated. Since the cells of the culture are diploid, it is thought that their reactions may indicate the reaction of somatic cells in the individual. However, as pointed out by a leading cytogeneticist, E.H.Y. Chu,[*] a number of factors can alter the rate of mutation in in vitro systems, including cell density, concentration of selective agents, and length of time that the cells are incubated. Further, the toxicity to the cells of the tested substances is not always known, and it may affect the sensitivity of cells that survive the toxic effects to alter mutation rates.

The effects of cell density and concentration of mutagens are especially important, since it is obvious that the density of cells in vivo (in the living organism) is enormously greater than that in tissue culture. Therefore, if density is a factor,

[*] E. H. Y. Chu, "Mutagenesis in Mammalian Cells," *Proceedings of the Twenty-Fourth Annual Symposium on Fundamental Cancer Research*, 1971.

results obtained in tissue culture could very well differ from results in vivo. Also, by the time a substance has been ingested or injected, its concentration when it reaches either somatic or germinal tissue has to be greatly altered.

Finally, studies of treated cells cannot reveal all mutations. Cytogenetic techniques reveal gross chromosomal damage, and some biochemical alterations resulting from mutation may be discerned. However, the majority of point mutations would not be detected.

MICROBIAL SYSTEMS

Many tests on the mutagenicity of various chemicals have been carried out on microorganisms, bacteria, and fungi. The value of these studies resides in the fact that all genes are composed of DNA. The effect of a substance on the DNA of bacteria may therefore indicate its effects on the DNA in cells of complex organisms. There is the further advantage that microbial systems allow scientists to discern mutation frequencies so low that they could not be picked up in *Drosophila,* much less mice. Nonetheless the reaction of single naked strands of DNA as found in a bacterium to mutagens *must* be different from reactions of the DNA-protein complex in true chromosomes.

Therefore, while results of mutagenicity testing in microorganisms may indicate biological effects of various chemicals on DNA, they cannot be considered true tests for the same substance in higher organisms. In this respect, studies on eukaryotes such as *Neurospora* may be better indicators of mutagenicity in man than bacteria.

STUDIES ON VARIOUS SPECIES

As has been stated before, *Drosophila* and mice have been extensively used to test mutagens. However, while they are diploid organisms with complex chromosome complements, the fact that mutation occurs at different rates in different species (Table 12.2) still prohibits total extrapolation of data from these organisms to man.

In addition, the specific locus test which has also been carried out using chemical mutagens reveals only the mutations at seven color loci. As pointed out before, other mutations which may occur would not be discerned by this method.

HOST-MEDIATED ASSAY

A technique called the *host-mediated assay* was developed in an attempt to circumvent the problem mentioned before on studies of microorganisms which involve differences that must exist in the metabolism of chemicals by the single-celled organisms and by large multicelled organisms, such as man. In this method, a mammal is injected with a bacterium or fungus and then treated with a chemical being tested for mutagenicity (Figure 15.9). After a period of time which varies according to the experiment, the microorganisms are recovered, and the rate of mutation in the injected microorganisms is compared to that in microorganisms which were administered the chemical directly. Any difference gives an indication of the extent to which the host animal has altered the chemical—that is, the chemical's ef-

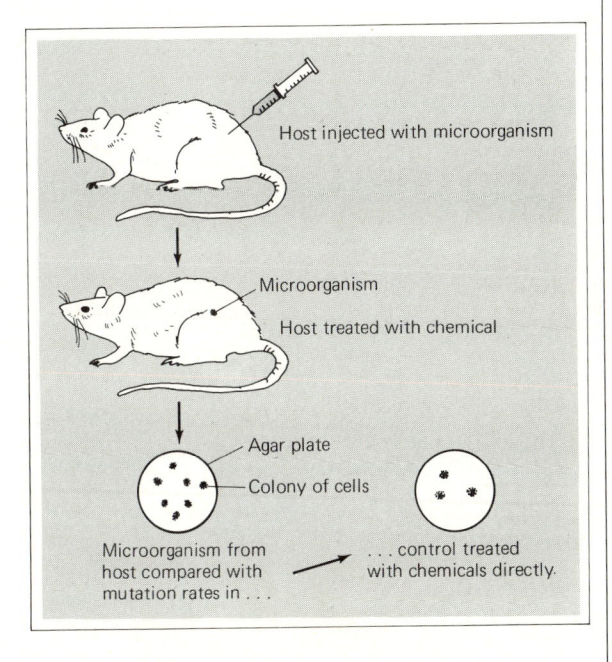

FIGURE 15.9
The *host-mediated assay*, a technique used to test the mutagenicity of chemicals on microorganisms after they are injected into and metabolized by a complex organism, such as a mouse.

fectiveness as a mutagen. Several substances have been found to be nonmutagenic in microorganisms but mutagenic in injected microbes in host mammals, such as cycasin.

Although the host-mediated assay technique does take into account the metabolic differences between unicellular and multicellular forms of life, there still remains the problem of extrapolating this data to man, who is much larger and more complex than experimental mammals and whose metabolism therefore may further change the effect of chemicals.

DOMINANT LETHAL ASSAY

Chromosomal aberrations resulting from treatment with x-rays are usually, as might be expected, dominant lethal mutations resulting in death at early stages of development and sterility and semisterility in some F_1 progeny. With this in mind, the dominant lethal assay was developed to test for the effects of chemicals in inducing such mutations (Figure 15.10).

Male mice or rats are treated with various doses of the chemicals being tested. They are then mated with untreated fe-

FIGURE 15.10
The dominant lethal assay, in which the mutagenicity of chemicals on males is tested. Males treated with the chemical are mated to untreated virgin mice. The litters are inspected for resorbed fetuses, which are presumed to be victims of dominant lethal mutations incurred in the sperm of the treated males.

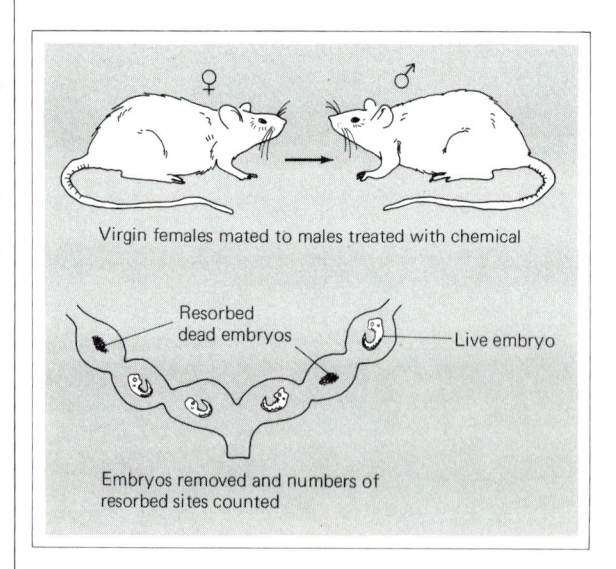

males at different times following the treatment. The females are then dissected at certain times in the gestation period and are studied for the numbers of offspring as compared with the numbers of corpora lutea (areas indicating the site of an ovulated egg) found in the ovary. The numbers of offspring are also compared with those produced by females mated with untreated males.

The drawback to this technique is that viable mutations and recessive mutations, which form the majority of mutation events, would not be detected. However, the dominant lethal assay has contributed to our knowledge, as have all the tests mentioned above. For example, using this test, scientists have shown that there is a definite difference between rats and mice in their sensitivity to DDT as a mutagen. Mice showed no loss of embryos, whereas rats did show a loss indicating lethal mutations.

Many of the substances that have been found to be mutagenic in different systems are also carcinogenic. The effects of chemicals in causing cancers is of great concern because of increasing evidence of cancers found in people who were exposed for long periods to certain chemicals. For example, a recent report* indicated that millions of people who were exposed to asbestos while working in shipyards during World War II may be in danger of developing a cancer of the lining of the chest and abdomen called *mesothelioma*. About 7 percent of shipyard workers have been found to succumb to mesotheliomas, and the incidence of this once rare disease is increasing. In each case, asbestos has been found in the lungs of the victims. It is disturbing that asbestos is known to be present in the air of most cities.

The teratogenic effects of chemicals were dramatically pointed up by the tragedy of the thalidomide babies (see Figure 4.12). The same problems that scientists have in determining mutagenicity of chemicals apply to studies of carcinogenesis and teratogenesis as well. Thalidomide had been tested on pregnant mice and found to cause no abnor-

OTHER BIOLOGICAL EFFECTS OF CHEMICALS

* R. Sherrill, "Asbestos, the Saver of Lives, Has a Deadly Side," *The New York Times Magazine,* Jan. 21, 1973, p. 12ff.

malities among the offspring. After the birth of the deformed babies, the drug was tested on other mammals and found to have teratogenic effects on rabbits similar to those found in the human babies.

LONG-RANGE EFFECTS OF CHEMICAL MUTAGENESIS

It is obvious that no one method exists that can be said to produce results valid to man. From past experience, it appears that new drugs and chemicals should be tested using a combination of tests on many different species. This is, however, easier said than done, for to test each drug and chemical that humans will be exposed to as thoroughly as it should be tested is too expensive and time-consuming for almost all laboratories.

At the 1972 meeting of the Environmental Mutagen Society, problems encountered in the production of a new chemical were discussed. The probability of a new product reaching the development stage was reported to be only 1 in 8000 because of the various tests the substance must pass for both effectiveness and safety. To reach commercialization, the chances are 1 in 36,000 at a cost of $2 million.

The fact that some industrial giants have initiated programs for screening chemicals for mutagenicity and other possible detrimental effects is encouraging. However, most scientists concerned about chemicals in our environment do not feel that there are enough of these programs, and that $2 million notwithstanding, not enough is being done to protect the public from possible harm from the enormous numbers of chemicals being released into our air, food, and water.

The conflict is great. On the one hand, there is no question that life has been eased and prolonged by the development of beneficial drugs and chemicals that are artificially introduced into our systems. On the other, because of the deficiencies in our systems of testing, some of these substances very likely will have detrimental effects in the long run.

Besides the difficulties we face in extrapolating experimental data from simpler forms of life to man, there exists the problem that mutagenic effects of chemicals now in our systems will not be expressed for many generations because most mutations are recessive. The same is true for carcinogenic potential, as illustrated by the cases of mesotheliomas some thirty years after most of the victims were heavily exposed to asbestos.

If, as some geneticists suspect, mutations are now being created in significant numbers by our exposure to chemicals and drugs, and this is proved in future generations by an increase in the number of genetic diseases and inherited anomalies, it will be too late to prevent what is occurring now in the genetic makeup of gametes of young adults. This is the most disturbing aspect of the problem.

Besides the prospect of increased numbers of progeny with inherited abnormalities, the increase in mutations is adding to our genetic load, rendering our species as a whole genetically weaker. In addition, with new techniques and medicines, we are also able to retain as reproductive individuals many who would normally have succumbed to various genetic conditions. Some of the inborn errors of metabolism serve as good examples of this. A child homozygous for phenylketonuria (pku), if not treated, would usually be severely retarded mentally and would not normally reproduce. Now we can prevent the disease from developing by special diets for pku homozygotes. These individuals, when they reproduce, are then passing the mutation to each of their children.

In short, man is capable of interfering with the processes of natural selection and the survival of the fittest. From the standpoint of genetic principles, this is harmful for man as a species. However, because man is also the only animal with a conscious sense of moral values, to do otherwise—to allow pku homozygotes, for example, to follow their natural path to mental degeneration—would be to counter the decent instincts which we possess, and such policies are anathema to many geneticists. Yet as others point out, our concern for such individuals may in the long run lead to man's extinction.

This is the situation now. Man as a biological organism faces an uncertain future. His evolutionary destiny is in question because of the increasing genetic load he must bear and because the environment to which he had become adapted through the evolutionary ages is changing at alarming speeds—by man's own hand.

However, our understanding of life processes has also increased with unbelievable swiftness. Recall that genetics as a science began only in this century. The techniques and discoveries made have been heralded as potentially of great use to solving many of the problems of man which we have discussed in this chapter. What these techniques are and the promise they hold for man will be our subject in the last chapter.

REFERENCES

Annual conferences are held by the Environmental Mutagen Society dealing with mutagenicity of various agents in our environment. A conference at which the various methods for assessing mutagenicity of chemicals were discussed:

Conference on Evaluating Mutagenicity of Drugs and Other Chemical Agents, 1970 Environmental Mutagen Society.

The following is a technical article:

Chu, E. H. Y., 1971: "Mutagenesis in Mammalian Cells," *Proceedings of the Twenty-fourth Annual Symposium on Fundamental Cancer Research.*

A review article on LSD, also quite technical:

Dishotsky, N. I., W. D. Loughman, R. E. Mogar, and W. R. Lipscomb, 1971: "LSD and Genetic Damage," *Science,* **172**:431–440.

A technical article on radiation mutagenesis:

Russell, W. L., L. B. Russell, and E. M. Kelly, 1958: "Radiation Dose Rate and Mutation Frequency," *Science,* **128**:1546.

A simply written book dealing with effects of radiation:

Wallace, B., and T. Dobzhansky, 1959: *Radiation, Genes, and Man,* Holt, Rinehart and Winston, Inc., New York.

An evaluation of the observations made by the Atomic Bomb Casualty Commission:

Miller, R. W., 1971: "Delayed Radiation Effects in Atomic Bomb Survivors," *Science,* **166**:569–574.

A good, easily comprehensible discussion of various environmental problems:

Commoner, B., 1966: *Science and Survival,* The Viking Press, Inc., New York.

An exposé of the use of radioactive waste as foundation for buildings:

Metzger, H. Peter, 1970: "Project Gasbuggy and Catch-85," *The New York Times Magazine,* Feb. 22, 1970, p. 26ff.

A review article on food additives:

Kermode, G., 1972: "Food Additives," *Scientific American,* **226**: 15–22.

A review article on a new type of nuclear reactor for generating electricity:

Creagan, R. J., 1973: "Boon to Society: The LMFBR," *Mechanical Engineering,* February 1973, pp. 12–16.

A report on the carcinogenic effects of asbestos:

Sherrill, R., 1973: "Asbestos, the Saver of Lives, Has a Deadly Side," *The New York Times Magazine,* Jan. 21, 1973, p. 12ff.

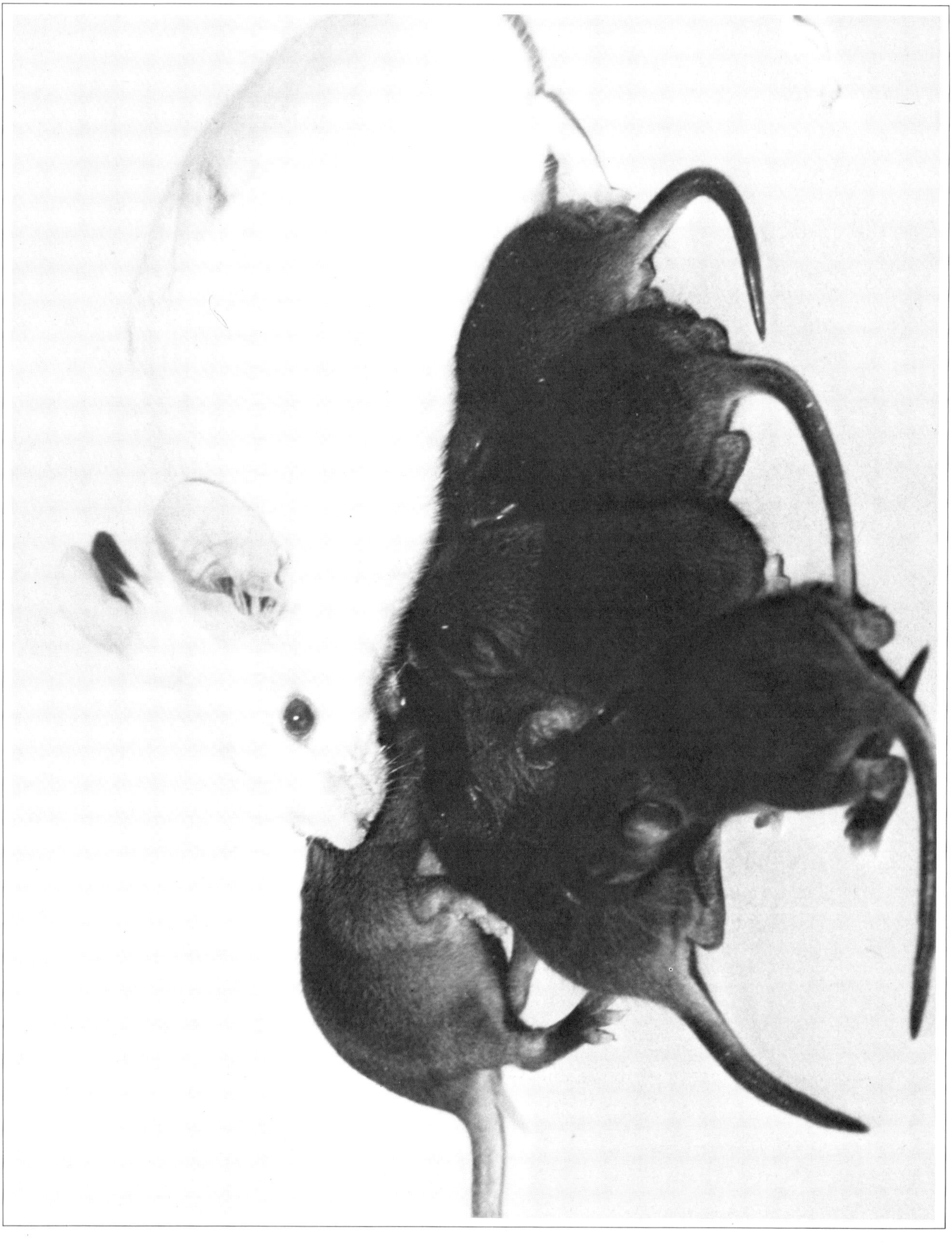

Chapter 16

Now and to Come

The work that has been done in this century on the elucidation of the principles and phenomena of genetics has brought us very close to an understanding of the basic mechanism of life processes. As exciting as recent accomplishments have been, though, most geneticists feel that the real impact of this young science is yet to be felt. Although our comprehension of cellular phenomena is far from complete (in every chapter we have mentioned some of the gaps in knowledge which abound in genetics), the successes that have been reached in molecular genetics lend to an optimism that, given time, many of these gaps will be bridged.

Implicit in such an attitude is that what we come to understand may some day be under our control. Since in genetics we deal with the basis of life itself, the significance of every new discovery is that as we approach our goal of complete elucidation, we approach the possibility of controlling life, with enormous potential to benefit man but also with the ever-present danger of abuse of such knowledge.

THE CONTROL OF LIFE?

"Frozen-thawed" mice and the albino foster mother who gave birth to them. [From D. G. Whittingham, Science, *vol. 74* (*cover photo*), *October 1972.*]

Now and to Come

Recognition of this potential has greatly concerned scientists of all fields. Many symposia have been held to deal with the subjects of euphenics, eugenics, and genetic engineering, the three main avenues by which man may control the destiny of an individual as well as our species in the future.*

*T. M. Sonneborn (ed.), *The Control of Human Heredity and Evolution,* The Macmillan Company, New York, 1965.

RESPONSIBILITY OF THE SCIENTIST

As was stated in Chapter 15, scientists' responsibility for potential uses of knowledge developed in the search for an understanding of natural phenomena extends beyond their own laboratories. More and more, scientists have come to recognize that it *is* their moral obligation to monitor the possible applications of their theories and techniques and to inform the public of possible consequences.

The use of atomic bombs at the end of the Second World War serves as a case in point. One wonders if today's scientific community would have allowed such a destructive force to be loosed upon crowded cities, although it must be pointed out that because of wartime restrictions, only a small number of scientists knew of the existence of nuclear weapons at that time. However, since then, wars in various parts of the world have been allowed to wreak havoc on the same scale of destruction with more conventional weapons; so, one must temper hope with the cynicism that man has evolved intelligence but *not* yet wisdom, and that for every new scientific development there is the real possibility that it can and would be turned against our world rather than used for its good.

What are some of the ways in which new techniques developed in various branches of genetics can be applied to man now, or hold potential to be developed for man and his world that will allow us to deal with increasing biological and medical problems that have been shown to have a genetic basis? This chapter will deal with three major avenues of approach to such problems: euphenics, eugenics, and genetic engineering.

EUPHENICS

EUPHENICS NOW

At the present time, *euphenics,* or the symptomatic treatment of genetic diseases, deals with the control of several inherited diseases, especially inborn errors of metabolism in which the missing or defective enzyme has been identified. One example of this, as mentioned before, is the condition known as *phenylketonuria,* or *pku,* determined by an autosomal recessive gene. Babies with this defect are unable to properly metabolize an amino acid, phenylalanine; the resulting chemical imbalance causes severe mental retardation.

Today it is possible to distinguish pku homozygotes from normal individuals by testing the urine of all newborn babies with ferric chloride. In affected children, the metabolic imbalance caused by the mutation will turn the urine green. Once such a child is detected, a diet free of phenylalanine is imposed, and the child can develop normally.

Although a number of inherited diseases can be treated in a similar manner, these constitute only a small fraction of known inherited diseases. For the most part, we have not identified the biochemical error. In other conditions, such as albinism, even though we know the metabolic block leading to the abnormality, we are powerless to correct it.

EUPHENICS IN THE FUTURE

One possible euphenic measure for the future would be to supply the known missing enzyme to individuals that would allow their cells to complete the required biochemical reaction. Some attempts to do this have been made without much success. Immunological difficulties are encountered, since the enzymes being supplied are antigenic and the body produces antibodies against them. However, as we mentioned in Chapter 9, new techniques in immunology that reduce the antigenicity of substances may play an important role here.

CURE FOR INHERITED ANEMIAS?

In Chapter 12, two anemias, Cooley's and Lepore, resulting from abnormally low hemoglobin production in individuals homozygous for mutations at the beta-chain locus, were discussed. Scientists studying these conditions hope that we may someday discover the factors regulating the beta gene activity and thereby increase the amount of beta chain synthesis. Furthermore, if the mechanism regulating hemoglobin synthesis in the fetus and the adult can be discerned, we might be able to cure lethal conditions, such as thalassemia major and sickle cell anemia, by suppressing synthesis of the abnormal beta chains and allowing fetal or gamma chains to be produced instead. In fact, individuals have been discovered whose β chain locus for some reason never becomes active. Such individuals function normally with fetal hemoglobin even as adults.

INCREASING IMPORTANCE OF GENETICS TO MEDICINE

The increasing numbers of human diseases that are being discovered to have a genetic basis lend great importance to the development of such euphenic measures. Three percent of all humans have hereditary diseases which are transmitted in a Mendelian fashion. Five percent more have "familial" traits which indicate a genetic basis but cannot be analyzed as Mendelian traits, and these numbers are increasing.

Two diseases which account for the deaths of hundreds of thousands a year, cancer and heart disease, are thought to have some heritable component. We have discussed at length the role of immunogenetics in cancer studies. Future work lies in discovering the nature of *resistance* to cancer, which also has a heritable component, and stimulating or creating the same condition in all individuals.

A recent study has shown that some forms of heart disease may be inherited as an autosomal dominant trait.* An understanding of the genetic basis would alert persons from families with an incidence of the disease to the possibility of incurring heart conditions and perhaps cause them to alter their diets and life habits accordingly. (For example, those who may inherit genes for heart trouble would then refrain from smoking and avoid high-fat diets.)

Related to heart diseases are the possible advances in the development of artificial organs, which may one day allow doctors to readily substitute vital artificial organs for natural ones without fear of immune reactions leading to rejection.

The rise in numbers of mutagens and/or carcinogens in our environment (see Chapter 15) has led geneticists to believe that these will continue to increase. Unless a thorough routine of screening for mutagenic effects is developed, science will be hard pressed to devise euphenic measures for each new disease condition in the future. One problem in the use

* L. K. Altman, "Study Links Some Heart Attacks to Genetic Factors," *The New York Times*, Nov. 16, 1972.

of euphenic measures is the existence of *phenocopies,* or conditions that mimic the effect of an inherited disease but actually are caused by other factors. To treat every case of albinism in the same manner, for example, may compound the situation, since albinism may be caused by blockage at different stages in the series of reactions leading to the production of pigmentation.

To restore affected individuals to normalcy by euphenic measures is the only compassionate goal for scientists to maintain, but to do so is to counter the forces of natural selection that are the basis for the evolutionary strength of a species. Pku homozygotes, for example, would normally not reproduce and transmit the harmful mutations to future generations. Selection against them would be total; however, when they do develop normally and produce progeny, every member of the next generation would be a carrier of the mutation, assuming that the spouse is normal. This has to add to our genetic load and to weaken the human species from the evolutionary point of view.

Should we then, as some believe, counsel such individuals against reproduction, or even legally forbid them from having children for the good of the species? And if we decide on this course for homozygotes of pku, should we not extend the same philosophy to those whose families have a history of heart trouble or schizophrenia? Such questions lead us to the area of eugenics.

EUGENICS

The increase or decrease of certain genetic constitutions by selective reproduction (or restraint of reproduction) is known as *eugenics,* an approach which more and more is making its impact felt in society. At the present time, the practice of eugenics has mostly taken the form of genetic counseling. As more and more inherited conditions are discovered, the importance of genetic counseling has become manifest, and its practice is growing.

GENETIC COUNSELING

The role of the genetic counselor, as was discussed in Chapter 1, is to inform concerned individuals of the nature of the mutant condition that concerns them. If it is inherited in a Mendelian fashion, then the probability of producing affected offspring can be calculated. The final decision for taking a risk, however, is entirely the responsibility of the individuals involved; it cannot be made by the counselor.

PRENATAL DIAGNOSIS OF INHERITED DISEASES

Because of recent advances in medical technology, the phenotype of the progeny can sometimes be predicted with certainty. A technique known as *amniocentesis* has been developed whereby a needle is inserted into the womb of a pregnant female and some of the amniotic fluid is withdrawn. Floating in the fluid are some of the cells of the embryo, which can then be cultured and studied for chromosomal abnormalities, as in families with a history of Down's syndrome, or tested for biochemical insufficiencies.

More than 30 different diseases can now be detected prenatally using amniocentesis.* Again, should an abnormal condition be detected, the decision for abortion is one which must be made by the parents, taking moral and legal aspects of the problem into consideration. In some cases, such as lethal syndromes, the decision is not as difficult as others which involve an abnormal but viable child.

*T. Friedman, "Prenatal Diagnosis of Genetic Disease," *Scientific American*, November 1971, pp. 34–43.

LAWS RESTRICTING MARRIAGES

Very little is now being done to impose restrictions on human reproduction. However, a few laws exist that prescribe who may or may not marry and/or reproduce. In Denmark, a law exists allowing for the sterilization of individuals whose IQ is 75 and below. Another law allows for sterilization on grounds of "a danger of hereditary affliction of progeny." A woman may be sterilized following a therapeutic abortion "provided the indication for this abortion is genetic and attributable to the mother."†

† J. Mohr, "Genetic Counseling," *Proceedings of the Third International Congress of Human Genetics*, Chicago, 1966.

In an article summarizing laws in the United States which prohibit certain types of marriages, M. Farrow and R. Juberg state:

A person may not marry a parent, a grandparent, child, or grandchild except in Georgia, where a man is not prohibited from marrying his daughter or grandmother. While all political units prohibit marriage between a person and a sibling, an aunt, or an uncle, their prohibitions vary considerably for other degrees of collateral relationship. The uncle-niece marriage is not prohibited in Georgia and among Jews in Rhode Island. Generally, marriage between persons with a coefficient of relatedness equivalent to first cousins or closer has been prohibited.*

* M. G. Farrow and R. C. Juberg, "Genetics and Laws Prohibiting Marriage in the United States," *Journal of the American Medical Association,* **209**: 534, 1969.

(From a geneticist's point of view, the exception for Jews in Rhode Island seems somewhat anachronistic!)

EUGENICS AND PEST CONTROL

Much eugenics *is* practiced by man on *other* forms of life, however. For example, one method of pest control which is gaining favor is the development of sterile insects that are released to mate with insects in the wild. This method has led to success in controlling the screwworm, a parasite of cattle in the southwestern United States. The advantage of such programs lies in controlling the organism without having to use chemicals which would destroy more than just the pest.

GERMINAL CHOICE FOR MAN

As for future possibilities, several techniques have been developed which could have implications for man. One is the freezing of embryos or reproductive cells of many higher organisms (including human sperm) in such a manner that they may be thawed and, in the case of gametes, used in artificial insemination. (See introductory illustration to this chapter.) This would allow couples to plan exactly on the age at which to conceive children, the time of year they wish to have them, and perhaps even allow their choice of germ cells. For example, although it is not yet possible to do so, scientists are working on methods of distinguishing X-bearing sperm from Y-bearing sperm. In the last years of his life, one of the most vigorous proponents of this method of eugenics for the betterment of man was the Nobel Prize winner Hermann Muller.

Muller advocated the establishment of "sperm banks" in which sperm:

. . . derived from persons of very diverse types but including as far as possible those whose lives had given evidence of outstanding gifts of mind, merits of disposition and character, or physical fitness. From these germinal stores couples would have the privilege of selecting such material, for the engendering of children of their own families, as appeared to them to afford the greatest promise of endowing their children with the kind of hereditary constitution that came nearest to their own ideals.*

* H. J. Muller, "Means and Aims in Human Betterment," in T. M. Sonneborn (ed.), *The Control of Human Heredity and Evolution*, The Macmillan Company, New York, 1965, p. 115.

This was an astonishing program to be advocated by such a prominent man of science. Genetically, there is no guarantee that the sperm or eggs from, say, musicians would contain all the genes that contribute to musical talent. And even if they did, the environmental effect on the development of the genetic potential for musicianship would be a factor that such banks could not hope to control.

Further, from the moral point of view, who is to decide what constitutes "merits of disposition and character"? As Dobzhansky has wryly observed, who would be the president of such a bank? However, even now we make decisions on the moral fitness of individuals. We have mental and penal institutions for those whom we consider unfit for participation in society. Muller would simply extend the limits of our judgment on who is or is not desirable.

TEST TUBE BABIES?

We now possess techniques that would make such a program, if desired, possible. Add to this the successful fertilization of human eggs in vitro that can be developed to the time of implantation—and which, if inserted into the uterus, give all indications of being able to develop normally to term—and you have the real prospect of test tube babies for the future.

CLONING AND A BRAVE NEW WORLD?

Furthermore, one can foresee the possibility of joining these methods with cloning techniques to bring eugenics one step closer to the "brave new world" envisioned by Aldous Huxley decades earlier. In Chapter 8 we described experiments by Briggs and King, and others, in which clones of amphibians had been developed through nuclear transplantation. All the frogs of a particular clone were genetically identical. The cloning of human cells in tissue culture is a well-known tool of cellular biologists (Figure 16.1).

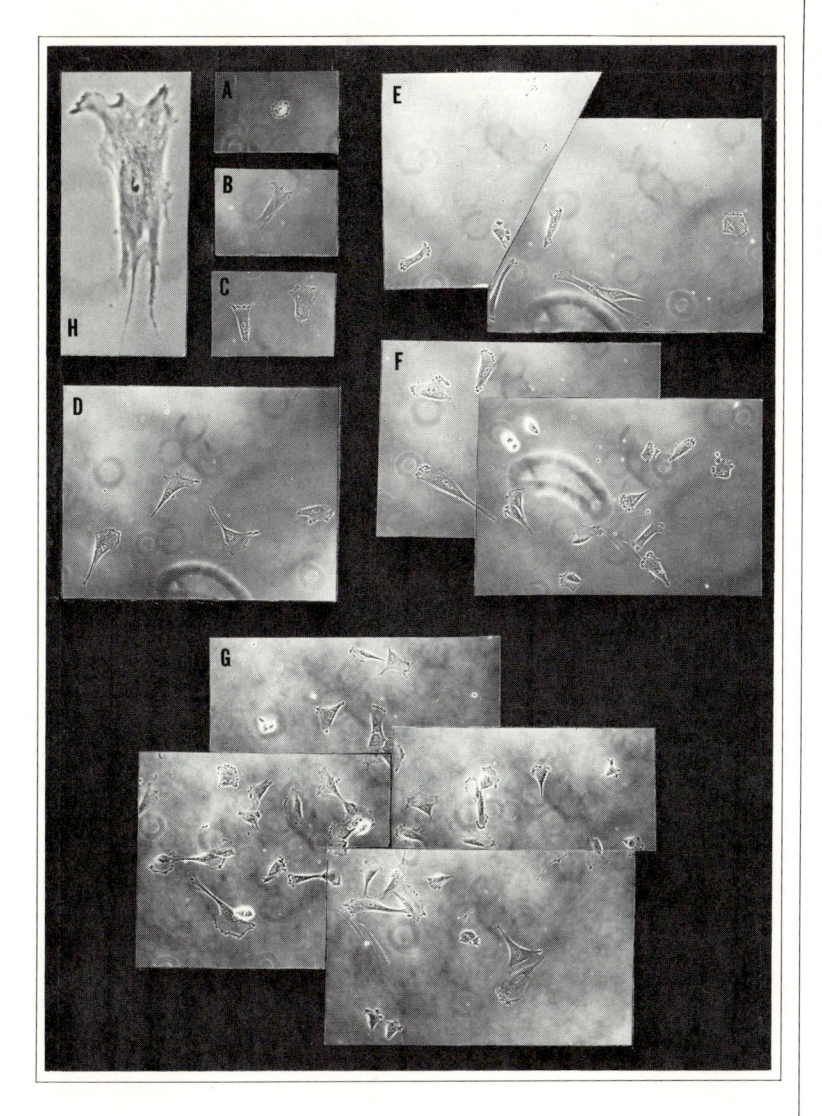

FIGURE 16.1
A series of photomicrographs of the formation of a clone of diploid human cells, all derived from the single cell in (A), through mitosis. (H) is an enlargement of the cell in (B). [Courtesy of Dr. P. DeMars, Medical Genetics, University of Wisconsin.]

Although the cells of these clones are not derived through nuclear transplants, since human egg cells are too small to manipulate in the same manner as frog eggs, they are also genetically identical. Someday cloning of cells from a zygote created in tissue culture by artificial insemination may be developed, and from one such egg cell, many identical individuals may be developed. If, as some geneticists have suggested, we may someday be able to screen germ cells for the desired combinations of genes and chromosomes, we would also be able to determine what types of individuals should be cloned. But one point should be made here: what may appear desirable now may not be so in the distant, or perhaps not so distant future, depending on our society and environment.

These and other methods of control over reproduction are at present repugnant to most people. Yet, given our rapidly changing world and the projected overcrowding of this finite planet, we cannot ignore the possibility that such methods may become necessary for the survival of man. That an astute, highly respected geneticist such as Muller was willing to subject himself to criticism by publicly avowing eugenic measures is in itself a sobering sign of our times.

Whereas it is the responsibility of scientists to inform the public of possible future developments, the informed public must then take steps either to prevent the developments or to deal with them responsibly. If reproductive freedom is to be preserved, for example, then birth control must be utilized today so that in the future we will not have the population explosion that would force governments to decide who shall reproduce, and when, and how often. [Toward this end, a survey in 1973 showed that the average number of children per family in the United States has dropped to 2.03, or below the number (2.1) necessary to maintain the present rate of population growth.]

Although individual freedom is a precious concept, we must decide whether the sacrifice of this freedom would be as distasteful as it might appear on the surface, if we could in the future determine the nature of an individual by controlling his genome and his environment to ensure that he will be a satisfied, happy, congenial person. Would this be so much worse than the competitiveness, aggression, and egocentricity leading to man's present inhumanity to man?

But because of man's history of turning potentially beneficial techniques against his fellow man, it is also possible that if reproduction were controlled, it might be used to develop populations of hostile, violent people. This possibility in itself may be reason enough to object to control over human reproduction.

These considerations are based on projected applications of some techniques which have been successful so far only in experimental animals. Whether they will ever be used on man remains for the future to determine. With all the problems that prevent geneticists from analyzing human genomes, such as the inability to identify where most genes lie on the chromosomes, it is difficult not to be skeptical that such control will be possible in the future. However, considering the advances that have been made in the last 20 years, perhaps we should heed the claims of the more optimistic geneticists and begin to plan for such programs even now.

GENETIC ENGINEERING

The work of microbial geneticists over the past 20 years has led to speculation that someday we may be able to alter man's genetic makeup. The production of new genes and the alteration of genomes, known as *genetic engineering,* has evoked great public interest because of projections of someday being able to create genes, beneficial mutations, and various other processes that constitute the creation of life itself. Many of these have already been accomplished using microorganisms.

ARTIFICIAL SYNTHESIS OF DNA

Based on the double helix concept of the structure of a gene, experiments have led to the synthesis, by Kornberg and his associates,* of DNA molecules that showed biological activity. Deoxyribonucleotides, enzymes, ions, and a piece of "primer DNA"—DNA from a virus—were mixed together to form new DNA molecules. These new molecules, when inserted into bacteria, caused the formation of new virus particles (Figure 16.2).

The implications of this experiment are profound. If it is possible to synthesize artificial DNA in a test tube, then it may also be possible to create specific genes which may then be inserted into cells to cause transformation. One can foresee the possibility of correcting mutant genes in this manner. For example, if the mutation for galactosemia can be identified, then a normal gene which codes for the missing enzyme preventing the homozygotes from metabolizing galac-

* A. Kornberg, *Enzymatic Synthesis of DNA,* John Wiley & Sons, Inc., New York, 1961.

FIGURE 16.2
Schematic illustration of the artificial synthesis of DNA. The artificial DNA has been shown to be biologically active because, when inserted into a bacterium, it directs the synthesis of new viral particles exactly like the virus from which the "primer" DNA was taken.

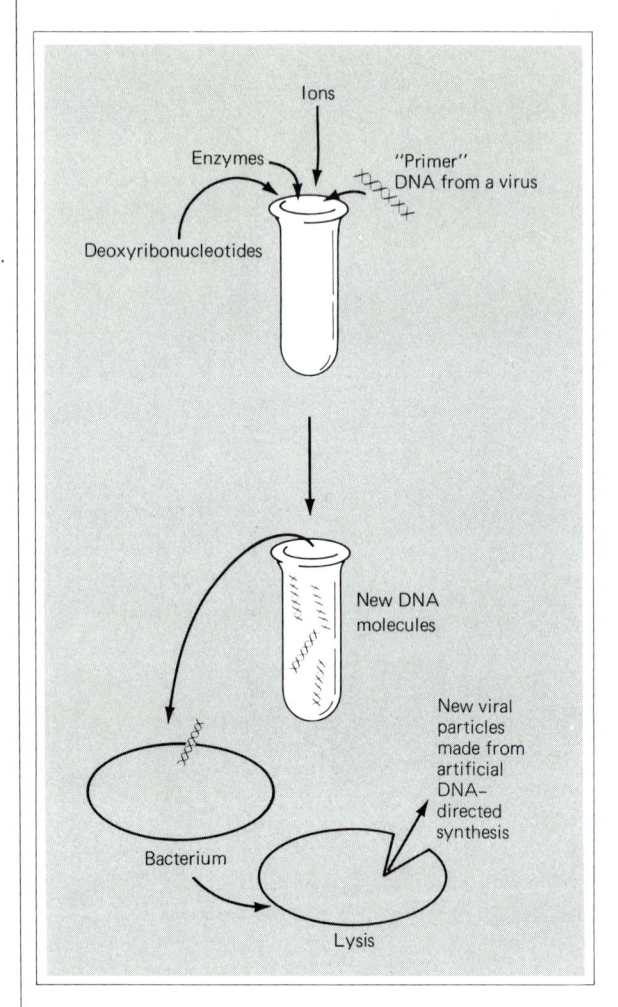

tose, a subunit of a sugar found in milk, could be synthesized and introduced into the individual's cells. This would supply the cells with the needed enzyme and would in effect be an artificially created back mutation. Some geneticists have suggested culturing mutant human cells and screening them for back mutations which could be reintroduced into the individual.

LAMBDA PHAGE AND THE LAC OPERON

A phenomenon in microorganisms which may serve in genetic engineering is *transduction,* the process by which a

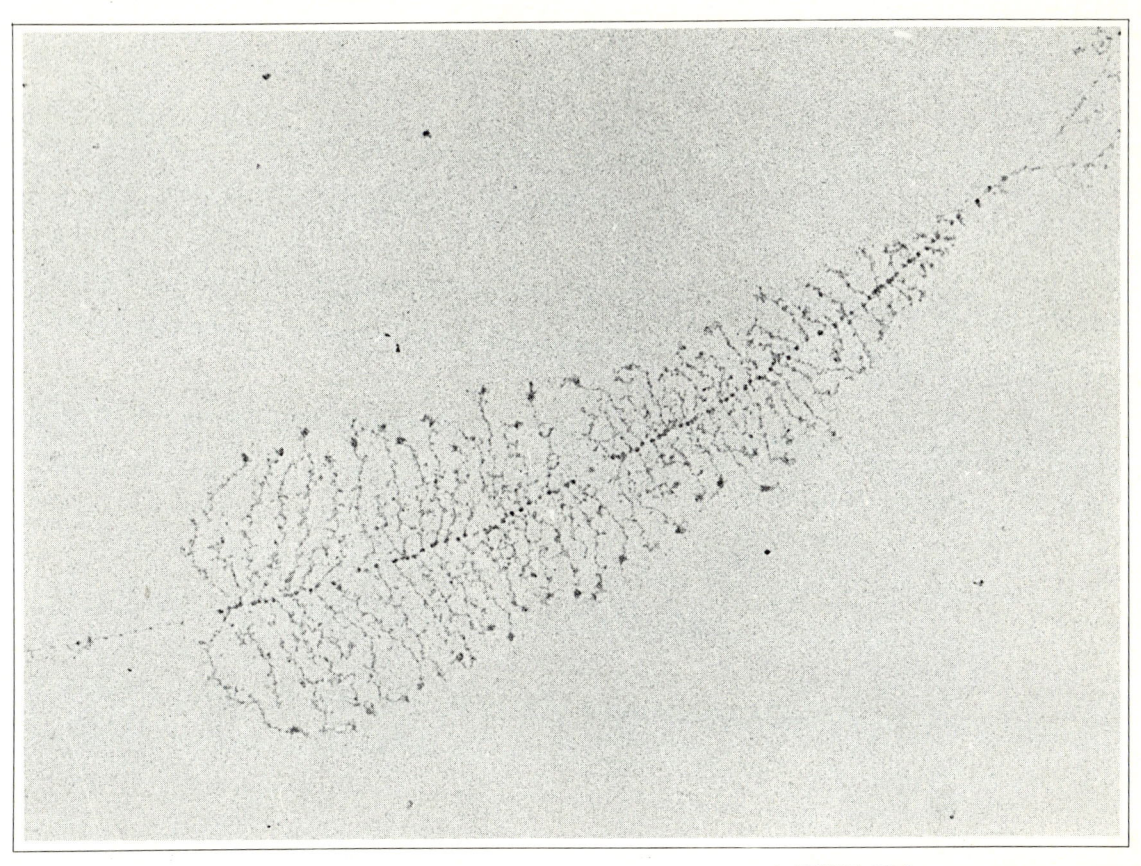

FIGURE 16.3
Electron micrograph of a single nucleolar gene isolated from the amphibian *Triturus*. The main axis is the DNA molecule from which project "matrix fibrils" believed to be transcribed RNA. [*Courtesy of O. Miller and B. Beatty, "Portrait of a Gene,"* Journal of Cellular Physiology, *vol. 74 supplement, pp. 222–232, 1969.*]

virus carries genetic material from one cell to another, causing the second host to be changed after incorporating the genetic material into its own genome. It has been found that some viruses are gene-specific with regard to transduction. For example, some derivatives of lambda viruses are known to transduce the lac operon in bacteria.

This fact was utilized by J. Shapiro and his co-workers at the Harvard Medical School in isolating some of the lac operon genes, including the promoter, operator, and one of the structual genes, β-galactosidase. Single genes had been visualized before (Figure 16.3), but this was the first time that a specific gene had been intentionally isolated and transferred from cell to cell. Figure 16.4 shows the single lac gene.

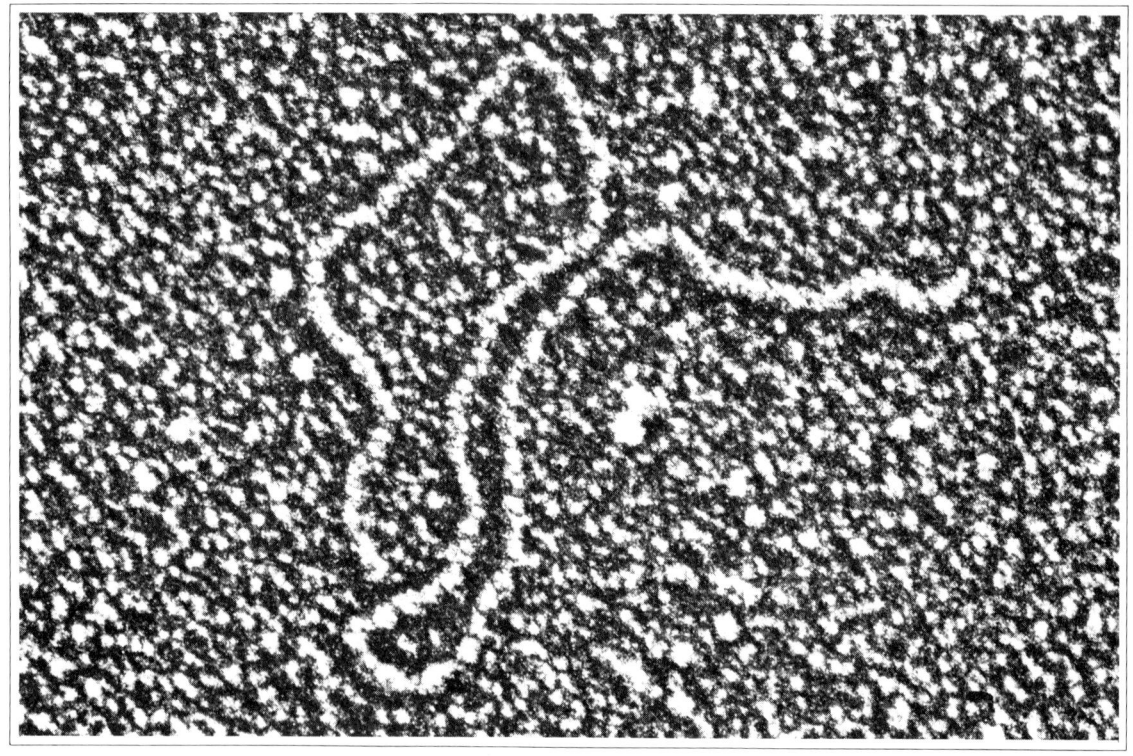

FIGURE 16.4
The single lac duplex. A duplex is formed by the joining of the two strands of DNA. [*Department of Microbiology and Molecular Genetics, Harvard Medical School.*]

One immediate possible application of the isolation of genes is to use specialized viruses to transduce normal genes into the cells of mutant individuals.

GENE INSERTION IN MAMMALIAN CELLS

C. R. Merril and his co-workers at the National Institutes of Health have been concerned with the treatment of galactosemia. They have treated human fibroblast cells from a galactosemia homozygote in tissue culture with phage bearing the *gal* operon from bacterial cells, and found what appeared to be enzyme activity in the fibroblasts which persisted through eight cell divisions. If true, and experimentation continues today, this would indicate that bacterial and bacteriophage DNA *can* enter human cells, and that at least some of the DNA can undergo replication and transcription in human cells.

P. I. Qasba and H. V. Aposhian at the University of Maryland School of Medicine have also shown that polyoma *pseudovirions* (particles containing viral protein coat and DNA from other cells) carrying DNA from mouse cells can insert this DNA into human cells in tissue culture systems. It is possible that DNA inserted in this manner may become incorporated into the chromosomes of the cell, although evidence for this has not been found.

In the future, viruses may be found or produced that transduce into humans different genes to be used for the correction of various mutant conditions resulting from mutations at those loci. In addition to the lambda virus, phages have been found that transduce the *tryptophan, maltose,* and *arabinose* operons from chromosomes of the bacteria *Escherichia coli* and *Salmonella.*

Here again, such techniques must never be allowed to be turned against man. The possibility of a genetic-bacterial-viral warfare conjures up visions of catastrophes of unbelievable proportions. If the consequences of nuclear warfare are abhorrent, possible results of the manipulation of genetic systems for hostile acts are many times more horrible.

THE FUTURE—WITH IMAGINATION

Knowing the base sequences of specific genes, as well as techniques for artificially synthesizing genes, gives rise to the tantalizing possibilities that we may someday alter the genetic makeup of man in more profound ways than have been imagined. One can envision, for example, the possibility that genes could be coded for the enzymes that allow the digestion of plants now inedible for man—such as grass or even wood and wood by-products—synthesized, and introduced into human systems. These enzymes would then allow man to subsist on diets of wood or grass or paper. As a molecular geneticist has observed, one could then say to a friend, with special meaning, "I certainly enjoyed your letter last week!" The benefit of such genetic engineering would lie in the elimination of famines.

It may also be in our power to alter various characteristics, not only physiological but structural as well, such that inhabitants of dry, hot climates might be intentionally endowed with adaptive traits similar to those found in the camel or kangaroo rat (see Chapter 13).

Still another area of great potential benefit to man would be to analyze and reproduce genetic differences between man and simpler forms of life, such as urodele amphibians, which

FIGURE 16.5
The results of experiments on regeneration of amputated limbs in the salamander *Ambystoma*. (A) Normal forelimb. (B) If the right forelimb is amputated at the shoulder and the hand is removed, and the remaining portion inserted into the back musculature, (C) the original stump regenerates into a normal limb, and the implant also regenerates a hand. [Adapted with permission of Macmillan Publishing Co., Inc., from Patterns and Principles of Animal Development, *by John W. Saunders. Copyright © 1970 by John W. Saunders, Jr.*]

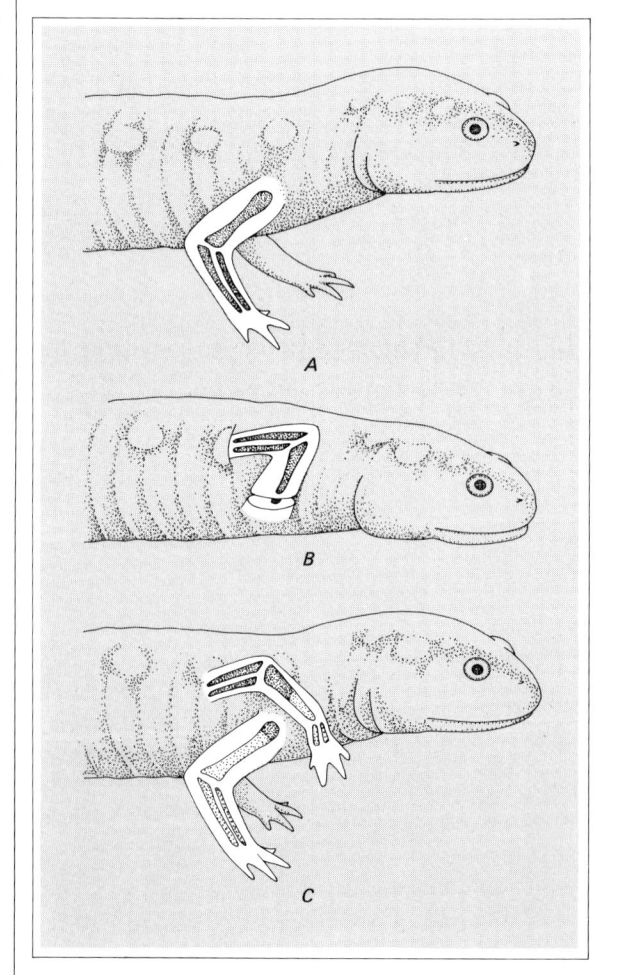

show remarkable properties of regeneration. For example, if a leg of a salamander is removed, it will be regenerated (Figure 16.5). If we could discover the differences between cells of such organisms capable of regeneration and those of man, which lack this capacity (and there is no question that such differences are ultimately genetic), then we may be able to encourage the regeneration of organs from an individual's own cells in tissue culture. Such organs would then replace worn and diseased organs without causing immunological reactions.

Most of these projections for the future are still only products of scientific imagination. Whether any of them will come true is a moot question. Although there is no doubt that gene synthesis and isolation have now been accomplished, bear in mind that they are mainly the products of experiments with microbial systems, in which the genetic material for the most part is naked strands of DNA. Recall, too, that bacteria and viruses are haploid. To extend the experiments to an organism of man's genetic complexity is not possible at the present time.

While it is easy to perceive the mechanism of gene transfer between microbes, it is far more difficult to envision the same process in the complex cells of man, in which chromosomes are composed not only of nucleic acids but also of proteins. To speak of genetic engineering in an organism in which most genes have not been identified and in which few of the autosomal genes that have been identified can be relegated to a particular chromosome (much less a particular locus on the chromosome or base sequence) is, at the least, premature.

To conjecture on the possible applications of genetics to man, as we have in this chapter, is, however, not merely an exercise in imagination. While it is true that the genetic problems of multicellular organisms are now seemingly impenetrable barriers, the brilliant successes attained so far bring all that we have postulated within the realm of possibility.

Such advances would directly or indirectly affect every human on earth one day. And when the possibilities are established, each one must be analyzed not just by the scientific community but by all people, as to whether they would lead to an improvement of man and his world. This task will be most difficult, requiring foresight of a quality that man has not to date exhibited. There is only one way that correct decisions can be approximated, and that is if there is knowledge on which the decisions can be based.

While the responsibility of scientists is to inform the public, the responsibility of the public is to remain informed and interested. There is no question that genetics holds much potential for the benefit of man in the future but since we are dealing with the basis of life itself, any mistakes we make may be literally fatal. Whether we use our knowledge for the good of man and his world will answer our question as to whether man has evolved not only intelligence but wisdom as well.

REFERENCES

The following are three meetings dealing with the potential applications of genetic advances in the future. All should be digestible by the layman with some genetics background:

"The Prospects of Gene Therapy," Fogarty International Center, National Institute of Health, Bethesda, Md. 1971.

Grobman, A. B. (ed.), 1970: *The Social Implications of Biological Education,* National Association of Biology Teachers, Philadelphia.

Sonneborn, T. M. (ed.), 1965: *The Control of Human Heredity and Evolution,* The Macmillan Company, New York.

Muller's paper on eugenics:

Muller, H. J., 1965: "Means and Aims in Human Betterment," in T. M. Sonneborn (ed.), *The Control of Human Heredity and Evolution,* The Macmillan Company, New York, pp. 100–122.

Other articles and references mentioned in this chapter:

Altman, L. K., 1972: "Study Links Some Heart Attacks to Genetic Factors," *The New York Times,* Nov. 16.

Farrow, M. G., and R. C. Juberg, 1969: "Genetics and Laws Prohibiting Marriage in the United States," *Journal of the American Medical Association,* **209**:534–538.

Friedmann, T., 1971: "Prenatal Diagnosis of Genetic Disease," *Scientific American,* November, pp. 34–43.

Kornberg, A., 1961: *Enzymatic Synthesis of DNA,* John Wiley & Sons, Inc., New York.

Mohr, J., 1966: "Genetic Counseling," *Proceedings of the Third International Congress of Human Genetics,* Chicago.

Qasba, P. K., and H. V. Aposhian, 1971: "DNA and Gene Therapy: Transfer of Mouse DNA to Human and Mouse Embryonic Cells by Polyoma Pseudovirions," *Proceedings of the National Academy of Sciences,* **68**:2345.

Merril, C. R., M. R. Geier, and J. C. Petricciani, 1971: "Bacterial Virus Gene Expression in Human Cells," *Nature,* **233**:398.

Appendix A

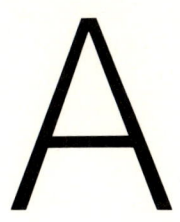

Answers to Chapter Problems

CHAPTER 1

2. 3 normal : 1 albino.
4. (a) None.
   (b) Half.
5. (b) $\frac{1}{4} \times \frac{1}{4} = \frac{1}{16}$
6. (b) $\frac{1}{4} \times \frac{1}{4} \times \frac{1}{4} = \frac{1}{64}$

CHAPTER 2

1. (A) Anaphase II of meiosis; (B) prophase I of meiosis; (C) metaphase of mitosis; (D) prophase II of meiosis; (E) prophase I of meiosis; (F) telophase of mitosis.

CHAPTER 3

2. (a) Half.
   (b) One-fourth.
3. All sons would be color-blind; all the daughters would be normal.
4. (a) One-fourth.

Answers to Chapter Problems

CHAPTER 4
1. (a) *Rr*; (b) *Rr* or *RR*; (c) *Rr*; (d) *Rr*; (e) *rr*; (f) *Rr*; (g) *Rr*.
2. (a) Autosomal dominant.
   (b) Any letter may be used. We shall use *T*. (X) *Tt*; (Y) *tt*; (Z) *tt*.
3. (a) Sex-linked recessive.
   (b) (A) *Ll*; (B) *LY*; (C) *lY*; (D) *LL*; (E) *Ll*.

CHAPTER 5
1. (b) Three-sixteenths.
2. Two different independently assorting pairs of genes are involved in pigment formation. If we assume one pair to be *aa* and another *cc*, the two albinos could have had the genotypes *AAcc* × *aaCC*. All the offspring would be dihybrids, *AaCc*, and therefore normally pigmented.
3. (a) Dominant gene.
   (b) The class of offspring that is homozygous dominant is missing, due to the fact that the mutant gene is lethal in the homozygous condition.
4. (b) All normal.
5. (a) (A) $\frac{CH}{Ch}$; (B) $\frac{Ch}{\rightarrow}$; (C) $\frac{Ch}{cH}$; (D) $\frac{cH}{\rightarrow}$; (E) $\frac{CH}{cH}$ or $\frac{CH}{Ch}$; (F) $\frac{ch}{\rightarrow}$
   (b) Crossing-over in one of the gametes of individual C.
7. 0.4*Rrss*, 0.4*rrSs*, 0.1*RrSs*, 0.1*rrss*.
9. (a) With independent assortment, the expected ratios of the progeny classes would be 1:1:1:1. The actual results show a decided deviation from this distribution.
   (b) 1*GgHh*:1*Gghh*:1*ggHh*:1*gghh*.

Appendix B

The Chi-Square Test

Because genetic experiments have usually relied on the analysis of data obtained from crosses of experimental plants and animals, it is important for geneticists to be able to determine when deviations from expected ratios are due to chance alone, or to some unexpected factor other than chance.

For example, in the tossing of a coin, one expects heads to turn up half the time and tails half the time; we thus say the chance for heads or tails is one-half. But if the coin is tossed only a few times, perhaps four, not surprisingly we may get heads three times and tails only once. To ascertain whether this deviation from the expected 2:2 ratio is due only to chance or perhaps to some defect of the coin, we can toss it many more times and would expect a closer and closer correlation with our expected ratio of 1:1.

But suppose we tossed three coins 50 times each and obtained the following ratios: 22:28, 30:20, 40:10. Which of these cases is due to chance alone, or are the deviations in any or all cases significant enough to warrant investigation into some other factor, such as the construction of the coins?

To translate this into genetics experiments, suppose we made a cross between one of Mendel's pea plants that is homozygous recessive and one that is heterozygous for the genes determining height. The cross would be $tt \times Tt$. Our expected ratio from this cross would be 1 Tt (or tall) to 1 tt (or short). Suppose the offspring from this cross in three different experiments show the ratios 22 tall to 28 short, 30 tall to 20 short, and 40 tall to 10 short.

The statistical tool which can be used to determine the "goodness of fit" of our actual results to our expected results is the *chi-square test*, which allows us to decide whether the deviations are due to chance alone or to some other factor which must then be investigated. A formula for calculating chi square (or χ^2) is

$$\chi^2 = \Sigma \left[\frac{(o-e)^2}{e} \right]$$

which, translated into words, means

Chi square = the sum of $\dfrac{\text{(observed results minus expected results)}^2}{\text{expected results}}$

In both the coin tosses and the crosses of pea plants mentioned above, the expected proportion for each class (of coin sides or tall and short offspring) is 50 percent, or 0.5. Let us use the chi-square formula to test the first set of results, 22:28, as shown in the accompanying table.

CLASS	OBSERVED RESULTS o	EXPECTED RESULTS e	DEVIATION $o-e$	$(o-e)^2$	$(o-e)^2/e$
Tall	22	25	-3	9	0.36
Short	28	25	3	9	0.36
					0.72

Having obtained the value of χ^2 as 0.72, we may now use a *table of chi square* which has been drawn up by statisticians and which we need not discuss more than to say that it allows us to determine whether a particular chi-square value can be expected to occur for a particular experiment due to chance alone. Since the chi-square value reflects the amount of deviation from expected ratios, the chi-square table in effect allows us to determine whether the deviations we have obtained are due to chance alone.

An abridged chi-square table follows.

	PROBABILITIES p								
DF	0.95	0.90	0.70	0.50	0.30	0.20	0.10	0.05	0.01
1	0.004	0.016	0.15	0.46	1.07	1.64	2.71	3.84	6.64
2	0.10	0.21	0.71	1.39	2.41	3.22	4.61	5.99	9.21
3	0.35	0.58	1.42	2.37	3.67	4.64	6.21	7.82	11.35
4	0.71	1.06	2.20	3.36	4.88	5.99	7.78	9.49	13.28
5	1.15	1.61	3.00	4.35	6.06	7.29	9.24	11.07	15.09

Source: Abridged from Fisher and Yates, *Statistical Tables for Biological, Agricultural and Medical Research,* Table IV, Oliver and Boyd Ltd., Edinburgh.

There are two aspects of the chi-square table that we must discuss briefly, *df* and the probability values from 0.95 to 0.01. The letters *df* stand for *degrees of freedom*. This term refers to the number of classes that can be said to be variable. Without going into sophisticated concepts of statistics, we may explain the degrees of freedom, which is always *one fewer than the number of classes involved* in the experiment, by using the analogy of putting on a pair of gloves. Although there are two gloves in a pair, there is only one choice that can be variable, the first choice: you can either put on a right glove or a left glove. Once the first choice has been made, and one of the gloves donned, then there can be no variation in the choice of the second glove.

The probability values refer to the proportion of times a particular experiment has a deviation that can be expected to be due to chance alone. The value of $p = 0.95$, for example, means that 95 percent of the time that a particular deviation from expected values occurs, that deviation is due to chance alone. Obviously, then, such a deviation would not be considered significant, and the results can be said to fit the expected numbers. Statisticians have established that whenever a chi-square value is equal to or less than a probability level of 0.05, that deviation is significant, for it means that only 5 percent or less of the time that the deviation occurs can it be expected to be due to chance alone.

To return to our experiment above: we have determined that the chi-square value is 0.72. Since we are dealing with two classes (the tall and short, or heads and tails), there is one degree of freedom. Looking at the chi-square table, we find that 0.72 for one degree of freedom falls between $p = 0.5$ and $p = 0.3$. The deviation is therefore not significant, and the ratio of 22:28 does fit the expected 1:1 ratio.

Try to establish the significance of the deviations obtained in the other two experiments using the chi-square test. (*Hint:* The ratio of 30:20 is not significant; however, the deviation found in the ratio of 40:10 is highly significant.)

The chi-square test may be used regardless of the number of classes involved. For dihybrid crosses, the results would involve the four phenotypic classes found in the classical 9:3:3:1 ratio. To apply the chi-square test to such experiments, the same steps are taken and the same formula is used. The only difference between the dihybrid cross and the one we discussed above is that there would be three degrees of freedom for the dihybrid cross results.

Appendix C

Determination of Linkage and Chromosome Mapping Using Crossover Frequencies

Let us use the experiment by Bateson and Punnett to analyze the distance between the *A* and *B* loci. Given our knowledge of the phenomena of linkage and crossing-over, we can make the following Punnett square, indicating which gametes are the parental types and which are the recombinants.

		Parental		Recombinant	
		AB	*ab*	*Ab*	*aB*
Parental	*AB*	*AABB*	*AaBb*	*AABb*	*AaBB*
	ab	*AaBb*	*aabb*	*Aabb*	*aaBb*
Recombinant	*Ab*	*AABb*	*Aabb*	*AAbb*	*AaBb*
	aB	*AaBB*	*aaBb*	*AaBb*	*aaBB*

Because of the law of dominance and recessiveness, the only *genotype* whose frequency can be exactly determined from the phenotype data is the homozygous recessive for both pairs of genes, *aabb*, which was found to be at a frequency of approximately $\frac{3}{14}$, or 0.21.

Since the law of probability we discussed in Chapter 1 states that the probability of two independent events occurring simultaneously is the product of their individual probabilities, that is, *ab* × *ab* = *aabb* = (*ab*)², then it stands to reason that the probability or frequency of occurrence of *ab* can be figured by taking the square root of *aabb*. Thus, if the frequency of *aabb* offspring is 0.21, then the frequency of the *ab* gamete from each parent is $\sqrt{0.21}$ = 0.46:

	AB	ab (0.46)	Ab	aB
AB				
ab (0.46)		aabb (0.21)		
Ab				
aB				

Since we are dealing here with a single pair of homologous chromosomes, one member of which carries both dominant genes, the other the recessives, it stands to reason that there are as many *AB* chromosomes as there are *ab* chromosomes, or also 0.46. Since all of the gametes should total 100 percent (or *AB* + *ab* + *Ab* + *aB* = 100 percent), then the number of recombinant gametes must equal 1.00 minus the parental types which add up to 0.92. In other words, the total number of *Ab* and *aB* gametes equals 1.00 − 0.92 = 0.08.

If crossing-over between the *A* and *B* loci occurs at a frequency of approximately 8 percent, we then say that the two genes are eight units apart, and the chromosome can be drawn in the following manner to indicate the distance between the *A* and *B* loci:

$$\overline{A - 8 - B}$$

The Punnett square can now be redrawn showing the frequencies of all gametes from both parents.

Note that we may now calculate the frequency of each genotype if we wish. For example, the frequency of *AAbb* individuals can be ex-

Determination of Linkage and Chromosome Mapping Using Crossover Frequencies

		Parental 0.92		Recombinant 0.08	
		AB 0.46	ab 0.46	Ab 0.04	aB 0.04
Parental 0.92	AB 0.46	AABB 0.212	AaBb 0.212	AABb 0.018	AaBB 0.018
	ab 0.46	AaBb 0.212	aabb 0.212	Aabb 0.018	aaBb 0.018
Recombinant 0.08	Ab 0.04	AABb 0.018	Aabb 0.018	AAbb 0.002	AaBb 0.002
	aB 0.04	AaBB 0.018	aaBb 0.018	AaBb 0.002	aaBB 0.002

pected to occur about $0.04 \times 0.04 = 0.0016$, or 0.002, to round off the figure.

The nail-patella syndrome and the ABO blood group loci in humans can serve as examples of how linkage data may assist geneticists in determining the genotypes of human offspring, including recombinant genotypes.

If we assume that the two loci are seven units apart, what types of progeny can be expected from a cross of two people whose genotypes are $I^A i Nn \times I^A I^B Nn$ and whose chromosomes are as shown in Figure C.1?

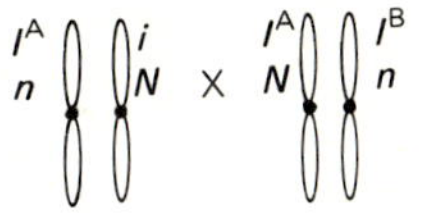

The Punnett square for this cross is shown below.

FIGURE C.1

		Parental 93%		Recombinant 7%	
		$I^A n$ 0.465	iN 0.465	$I^A N$ 0.035	in 0.035
Parental 93%	$I^A N$ 0.465	$I^A I^A Nn$ 0.216	$I^A iNN$ 0.216	$I^A I^A NN$ 0.016	$I^A iNn$ 0.016
	$I^B n$ 0.465	$I^A I^B nn$ 0.216	$I^B iNn$ 0.216	$I^A I^B Nn$ 0.016	$I^B inn$ 0.016
Recombinant 7%	$I^A n$ 0.035	$I^A I^A nn$ 0.016	$I^A iNn$ 0.016	$I^A I^A Nn$ 0.001	$I^A inn$ 0.001
	$I^B N$ 0.035	$I^A I^B Nn$ 0.016	$I^B iNN$ 0.016	$I^A I^B NN$ 0.001	$I^B iNn$ 0.001

Determination of Linkage and Chromosome Mapping Using Crossover Frequencies

The percentage figure within each progeny square is the probability for that genotype.

A simpler and more direct way of determining crossover frequencies would be the use of a testcross. Since, by definition, one of the parental genotypes of a testcross is homozygous recessive, the genotypes of all gametes produced by the hybrid parent would be immediately reflected in the phenotypic ratio of the offspring. The progeny ratio, in other words, would allow a direct assessment of the map distance between genes.

For example, a testcross between dihybrids (with two pairs of linked genes in the coupling phase) and homozygous recessive individuals would involve chromosomes shown in Figure C.2.

If 90 percent of the offspring manifested the genotypes, *DdEe* and *ddee*, involving parental combinations of genes, and 10 percent *Ddee* and *ddEe*, which would result from recombinant gametes, then we can assume the genes *D* and *E* are 10 map units apart from one another.

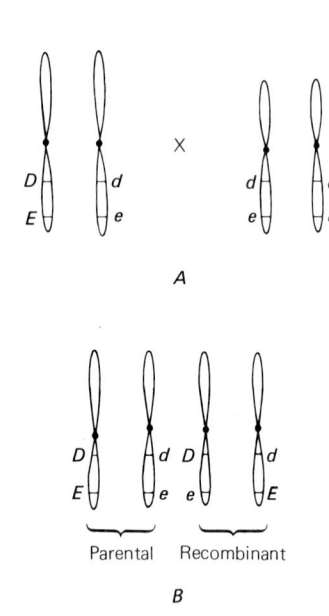

FIGURE C.2

Appendix D

The Chemical Nature of Mutation

Our understanding of the chemical nature of genes makes it easier to understand the changes which result in mutation. Since the normal function of a gene depends on the sequence of nucleotides, it is clear that changes in the normal sequence would result in mutation.

What are some of the ways in which changes in nucleotide sequence can be brought about? One was postulated by Watson and Crick, following their proposal for the structure of the DNA molecule, called *tautomerization*. Basically, tautomerization involves the shifting of hydrogen bonds from one part of a base to another part. This shift results in abnormal pairing during replication. Figure D.1 illustrates the difference between adenine and its tautomeric form, which allows it to pair with cytosine instead of thymine. As DNA replication continues, the tautomer causes the following substitution of a G–C pairing for the normal A–T:

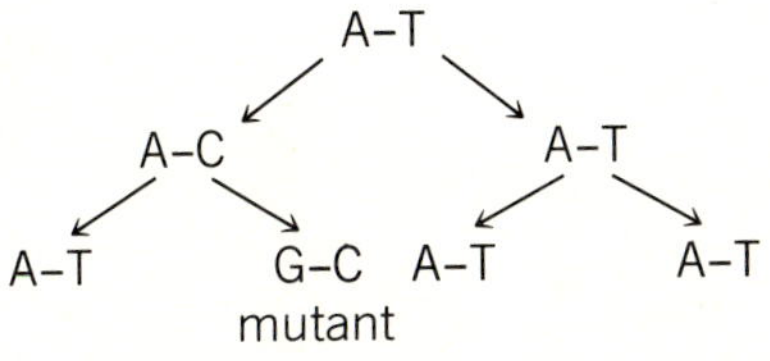

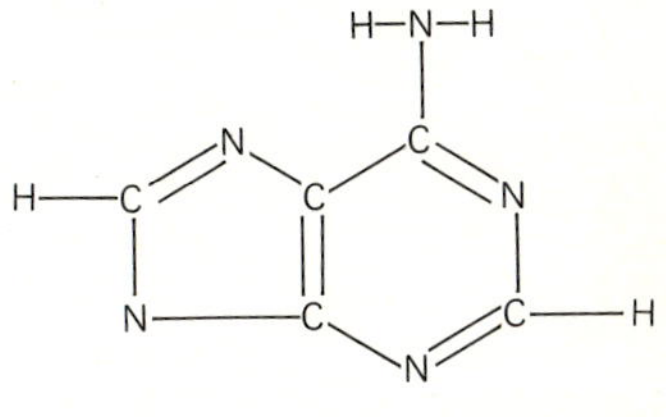

```
           A-T
          ↙   ↘
       A-C     A-T
      ↙  ↘   ↙  ↘
    A-T  G-C A-T A-T
       mutant
```

TAUTOMERIZATION

A

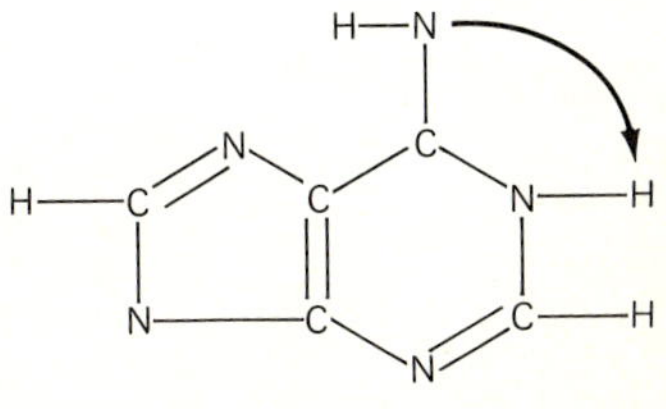

B

FIGURE D.1

The DNA containing the G–C pair where normally there is an A–T would be mutant, since the sequence change would result in the substitution of a new amino acid for one normally in that position in the polypeptide for which the gene is coded.

BASE ANALOG

Another possible cause of mutation lies in the ability of *base analogs* to pair occasionally with the nitrogenous bases found in DNA. Base analogs have almost the same chemical makeup as the bases normally found in DNA, with slight differences in one part of the molecule. Figure D.2 illustrates the chemical structure of thymine and of its base analog, 5-bromouracil.

Because of its similarity to thymine, 5-bromouracil can pair with adenine, and occasionally with guanine. The following sequence of events can then occur with a series of replications:

```
                         A–T
                  ┌───────┴───────┐
                A–T              A–Br
              ┌──┴──┐          ┌──┴──┐
            A–T    A–T        A–T   G–Br
           ┌─┴─┐  ┌─┴─┐      ┌─┴─┐  ┌─┴─┐
          A–T A–T A–T A–T  A–T A–T G–C A–Br
              Normal                Mutant
```

Experiments have shown that base analogs such as 5-bromouracil are mutagenic if present during DNA replication. Errors that occur during replication due to substitution of base analogs or to tautomeric shifts are sometimes referred to as *copy errors*.

FIGURE D

OTHER SOURCES OF MUTATION

Some mutation-causing agents, or mutagens, such as nitrogen mustard, directly affect the structure of nucleotides found in DNA. Nitrogen mustard, for example, adds a chemical group to guanine, changing it to an analog of adenine, which is therefore able to pair with thymine. Nitrous acid, another mutagen, changes cytosine into uracil, which as you know, pairs with adenine. In both cases, replication results in new DNA strands containing abnormal sequences of bases. Yet other mutagens, such as acridine dyes, appear to cause the insertion or deletion of nucleotides.

As you can see, another of the academic by-products of the elucidation of the double helix by Watson and Crick is the clarification of the chemical nature of mutations. With this knowledge it may become possible to correct mutant genes known to possess such chemical changes.

Appendix E

Glossary of Terms

Acute dosage: Exposure to a large amount of radiation at one time. Acute dosages have been found to be more mutagenic than chronic dosages.

Alleles: A pair of genes located at corresponding positions on a pair of homologous chromosomes; alleles interact to determine a particular trait.

Amino acids: Molecules which form polypeptides when linked together in a specific sequence.

Aneuploidy: The addition or loss of individual chromosomes leading to abnormal chromosome constitutions.

Antibody: A group of proteins produced by white blood cells in response to the presence of antigens in the bodies of complex organisms.

Anticodon: A sequence of three nucleotides on one of the loops of *tRNA* which is complementary to a codon on *mRNA*.

Antigen: Foreign factors which elicit the production of antibodies in an organism.

Autoimmune diseases: Diseases in a complex organism believed to result from the individual's immune system reacting to his own cells as antigens.

Autosomal traits: Traits determined by genes on autosomes.

Autosomes: Chromosomes other than the sex chromosomes.

Bacteriophage: Virus which parasitizes bacteria.

Balanced polymorphism: The transmission of two different alleles at equal rates to the next generation, maintaining an equal frequency of the two alleles in the population.

Base analogs: Molecules similar in structure to nitrogenous bases found in nucleic acid, which, when they are substituted for the normal bases, result in abnormal base pairings leading to mutations (see Appendix D).

Carcinogens: Agents known to precipitate the development of cancer.

Causal analysis: The method used in tracing development from later to earlier stages by studying related developmental events in complex organisms.

Centromere: Region of constriction of chromosomes which holds sister chromatids together and the area where chromosomes are attached to spindle fibers during cell division.

Chiasma: Configuration made by chromatids of homologs overlapping one another during tetrad formation of meiosis, when genetic crossing-over is believed to occur.

Chromatid: A duplicated half of a chromosome during cell division which is bound to the other sister chromatid by the centromere.

Chronic dosage: A number of exposures to lesser amounts of radiation.

Cistron: The smallest unit of DNA that codes for a polypeptide. Is synonymous with structural gene.

Cloning: The technique of developing large numbers of genetically identical cells or organisms.

Codominance: The situation in which traits determined by both alleles of a hybrid are fully expressed; there is no masking effect or dilution of phenotype compared to the homozygotes for each allele.

Codon: A sequence of three nucleotides on the *mRNA* which codes for a particular amino acid.

Coefficient of consanguinity: The probability that an individual would be homozygous for the same allele transmitted to his parents from a common ancestor.

Concordance: The agreement in manifestation or lack of manifestation of a trait in twins.

Conjugation: A form of sexual reproduction in simple organisms, such as bacteria, in which the cells lie side by side and DNA from one cell enters the other.

Consanguineous marriages: Marriages between related individuals.

Copy errors: Mistakes in base pairing made during DNA replication, due to the presence of tautomers, base analogs, or other mutagens.

Correlation coefficient: Degree of similarity in the expression of a trait in twins.

Coupling phase: Genetic makeup of the chromosomes of a dihybrid for two pairs of linked genes such that the two dominant genes are both on one homolog and the recessive genes are on the other homolog.

Creationism: Theory of the appearance and evolution of life on earth due to a creative force or supernatural power other than biological factors.

Crossing-over: Process by which chromatids of homologs exchange portions of genetic material; usually during tetrad formation of meiosis.

Darwinism: Theory of evolution based on natural selection and the survival of the fittest, as proposed by Charles Darwin.

Degenerate code: A code system in which more than one triplet sequence of nucleotides codes for a particular amino acid.

Deletion: A chromosomal abnormality in which a piece of the chromosome is missing, or a mutation in which a portion of a gene is missing.

Dihybrid: An individual heterozygous for two pairs of genes.

Diploid number: The number of chromosomes found in cells other than mature gametes; the species number of chromosomes; sometimes referred to as *2n*.

Dispermy: Fertilization of a single egg by two sperms, resulting in the presence of an extra set of chromosomes in the zygote.

Dizygous twins: Twins who develop from two eggs; these are called *fraternal,* not identical, twins.

DNA: Deoxyribonucleic acid, the chemical makeup of genes.

Dominance and recessiveness: A Mendelian law referring to the situation in a hybrid in which the trait determined by one allele masks the trait determined by the second allele. The trait that is expressed is dominant; the one that is masked is recessive.

Down's syndrome: Mutant condition resulting from chromosomal trisomy or translocation, characterized by severe mental retardation. Also known in the past as mongoloid idiocy.

Duplication: Aberration of chromosome structure in which an extra piece of the chromosome is present or an extra sequence of nucleotides exists within a gene.

Dyad: A doubled chromosome following anaphase I of meiosis. It consists of sister chromatids held together at the centromere.

Enzymes: Proteins which function as biological catalysts for chemical reactions that occur in the cell.

Episomes: Extrachromosomal genetic factors found in cells which differ from complete viruses because of the absence of a protein coat.

Epistasis: The masking of a trait determined by one pair of genes by a trait determined by a different pair of genes.

Euchromatin: Lightly staining portions of chromosomes.

Eugenics: The program of selective reproduction or prohibition of reproduction by individuals with specific genetic constitutions.

Euphenics: The symptomatic treatment of inherited diseases.

Fitness: From the evolutionary point of view, the ability of an organism to adapt to its environment and to reproduce young which, in turn, are adapted and reproduce.

Gamete: Synonymous with *germ cell* and *sex cell;* the cell which takes part in fertilization and development of a new organism. Only gametes undergo meiotic division.

Gammaglobulins: A group of proteins in the plasma of blood involved in immune reactions; known also as antibodies.

Gene: Classical unit of inheritance.

Genetic code: The sequences of three nucleotides in *mRNA* (which are complementary to triplet sequences in DNA) that code for specific amino acids. There are sixty-four triplets in the code.

Genetic drift: A random change in gene frequency from one generation to another in a population.

Genetic engineering: The intentional alteration of genetic constitutions by the substitution or addition of new genetic material.

Genetic load: The number of unexpressed detrimental mutations being carried in the genomes of a population.

Genome: The total genetic constitution of an organism.

Genotype: The specific genetic constitution of an organism; pertains to specific traits under study.

Germinal: Pertaining to gametes. A germinal mutation would be one in the germ cells.

Haploid: The number of chromosomes found in mature gametes of sexually reproducing organisms; one-half the species number of chromosomes; sometimes referred to as the *n* number of chromosomes.

Hemizygous: A term pertaining to males who possess no apparent alleles on the Y chromosome to genes on the X chromosome. Having only one copy of the X-linked genes, males can be neither homozygous nor heterozygous.

Hermaphrodite: An individual who possesses both ovarian and testicular tissue in gonads, which, however, are usually not functional.

Heterochromatin: Darkly staining portions of chromosomes, believed due to high degree of coiling.

Heterozygotes: Individuals possessing different alleles for a particular trait. This term is synonymous with *hybrid*.

Histocompatibility loci: A large group of genes in complex organisms which determine an individual's specific tissue antigens on the surface of all his cells.

Holandric gene: Gene found on the Y chromosome. As yet, no holandric genes have been clearly identified.

Homologous chromosomes: Pairs of chromosomes in sexually reproducing organisms which carry alleles and are physically alike in shape, length, and position of centromeres. Each member of a pair is called a homolog. Each parent contributes one of each pair of homologs to their progeny.

Homozygotes: Individuals which are "true-breeders" when crossed with each other because they possess two of the same alleles for a particular trait.

Hybrid vigor: Increased fitness of individuals due to heterozygosity; sometimes referred to as heterosis.

Immune reaction: The response of an organism to the presence of antigens by producing antibodies or by the phagocytic (actual consumption of foreign cells) activity of white blood cells.

Inbreeding: The process of breeding related individuals, which results in increased homozygosity and therefore uniformity of phenotype among individuals of an inbred strain.

Incomplete dominance: A system in which the interaction of alleles in a hybrid results in a phenotype which is intermediate between the phenotype of homozygous dominant and homozygous recessive individuals.

Induced mutation: Mutation as a result of man-made factors.

Induction (embryonic): The process by which one group of embryonic cells causes another group of cells to develop.

Induction (molecular): The activation or derepression of a gene which allows the synthesis of enzymes by a cell.

Interferon: Protein produced by cells in response to viral infection, which prevents replication of viruses.

Inversion: A reversal in the order of genes on a chromosome, or in the order of nucleotides within a gene.

Isozymes: Enzymes which catalyze the same chemical reactions but which exist in slightly different molecular forms.

Karyotype: A composite picture of an individual's chromosomes, made by taking a photomicrograph of specially prepared cells and then cutting out the chromosomes and matching them.

Klinefelter's syndrome: Aberrant male phenotype resulting from XXY constitution of sex chromosomes, leading to sterility and other abnormalities.

Lamarckism: Theory of the inheritance of acquired characteristics as proposed by Lamarck, suggesting the ability of living organisms to transmit traits acquired in response to environmental factors.

Linkage group: All of the genes linked together on one particular chromosome.

Lyon hypothesis: Theory proposed by M. Lyon that one of the X chromosomes in females is inactivated during early development.

Lysis: Process of breaking open cells, as in viral infections.

Lysogenic: Term referring to bacteria which host viruses that do not reproduce.

Meiosis: A special kind of cell division occurring only in germ cells of sexually reproducing organisms, which results in mature germ cells containing one-half the species number of chromosomes, or the haploid number.

Messenger RNA (mRNA): Carries the genetic message to the ribosome, where the message is translated into a protein molecule.

Metastasis: The spreading of malignant cells through the body of an organism.

Mitosis: Process of cell division resulting in daughter cells which are genetically identical to the original cell.

Monohybrids: Individuals heterozygous for one particular pair of alleles.

Monosomy: Presence of a single copy of a chromosome in cells of a normally diploid organism.

Monozygous twins: Identical twins developed from one egg.

Mosaicism: The presence of cells of different genetic constitutions in the same individual, e.g., cells with XX or XY chromosomes in some hermaphrodites.

Multiple alleles: The existence in a population of a number of alleles determining variations of a particular trait, such as eye color; however, only two of these are present in any one individual.

Mutagens: Factors capable of causing mutation.

Mutation: A permanent, heritable change involving a gene (point mutation) or chromosomes (chromosomal mutation).

Muton: The smallest genetic unit that can mutate.

Nondisjunction: Aberration of cell division in which chromosomes or chromatids that normally separate during anaphase remain together, resulting in daughter cells with too many or too few chromosomes.

Nonsense codons: Codons which do not code for an amino acid; they are thought to terminate messages.

Nuclear transplantation: Technique of transferring nuclei from the cells of an organism to enucleated egg cells.

Nucleocapsid: Inner core of some animal viruses formed within the cell during reproduction of virus particles, such as influenza viruses.

Oncogene theory: Theory which states that cancer-causing viral genes became incorporated into the DNA of man ages ago during his evolution; cells become malignant if the correct combination of oncogenes is present.

Oogenesis: The process of maturation of female germ cells.

Operator gene: In bacteria, the gene that controls transcription of structural genes.

Operon: In bacteria, consists of the operator, promoter, and structural genes, which determine the presence of a set of enzymes that catalyze a particular set of chemical reactions.

Parthenogenesis: The development of organisms from unfertilized eggs.

Pedigree chart: Symbols used to depict the phenotype and, sometimes, genotype of individuals in many generations of human families; used as a tool by geneticists to follow the transmission of an inherited condition.

Phagocytes: White blood cells which react to antigens and foreign cells by engulfing and consuming them.

Phenocopy: A condition caused by environmental factors which mimics a condition known to be genetically determined.

Pleiotropy: The occurrence of a syndrome of diverse effects resulting from the mutation of a single gene, e.g., the sickle cell anemia syndrome.

Polar body: Nonfunctional cell resulting from unequal division during the maturation of egg cells.

Polygeny: The determination of a trait by several pairs of genes with additive effects on the phenotype.

Polymorphism: The presence of two or more allelic forms of a gene in a population.

Polypeptides: Chains of amino acids which will assume further structural configurations to become functional proteins.

Polyploidy: Presence of extra sets of chromosomes.

Polysome: Cluster of ribosomes formed during protein synthesis.

Polytenization: Formation of giant chromosomes by replication of DNA without separation of chromatids; found primarily in insects.

Position effects: The inactivation of genes by neighboring genes.

Potentiation: The effect of increasing mutagenicity of a substance when administered together with other substances.

Proband: The individual who first brings an inherited trait to the attention of geneticists; usually singled out in pedigree charts.

Promoter: Part of an operon situated next to the operator gene. The site for the attachment of enzyme molecules necessary for transcription.

Prophage: Term referring to dormant viruses during the temperate phase of their life cycle.

Protovirus theory: Theory stating that the presence of pieces of viral genetic information in a cell as a result of separate viral infections can lead to cancer if the correct combination of genes is present.

Provirus theory: Theory of transformation of normal cells to cancer cells by the infection of cancer viruses, whose nucleic acids cause the transformation.

Pseudovirions: Viruslike particles, experimentally produced, which contain DNA from complex organisms, such as mice, surrounded by a virus protein coat.

Races: Subgroups in man who share certain genes in common at higher frequency than other groups.

Random assortment: Mendelian law stating that genes determining different traits will be transmitted randomly to the next generation; true only of genes located on different pairs of chromosomes. This results in a random assortment of gene combinations, leading to variability in the offspring.

Recombinant gametes: Gametes with new gene combinations on chromosomes, formed as a result of crossing-over.

Recon: The smallest genetic unit that can undergo crossing-over.

Reductional division: The first division of meiosis, resulting in daughter cells containing one representative of each pair of homologous chromosomes.

Regulator gene: In bacteria, the gene which produces a repressor substance that binds to the operator gene.

Repair mechanisms: Innate ability of cells to repair some forms of genetic damage which otherwise would result in mutation.

Replication: The synthesis of new DNA molecules during cell division.

Repressor: A substance produced by the regulator gene, which binds to the operator gene.

Repulsion phase: Genetic makeup of the chromosomes of a dihybrid for two pairs of linked genes such that the dominant allele of one pair and the recessive of the other are linked on the same homolog.

Ribosome: Cell organelle which serves as the site for protein synthesis.

RNA: Ribonucleic acid, the genetic material of some viruses and the chemical makeup of molecules important for protein synthesis in all cells.

Secondary nondisjunction: Nondisjunction occurring in cells of already trisomic individuals.

Segregation: Mendelian law stating that each of a pair of alleles in an individual must be transmitted separately to the next generation.

Selection: The differential perpetuation of genes from one generation to another.

Sex chromosomes: Usually one pair of chromosomes which carry, among others, genes determining the sex of the individual.

Sex-limited traits: Traits determined by autosomal genes whose expression differs in males and females because of the influence of hormones; e.g., balding is a sex-limited trait.

Sex linkage: Pertains to all genes found on the X chromosome. Synonymous with *X linkage*.

Sibship: The siblings of one generation in a family.

Somatic: Pertaining to body cells other than gametes. A somatic mutation is one occurring in a body cell rather than a sex cell.

Species: Populations that are genetically so different that they can no longer interbreed.

Spermatogenesis: Process of maturation of male germ cells.

Spontaneous mutation: Mutation which arises from natural causes.

Structural gene: In bacteria, the gene that contains the coding for a specific polypeptide.

Synapsis: Process of pairing of homologous chromosomes during prophase I of meiosis.

Tautomers: Alternate forms of a substance which differ slightly in their molecular structure.

Temperate phase: Part of the life cycle of viruses during which time they remain dormant in the host cell.

Teratogens: Agents which cause deformities in developing fetuses.

Tetrad: A cluster of four chromatids formed by duplicated homologs lying parallel to one another, following synapsis.

Tolerance: An immunological phenomenon in which an organism does not react to the presence of antigens.

Transcription: The synthesis of new RNA molecules using DNA as a template.

Transduction: Process of virus-mediated transfer of genetic material from one host cell to another.

Transfer RNA (tRNA): RNA molecules which carry amino acids to mRNA on ribosomes to be bonded together into a polypeptide.

Transformation (genetic): Genetic changes caused in a cell by the introduction of genes from another cell.

Transformation (neoplasia): Changes in cell characteristics following infection by oncogenic viruses.

Translation: The process by which the message of the DNA encoded in mRNA is converted into a specific polypeptide at the ribosome.

Translocation: Chromosomal aberration in which a piece of one chromosome breaks off and attaches itself to another chromosome.

Transsexual: An individual who is usually normal in genetic constitution but who undergoes surgery for sex change because of psychological problems.

Trisomy: Presence of three copies of a chromosome in cells of a normally diploid organism.

Turner's syndrome: An aberrant female phenotype usually resulting from monosomy of the X chromosome. Its most common trait is gonadal dysgenesis (abnormal gonads leading to sterility).

Vegetative phase: Part of the life cycle of viruses, during which time they reproduce and cause the cell to rupture.

X-linkage: Linkage of genes on the X chromosome.

XXX syndrome: Trisomy of the X chromosome with variable effects on the phenotype of the female individual.

XYY syndrome: Presence of an extra Y chromosome resulting in unusually tall males; has been referred to as the *criminal syndrome* in the past, but the association with aberrant behavior patterns has not been clearly established.

Zygote: A fertilized egg cell.

Index

Index

ABO blood group, 173–175
  genetic basis of, 174–175
  immune basis of, 173–174
ABO incompatibility, erythroblastosis fetalis and, 178–179
Abortive infection, defined, 199
Achondroplasia, 243–244
Acquired characteristics, theory of the inheritance of (Lamarckism), 286–288
Acquired immunity, defined, 172–173
Acute doses of x-rays, 310
Adaptation of living organisms, 266–270
Albinism, 94, 95, 128–129, 333
Alkaptonuria, 127–128
Alleles, defined, 13, 38–39
Allelism, concept of, 13–14
Amino acids:
  described, 110–111
  quantity of, 137
  (See also Protein synthesis; Proteins)

Amish, genetic defects among, 72–73
Anaphase, mitosis, 32
Anaphase II, 37
Anemia:
  Cooley's, 240–241, 331
  cures for, 331–332
  Lepore's, 331
  pyruvate kinase hemophylic, 72
  sickle cell, 237–239, 331–332
Aneuploidy, defined, 214
  (See also Down's syndrome; XYY syndrome)
Anthropocentrism, 284
Antibodies:
  circulating: blood groups and resistance to disease and, 173–180
  immunity to disease organisms and, 180
  origin of, 175
  defined, 172
Antibody molecules:
  genetic determination of, 185–187

Antibody molecules:
  physical nature of, 184–187
Anticodon, defined, 132, 138
Antigen-antibody reactions,
  specificity of, 185
Antigens, defined, 172
  (See also Histocompatibility
  loci)
Aposhian, H. V., 343
Arabinose operon, 343
Artificial immunity, defined, 172
Atomic Energy Commission (AEC),
  314–316
Auerbach, Charlotte, 317
Autoimmune diseases, 189
Autosomal dominance in pedigree
  charts, 67–68
Autosomal recessiveness in
  pedigree charts, 69–70
Autosomal traits, 56
Autosomes, defined, 52, 53
Avery, D. T., 114

Back mutations, defined, 244–245
Background radiation, 306
Bacteriophages (phages), 194–198
Balbiani rings, 155
Banker, B. Q., 165
Barr bodies, X chromosomes and,
  159–160
Base pairing, specificity of,
  119–121
Bateson, W., 98, 99
Beadle, G., 125–127, 236
Beagle, H.M.S. (ship), 257–258
Behavior, genetics and, 300–302
Benzer, S., 247
Bivalents, defined, 35
Bleeder's disease (hemophilia), 58, 80
Blood theory of heredity, 6
Blood types (blood groups):
  and circulating antibodies and
    resistance to disease,
    173–180
  codominance in, 91
  determining, 14
  legal implications of typing,
    179–180
Borlaug, Norman, 74

Boveri, Theodor, 33, 34, 149, 159
Briggs, R., 149–151, 336
Bryan, William Jennings, 284
Burnet, Sir Macfarlane, 187, 188

Cancer, 193–194, 199–210
  agents causing, innate resistance
    and, 204–207
  changes in cell membrane of
    transformed cells, 201–202
  chromosome abnormalities and,
    208–209
  difficulties in curing, 204
  immune mechanism and, 172
  theories on origin of, 207–209
  as transformation in animal cells,
    defined, 199–200
  unrestrained and infinite growth
    in, 200–201
  viruses in human, 202–204
Carcinogens, 204–207, 308
  mutagens as, 323
Catastrophism, 284
Causal analysis of gene action in
  development, 166
Cell division (see Meiosis; Mitosis)
Cell fusion technique, 102
Cell-mediate immune reaction,
  180–184
Cell membranes:
  changes in, of transformed cells,
    201–202
  defined, 29
Cells, described, 27–30
Cellular complexity, 161–162
Centrioles, defined, 29
Centromere, defined, 30
Characteristics, theory of the in-
  heritance of acquired,
  286–288
Chase, M., 115, 116
Chemical mutations, 317–326
  chemical nature of mutations,
    359–360
  determination of, 319–323
  long-range effects of, 324–326
Chiasmata, defined, 98
Chromatin:
  defined, 28
  in prophase, 30

Chromosomal mutations, 213-233
 aberrations of chromosome number, 214-229
 defined, 214
 structural abnormalities of chromosomes, 229-232
Chromosomes:
 cancer and abnormalities in, 208-209
 defined, 28
 DNA of microorganisms versus, 141-142
 evolution of, 270-271
 identification of gene loci on, 142-143
 intrachromosomal divergence, 271-274
 mapping: by crossing-over data, 100
  determination of linkage and chromosome mapping using crossover frequencies, 355-358
  in mammals, 100-103
 number of, in an organism, 30
 progressive changes in, during development, 154-157
 proteins and nucleic acids in, 109-112
 qualitative differences in, 33
 sex (see Sex chromosomes)
 species-specific number of, 34
 (See also Meiosis; Mitosis)
Chronic doses of x-rays, 310
Chu, E. H. Y., 319
Cis arrangement, defined, 97
Cis-trans positions in linkage, 97-98
Clonal selection theory, 186-187
Clone, defined, 151
Cloning:
 described, 149-151
 future uses of, 336-339
Codominance, 89-92
 defined, 91
Codominant alleles, defined, 14
Codon, defined, 137
Coefficient of consanguinity, described, 70-73
Color blindness, 56-58
Coloration, adaptation of, 266-268
Commoner, Barry, 314

Compatibility (transplants), 171
Concordance in twin studies, 298
Conjugation, defined, 142
Consanguineous marriages, 69-81
 coefficient of consanguinity and, 70-73
 inbreeding versus hybrid vigor and, 73-74, 76, 80
Control of life processes, implications of, 122, 329-330
Cooley's anemia (thalessemia), 240-241, 331
Correlation coefficient in twin studies, 298
Correns, Karl E., 22
Cosmic rays, 306
Couloumbre, A., 165
Coupling phase, defined, 97
Creagan, R. J., 317
Creationism, 284-286
Cri du chat syndrome, 231
Crick, Francis, 117
Criminal syndrome (XYY syndrome), 223-225
Crossing-over, 98-103
 chromosome mapping by, 100
 defined, 98
 determination of linkage and chromosome mapping using frequencies of, 355-358
Cytoplasm:
 in egg cell to be fertilized, 43
 centrioles of, 29
 defined, 28

Danforth's short tail gene (*Sd*) 167
Darrow, Clarence, 284
Darwin, Charles, 257-260, 288
Darwinism, 257-274
 evidence supporting, 263-274
 (See also Evolution)
Daughter cells, defined, 27
DDC (drug), 158
Degenerate code, 138
Deletion, defined, 230-231
Deoxyribonucleic acid (see DNA)
Dermatoglyphic patterns, 215
Descent of Man, The (Darwin), 260
Developmental genetics, 149-157

DeVries, Hugo, 22
Dihybrids, defined, 16
Diploid number, defined, 34
Diploidy, estimation of recessive mutations and, 244
Division in mitosis, 32
Division I in meiosis, 38
Division II in meiosis, 78
DNA (deoxyribonucleic acid):
  artificial synthesis of, 339–340
  bases in, 137
  cellular complexity and, 161–162
  chemical mutagenesis and, 320
  in chromosomes, 109–112
  discovery of, 112–117
  in gene regulation, 145–148
  and identification of gene loci on chromosomes, 142–143
  of microorganisms versus chromosomes, 141–142
  puffing and, 155–156
  repetitive, 241–243
  RNA compared with, 129–132
  significance of, 116–117
  structure of, 117–121
    double helix, 117–119
    specific base pairing, 119–121
  variability of, 136–137
  in viruses, 194–199, 203, 205, 207
  (See also Replication; Transcription)
Dominance:
  codominance, 89–92
    defined, 91
  incomplete, 89–92
    defined, 89–90
Dominance and recessiveness, law of, 8–14
Dominant genes, mutation of, 243–244
Dominant-lethal assay technique, 322–323
Double helix, described, 117–119
Down's syndrome (mongoloid idiocy):
  described, 215–218
  frequency of, 45
Drugs, mutagenicity of, 318
Duplication:
  defined, 230–231

Duplication:
  gene, 241–243
Dyads, defined, 36

Ecology, 276
Ecosystems, 276, 290
Egg:
  age effect on formation of, 44–45
  divisions forming, 42–43
  mRNA in, 135–136
  second meiotic division and fertilization of, 45
Ehrenstein, G. von, 140
Electrophoresis technique, 238
Ellis-van Creveld syndrome, 72
Embryology, 263
Embryonic cells, genetic potential of, 149–151
Embryonic induction, 162–165
Embryos, similarities between, among species, 263, 265
Endoplasmic reticulum, defined, 29
Environment:
  course of evolution influenced by, 276
  fulfillment of genetic potential and, 255
  man and, 289–292
Enzymes:
  function of, 110
  genes and, 125–129
Episomes, defined, 198
Epistasis, defined, 93–94
Equational division, defined, 38
Equator of spindle, defined, 31
Erythroblastosis fetalis, 176, 178–179
Essentialism, 284
Euchromatin, defined, 160
Eugenics, 333–339
Euphenics, 330–333
Evolution, 255–303
  chromosomal, 270–271
  evidence supporting Darwinian theory of, 263–274
  factors influencing course of, 274–277
  man and, 283–303
  molecular, 274
Evolutionary fitness, 261–262

Farrow, M., 334–335
Ferguson-Smith, M. A., 55
Flagellum, defined, 42
Food additives, 317–318
Fossils, 263

Galactosemia, 128
Gametes, defined, 34
Gametogenesis:
    completion of, 41–44
    X and Y chromosomes during, 53–54
    (*See also* Meiosis)
Gammaglobulins, defined, 184
Garrod, Sir Archibald, 127–128
Gene action:
    causal analysis of, in development, 166
    heterochromatin effects on, 159–161
    hormones and, 157
    problems of, in complex organisms, 140–143
    regulation of, 145–169
        in complex organisms, 157–161
        in lac operon, 145–148
        redefinition of gene, 148–149
Gene duplication, 241–243
Gene insertion in mammalian cells, 342–343
Gene loci, identification of, on chromosomes, 142–143
Genes, 109–169
    defined, 148–149
    enzymes and, 125–129
    identification of genetic substance, 112–117
    (*See also* DNA)
    mechanism of gene-determined protein synthesis, 129–136
    significance of molecular nature of, 122
Genetic code, 137–140
    universality of, 140
Genetic counseling, 333–334
    pedigree charts used in, 82–83, 85
Genetic drift affecting Hardy-Weinberg law, 252
Genetic engineering, 339–343

Genetic load, defined, 248
Genetic message:
    termination of, 139–140
    translation of, 131–133
    (*See also* Transcription)
Genetic mosaicism, 226
Genetic polymorphism, defined, 179
Genetics:
    defined, 3
    developmental, 149–157
    laws of, 6–18
        law of dominance and recessiveness, 8–14
        law of random assortment (*see* Random assortment, law of)
        law of segregation (*see* Segregation, law of)
    molecular: center of study in, 110
        implications of, 116–117
    population, 248–252
        defined, 249
    value of microorganisms in, 116
Genotype, defined, 18
Geologic ages, 264
Germinal mutations, defined, 214
Glossary, 361–371
Gonadal dysgenesis (Turner's syndrome), 55, 218–221
Goodness of fit, defined, 21
Griffith, F., 113–115
Gross mutations (*see* Chromosomal mutations)

H-2 locus, identified, 193
Hamerton, J., 208–209
Haploid number, defined, 34
Hapsburg lip, 6
Haptoglobins, 242
Hardy-Weinberg law, 249–252
Harvard Educational Review (journal), 297
Heart disease, 332
Heber, R., 299
Hemizygous, males as, 55
Hemoglobin:
    abnormalities in, 237–241
    gene regulation in, 152–154
Hemophilia (bleeder's disease), 58, 80
Hemophilia B, 72

Heredity in man, Mendelian, 63–87
  (See also Inheritance)
Hermaphrodites, 226–227
Hershey, A. D., 115, 116
Heterochromatin, effects on gene action, 159–161
Heteropyknotic, defined, 160
Heterosis (hybrid vigor), inbreeding versus, 73–74, 76, 80
Heterozygotes, 12, 57
Histocompatibility loci, 182
Histones, defined, 158
Historical theories on origin of life, 4–6
  (See also Evolution)
Holandric gene, defined, 55
Homologs, defined, 35
Homozygotes (homozygous), defined, 12
Homuculus, defined, 5
Hormones, gene action and, 157
Host-mediated assay technique, 321–322
Huebner, R., 207
Hugon, Daniel, 225
Human races, 292–300
  genetic basis of, 293–296
  IQ and, 297–300
  religions and, 296–297
Huntington's chorea, 82
Huxley, Aldous, 336
Hybrid vigor (heterosis), inbreeding versus, 73–74, 76, 80

Immune reactions, 171–191
  cell-mediated reaction, 180–184
  and circulating antibodies: blood groups and resistance to disease and, 173–180
  immunity to disease organisms and, 180
  origin of, 175
  defined, 171–173
  physical nature of antibody molecule, 184–187
  recognition of self and, 187–189
Immunological memory, 189–190
Immunology, 171
Inborn Errors of Metabolism, The (Garrod), 128

Inbred strains, defined, 74
Inbreeding, hybrid vigor versus, 73–74, 76, 80
Incomplete dominance, 89–92
  defined, 89–90
Induced tolerance, defined, 188
Inducer substances, 146
Induction, defined, 162
Industrial melanism, 267
Infinite growth of transformed cells, 200–201
Ingram, Vernon, 239
Inheritance:
  patterns of, 3–25
  physical basis of, 27–49
  polygenic, 92–93
Inhibition, contact, defined, 201
Innate immunity, defined, 180
Insertion, defined, 235–236
Instinct, 300
Intelligence quotient (IQ), human races and, 297–300
Intercellular interaction in development, 162–165
Interferons, described, 205–206
Intermediate repetitive DNA, 242
Interphase in mitosis, 30
Interphase I in meiosis, 34–35
Interphase II in meiosis, 38
Intrachromosomal divergence, 271–274
Inversion, defined, 231–232
In vitro studies of human cells, 319–320
IQ (intelligence quotient), human races and, 297–300
Isozymes, defined, 151

Jacob, F., 129, 142, 145, 148, 157, 198
Jensen, A. R., 297, 299
Jones, K. W., 271
Juberg, R., 334–335

Karyotypes, described, 51
Kessler, S., 223
Kettlewell, H. B., 267
Khrushchev, Nikita, 288
Kim, S. H., 132–133
Kindred, defined, 67–68

King, T., 149-151, 336
Klinefelter's syndrome, 58, 222-225

Lac operon, 145-148, 161, 340-342
Lactate dehydrogenase (LDH), 151-154
Lamarck, Jean Baptiste, 286
Lamarckism (theory of the inheritance of acquired characteristics), 286-288
Lambda phage, 340-342
Landsteiner, K., 173
LDH (lactate dehydrogenase), 151-154
Lepore hemoglobin (HG lepore), 240
Lepore's anemia, 331
Lerner, I. M., 288
Lethal genes, 103-104
Limb-girdle muscular dystrophy, 72
Linkage:
  described, 96-98
  determination of, and chromosomes mapping using crossover frequencies, 355-357
Linkage group, defined, 96
Lipoprotein, defined, 202
Lippmann, F., 140
Little, C. C., 182, 193
Lyell, Charles, 258, 260
Lyon, M. F., 159
Lyon hypothesis, 159-161
Lysenko, Trofim D., 288
Lysergic acid diethylamide (LSD), 318
Lysis:
  defined, 195
  virus caused, 199
Lysogenic, defined, 196
Lytic phase of bacteriophages, 195-196

McCarty, M., 114
McKusick, Victor, 70, 72
Macleod, C. M., 114
Malthus, Thomas, 258, 276
Maltose operon, 343

Man:
  and his environment, 289-292
  human races, 292-300
    genetic basis of races, 293-296
    IQ and, 297-300
    religions and, 296-297
  Mendelian heredity in, 63-87
  origins of, 288-289
Man-made radioactivity, 308-317
Map units, defined, 100
Mapping chromosomes:
  by crossing-over data, 100
  determination of linkage chromosome, using crossover frequencies, 355-358
  in mammals, 100-103
Markert, C., 151
Marriages:
  consanguineous, 69-81
    coefficient of consanguinity and, 70-73
    inbreeding versus hybrid vigor and, 73-74, 76, 80
  laws restricting, 334
Mayr, Ernst, 283
Medawar, P. B., 283
Meiosis, 34-38
  anaphase I in, 36
  anaphase II in, 37
  division I in, 38
  division II in, 38
  interphase I in, 34-35
  interphase II in, 38
  metaphase I in, 35
  metaphase II in, 37
  mitosis compared with, 46, 47
  nondisjunction in, 45
  prophase I in, 35
  prophase II in, 37
  random assortment law and, 39-41
  segregation law and, 38-39
  telophase I in, 38
  telophase II in, 38
Mendel, Gregor, 89, 257
  influence of, 122
  laws of genetics and, 6-18
  Lysenko and, 288
Mendelian heredity in man, 63-87
  analytical difficulties in man, 63-66

Mendelian traits:
  defined, 63
Mendel's laws:
  chromosome movement and, 34
  rediscovery of, 22
  (*See also* Genetics, laws of)
Merril, C. R., 342
Mesothelioma, 323
Messenger ribonucleic acid (*see* mRNA)
Metabolism, inborn errors of, 127–129
Metaphase in mitosis, 32
Metaphase I in meiosis, 35
Metaphase II in meiosis, 37
Metastasis, defined, 201
Microbial systems, 320
Microscopes, effects of development of, 4–5
Migeon, B. R., 102
Migration affecting Hardy-Weinberg law, 252
Miller, C., 102
Mitosis, 30–33
  anaphase in, 32
  division in (completion), 32
  genetic continuity and, 45–48
  interphase in, 30
  meiosis compared with, 46, 47
  metaphase in, 32
  prophase in, 30–31
  telophase in, 32
Modifier genes, described, 94–96
Molecular evolution, 274
Molecular genetics:
  center of study in, 110
  implications of, 116–117
Mongoloid idiocy (*see* Down's syndrome)
Monkey Trial, 284
Monod, Jacques, 129, 145, 148, 157
Monosomic, defined, 215
Monosomic conditions (*see* Turner's syndrome)
Monozygous twins, defined, 46
Moos, R. H., 223
Morgan, T. H., 96
Mortality rate of offspring of consanguineous marriages, 71

mRNA (messenger RNA):
  genetic code and, 138
  in protein synthesis, 131–135
  transcription of, 139
Müller, Hermann, 308, 335–336, 338
Multicellular organisms, levels of organization of, 161–168
Multiple allelism, defined, 13–14
Multiple-germ-line theory, 186
Muscular dysgenesis (*mdg*), 165–167
Mutagenesis (*see* Mutations)
Mutagens, defined, 305
Mutation rates, determination of, 243–248
Mutations, 305–327
  affecting Hardy-Weinberg law, 252
  chemical, 317–326
    chemical nature of mutations, 359–360
    determination of, 319–323
    long-range effects, 324–326
  chromosomal, 213–233
    aberrations of chromosome number, 214–229
    defined, 214
    structural abnormalities of chromosomes, 229–232
  course of evolution influenced by, 274
  defined, 55, 213–214
  in evolutionary change, 256
  point, 235–248
    defined, 235
  radiation and, 305–317
  rates of, determination of, 242–248
  spontaneous, in inbred strains, 74

Natural immunity, defined, 175
Natural selection, theory of, 257–260
  (*See also* Evolution)
Nature versus nurture, 298–300
Neel, James V., 251–252, 279
Neo-Darwinism, defined, 261
Neutral mutations, defined, 245

Nilsson-Ehle, H., 92
Nitrogenous bases:
  in nucleic acids, 112
  as variable component of DNA, 119
Nobel prizes, 119
Nondisjunction:
  defined, 45
  in Down's syndrome, 217–218
Nonsense codons, 140
Nuclear energy:
  for peace, 315–317
  radiation from, 311–315
Nuclear membrane:
  changes of, in transformed cells, 201
  defined, 28
Nuclear transplantation, 149–151
Nucleic acids (see DNA: RNA)
Nucleocapsid, defined, 199
Nucleolus, defined, 28
Nucleotide, defined, 112
Nucleus, defined, 28

Oncogenes theory, 207
Oncogenic, defined, 203
One-gene-one-enzyme hypothesis, 126–127, 148
Oogenesis:
  defined, 42
  spermatogenesis differentiated from, 44–45
Operator gene, defined, 146
Operon, defined, 146
Organ systems, interaction between developing, 165
Organic compounds, defined, 110
Origin of Species, The (Darwin), 260
Out-crossing (hybrid vigor), inbreeding versus, 73–74, 76, 80

Parasitic birds, adaptations of, 266
Parental types, defined, 98
Parthenogenesis, sex determination in, 59
Passive immunity, defined, 172–173
Pauling, Linus, 237–238

Pedigree chart, 66–70
  analysis of, 67–80
  consanguineous marriages and, 69–81
  coefficient of consanguinity and, 70–73
  inbreeding versus hybrid vigor and, 73–74, 76, 80
  in genetic counseling, 82–83, 85
  purpose of, 66
  symbols used in, 66–67
Penrose, L. S., 215
Peptide bond, defined, 110–111
Pest control, eugenics and, 335
Phages (bacteriophages), 194–198
Phagocytes, defined, 172
Phenocopies, defined, 83, 333
Phenotypes, defined, 18
Philadelphia chromosome, 208, 231
Phosphoric acid in nucleic acids, 112
Photoreactivation, defined, 245
Physiological adaptation, 268, 270
Plasma cells, defined, 175
Pleiotropy, defined, 166
Point mutations, 235–248
  chemical nature of, 359–360
  defined, 235
  determination of, 319–323
Polar body, defined, 42
Polygenic inheritance, 92–93
Polymorphism affecting Hardy-Weinberg law, 251–252
Poly I:C, 206
Polynucleotide, defined, 112
Polypeptide:
  completion of, 133–136
  defined, 110–111
Polyploidy, 227–229
  defined, 214
Polysomes, defined, 131
Polytenization, defined, 155
Population explosion, 277
Population genetics, 248–252
  defined, 249
Potentiation, 318–319
Preformation theory, 5
Prenatal diagnosis of inherited diseases, 334

Primary nondisjunction, defined, 217
Primary structure of protein,
    defined, 111
Probabilities, laws of, prediction of
    offspring by, 18–22
Proband, defined, 67
Productive infection, defined, 199
Promoter region, defined, 146
Prophage, defined, 196
Prophase in mitosis, 30–31
Prophase I in meiosis, 35
Prophase II in meiosis, 37
Propositus (proposita), defined, 67
Protein synthesis:
    cell producing correct protein,
        136–137
    mechanism of gene-determined,
        129–136
Proteins:
    in chromosomes, 109–112
        nature and function of proteins,
            110–111
    progressive changes in, during
        development, 151–154
    variability of, 136–137
Protovirus theory, 207–208
Provirus theory, 207
Pseudovirions, 343
    defined, 198
Puffing, defined, 154–157
Punnett, R. C., 12, 98, 99
Punnett square, described, 12–13
Pyruvate kinase hemophilic anemia,
    72

Qasba, P. I., 343
Quantitative traits in polygenic
    inheritance, 93
Quaternary structure of proteins,
    defined, 111

Races (see Human races)
Radiation mutagenesis, 306–317
Random, evolutionary change as,
    262–263
Random assortment, law of, 15–18
    linked genes and lack of, 96–97
    phenotypic and genotypic ratios
        on basis of, 57

Random assortment:
    physical basis for, 39–41
Recessive genes, mutation of,
    244–248
Recombinant types, defined, 98
Reduction division, defined, 36
Regulator genes, 157–158
    defined, 146
Regulator substances, 158–159
Rejection (transplants), 171
Repair mechanisms in cells, 245,
    247
Repetitive DNA, 241–243
Replication, defined, 121
Repressor, defined, 146
Reproductive isolation, defined, 271
Repulsion phase, defined, 97
Resistance transfer factors (RTF),
    198
Reverse transcriptase, 203
Rh blood types, 175–178
    genetic basis of, 176
    incompatibility, techniques to
        minimize, 176–178
Ribonucleic acid (see RNA)
Ribosomes, defined, 29
Rickets, 293
RNA (ribonucleic acid):
    DNA compared with, 129–132
    mRNA: genetic code and, 138
        in protein synthesis, 131–135
        transcription of, 139
    puffing and, 155–156
    tRNA: genetic code and, 138
        in protein synthesis, 131–135
        termination of message and,
            140
    in viruses, 194, 199, 203–205, 207
Rous, Peyton, 202–203
Royal families, 80, 81
RTF (resistance transfer factors),
    198
Russell, W. L., 309

Scarr-Salapatek, S., 299
Schmidt-Nielsen, Knut, 270
Schull, William J., 251–252, 279
"Science Framework for California
    Public Schools," 284–285
Scientists, responsibility of, 330

Index

Scopes, John, 284
Secondary nondisjunction, defined, 218
Secondary structure of proteins, defined, 111
Segregation, law of, 14
  physical basis for, 38–39
Selection affecting Hardy-Weinberg law, 250–251
Self, immune reactions and recognition of, 187–189
Semiconservative replication, defined, 121
Sex chromosomes:
  aberrations in, 218–226
  determining sex, 51–61
  in other organisms, 59
  function of, 51–54
  (See also X chromosome; Y chromosome)
Sex-linked traits, 58–59
Sex ratio of offspring, possible effects of x-rays on, 311
Sexes, why only two, 59
Sexual reproduction, course of evolution influenced by, 277
Shapiro, J., 341
Sibship, defined, 66
Sickle cell anemia, 237–239, 331–332
Sister chromatids, defined, 30–31
Smithies, O., 242
Snell, G., 182, 193
Soluble RNA (see tRNA)
Somatic mutations, defined, 214
Somy (suffix), use of, 214–215
Specific locus test, 309
Specificity of base pairing, 119–121
Speck, Richard, 223
Spermatids, defined, 41
Spermatogenesis:
  oogenesis differentiated from, 44–45
  X and Y chromosomes following, 53–54
Spindle, mitotic, defined, 31
Spontaneous generation theory, 4–5
Spontaneous mutations, defined, 306
sRNA (see tRNA)
Structural genes, defined, 145
Sugar in nucleic acids, 112

Sutton, W. S., 34
Synapsis, defined, 35

Tatum, E., 125–127, 236
Tautomerization, 359–360
Telophase in mitosis, 32
Telophase I in meiosis, 38
Telophase II in meiosis, 38
Temin, H., 207
Temperate phase of bacteriophages, 196–197
Teratogens, 308
Tertiary structure of proteins, defined, 111
Test tube babies, 336
Tetrad, defined, 35
Thalessemia (Cooley's anemia), 240–241
Thalassemia major, 241
Thalassemia minor, 241
Thalidomide babies, 83, 323–324
Todaro, G., 207
Transcription, described, 131
Transduction, described, 195–196, 340–341
Transfer ribonucleic acid (see tRNA)
Transformation:
  defined, 115
  through viral infection, 199–200
  (See also Cancer)
Transformed cells, mutations and, 213–214
Transfusion technique, 173
Translation, described, 131–133
Translocation, defined, 229
Translocation carrier, defined, 229
Transplants, cell-mediated immune reaction to, 180–184
Transsexuals, 226–227
Trisomy 21 (see Down's syndrome)
tRNA (transfer ribonucleic acid):
  genetic code and, 138
  in protein synthesis, 131–135
  termination of message and, 140
Tryptophan operon, 343
Tschermak von Seysenegg, Gustav, 22
Turner's syndrome (gonadal dysgenesis), 55, 218–221
Twin studies, 298

Unrestrained growth of transformed cells, 200–201

Variability:
  in evolutionary change, 256
  in populations, 260–261
Variegated position effects, defined, 161
Vavilov, N. I., 288
Vegetative stage, defined, 195
Vestigial organs, 279
Victoria (Queen of England), 80
Viruses, 193–199
  animal, 199
  bacteriophages, 194–198
  characteristics of, 194
  in human cancers, 202–204
    (See also Cancer)
  viruslike genetic particles, 198

Wallace, Alfred Russel, 260, 261
Watson, James, 117
Watson-Crick double helix, 117–119
Widow's peak, 57–58
Wollman, E. L., 142, 198

X chromosome:
  Barr bodies and, 159–160
  function of, 52–53
  genes linked with, 143
  and genetics of Y chromosome, 55–56
  X-linked traits, 56–58
XO (Turner's syndrome), 55, 218–221
X-rays, 309–311
XXX superfemales, 223
XXY (Klinefelter's syndrome), 58, 222–225
XYY syndrome (criminal syndrome), 223–225

Y chromosome:
  function of, 52–54
  genetics of, 55–56

Zygotes:
  cell division in, 45–48
  defined, 27

Catalog

If you are interested in a list of fine Paperback
books, covering a wide range of subjects
and interests, send your name and address,
requesting your free catalog, to:

McGraw-Hill Paperbacks
1221 Avenue of Americas
New York, N.Y. 10020